Bicuspid Aortic Valve Syndrome

Gaetano Thiene • Chiara Fraccaro
Giuseppe Tarantini • Cristina Basso
Editors

Bicuspid Aortic Valve Syndrome

A not so Benign Congenital Heart Disease

Springer

Editors

Gaetano Thiene
Cardiovascular Pathology Unit,
Department of Cardiac, Thoracic,
Vascular Sciences and Public Health,
School of Medicine
University of Padua
Padua, Italy

Giuseppe Tarantini
Cardiology Clinic, Department of
Cardiac, Thoracic, Vascular Sciences
and Public Health, School of Medicine
University of Padua
Padua, Italy

Chiara Fraccaro
Cardiology Clinic, Department of
Cardiac, Thoracic, Vascular Sciences
and Public Health, School of Medicine
University of Padua
Padua, Italy

Cristina Basso
Cardiovascular Pathology Unit,
Department of Cardiac, Thoracic,
Vascular Sciences and Public Health,
School of Medicine
University of Padua
Padua, Italy

ISBN 978-3-031-85588-7 ISBN 978-3-031-85589-4 (eBook)
https://doi.org/10.1007/978-3-031-85589-4

This Springer imprint is published by the registered company Springer Nature Switzerland AG
The registered company address is: Gewerbestrasse 11, 6330 Cham, Switzerland

Foreword

Bicuspid aortic valve (BAV), while being the most frequent congenital heart disease, has alternated between being considered a curiosity, clinically suspected on the basis of an ejection click, to being feared as the angel of death, bringing about a heavy burden of endocarditis and aortic dissection. Like all things, it took time and consistent work to understand BAV's true clinical place, to be considered without fear but with the attention that it deserves to protect the patients, to treat the complications in time, and with the adequate measures to ensure that patients live their life with the full extent of their life expectancy and the full measure of quality of their life. Our early professional life (when no imaging was available besides chest X-ray) was centered around hemodynamic valve diseases, the high risks incurred by these patients without and with surgical interventions and cared little about abnormalities that did not threaten patients immediately. While the pathologists warned us about the importance of BAV, it is only with rapidly improving imaging starting with echocardiography that we evolved. In our own career, contributions to understanding of BAV were slow to come but illuminated our clinical practice before informing the guidelines for the management of valve diseases. Thus, with the rising interest in BAV, there is great accumulation of knowledge but with dispersion throughout abundant literature.

Gaetano Thiene, a luminary of cardiac pathology, spent a rich career providing scientific facts for the education of clinicians and improving clinical practice. He has gathered his partners, Chiara Fraccaro, Giuseppe Tarantini, and our friend Cristina Basso to edit a sum on BAV, calling upon multiple teams with experience in all matters regarding BAV. All aspects of BAV are addressed, starting from BAV embryology detailing how the normal and abnormal aortic valves are formed to the congenital abnormalities associated, to the anatomy of the aortic valve, so simple and so complicated to understand and to the morphologic classification of BAV so important to have a common language. The genetic basis of BAV is a teaser to wet our interest. The clinical complications, valve dysfunction, endocarditis, and aortopathy are then fully detailed, followed by imaging, echocardiography, MRI, and CT-angio. This full panorama is completed by chapters on surgical and interventional treatments of BAV. This extensive examination of BAV in all its aspects is impressive by its comprehensiveness but is also unique by the magnitude of illustrations that allowed us, as educated as we thought we were, to quickly acquire new understanding of BAV. Thus, there is new understanding to be gotten on BAV for all audiences, from the nurse to the medical student to the young colleague to the graying professors of medicine. This didactic enterprise is absolutely warranted, because BAV needs our attention and badly.

BAV represents a large burden of carriers and requires considerable scientific work. Solid epidemiological work on all forms of BAV remains needed with the corollary on better understanding of genetics with the appropriate animal models defining all genes involved in the genesis of the normal and abnormal aortic valve and the interferences that yield abnormal aortic valves. Clinically, it is crucial that we all speak the same language and phenotype these valves very accurately and the new nomenclature is key in this process. We also need to understand how to comprehensively image the BAV, normally functioning and abnormal, and the pathological consequences when complications occur. Therapeutically, we know that we tend to intervene too late and for the purpose of early treatment, it is indispensable that valve repair expands its performance, but it is also essential that we understand better the roles and pitfalls of interventional treatment. Considerable tasks await us, and we can be grateful to Gaetano Thiene, to Christina Basso and the whole team for enhancing our understanding of BAV so extensively.

Good Reading!

Minneapolis Heart Institute
Minneapolis, MN, USA
Maurice Enriquez-Sarano

Mayo Clinic
Rochester, MN, USA
Hector I. Michelena

Preface

Bicuspid aortic valve (BAV) is the most frequent congenital heart disease, with an incidence of approximately 1%. It can be silent and associated with normal valve function. However, a series of complications may occur with time: valve incompetence, valve stenosis by dystrophic calcification, infective endocarditis, progressive dilatation of the ascending aorta, aortic dissection, and sudden death.

The problem of BAV is not just the number of the cusps but also the aortic wall. Severe degenerative changes (elastic fiber fragmentation, smooth muscle cell death, and mucoid extracellular matrix accumulation) are observed in the aortic wall of BAV patients, accounting for progressive aneurysmal dilatation of the ascending aorta, valve incompetence, and wall dissection. The link between valve and aortic wall pathology finds most probably an explanation in the embryology of the arterial pole, since neurocrestal cells play a role in the development of the ascending aorta, aortic arch, and semilunar valves. BAV has a significant genetic component as to require screening of first-degree relatives.

Padua, Italy Gaetano Thiene
Padua, Italy Chiara Fraccaro
Padua, Italy Giuseppe Tarantini
Padua, Italy Cristina Basso

Contents

Part I

Epidemiology, Embryology, Anatomy and Genetics

1 Introduction: History and Epidemiology

Gaetano Thiene, Chiara Fraccaro, Giuseppe Tarantini, and Cristina Basso

Bicuspid aortic valve (BAV) (Fig. 1.1) is the most frequent congenital heart disease with an incidence of 0.5–1.5%, doubling the remaining ones.

It is usually silent, normally functioning in infancy and childhood. A series of complications occur throughout the natural history: dilatation of the ascending aorta with valve incompetence, dystrophic calcification with valve stenosis, infective endocarditis, aortic dissection, and sudden death. The term "BAV syndrome" is considered more appropriate, like "Marfan syndrome."

Indeed, the problem of BAV is not only the number of cusps but also the aortic wall, which is involved by Erdheim's medionecrosis aortae idiopatica [1] (elastic fibers disruption, smooth muscle cells death, cysts of proteoglycans within the extracellular matrix) with progressive dilatation of the aorta (Fig. 1.2). The cause lies in the embryology of the arterial pole, where neurocrestal cells play a fundamental role, both in the development of semilunar cusps and in the wall of ascending thoracic aorta and aortic arch. The frequent association of BAV in aortic arch coarctation (~40%) gives support to this theory [2, 3].

Prevalence of BAV in necropsy studies is reported in Table 1.1 [4, 5, 7–9, 11, 12]. Osler was the first in 1886, with 1.2% [4].

Prevalence of BAV was estimated also in vivo with echocardiography studies by Basso et al. [13] and by Nistri et al. [14] Table 1.2 (Fig. 1.3).

1.1 History

Leonardo Da Vinci, working in 1510 with Marco Antonio Della Torre (who had graduated at the University of Padua and then became Professor of Anatomy and Surgery at the University of Milan in Pavia), observed cases of BAV at autopsy and stated that an aortic valve with two or four cusps works worse than tricuspid valve (Fig. 1.4). He compared the aortic valve to a gate in a river (Fig. 1.5).

In 1844 Paget first brought the attention of BAV to develop disease [15]. In 1866, Peacock noted the tendency of these valves to develop stenosis and incompetence [16].

In his large experience, Osler reported also 18 cases with infective endocarditis [4].

G. Thiene (✉) · C. Basso
Cardiovascular Pathology Unit, Department of Cardiac, Thoracic, Vascular Sciences and Public Health, School of Medicine, University of Padua, Padua, Italy
e-mail: gaetano.thiene@unipd.it; cristina.basso@unipd.it

C. Fraccaro · G. Tarantini
Cardiology Clinic, Department of Cardiac, Thoracic, Vascular Sciences and Public Health, School of Medicine, University of Padua, Padua, Italy
e-mail: chiara.fraccaro@unipd.it; giuseppe.tarantini.1@unipd.it

G. Thiene et al. (eds.), *Bicuspid Aortic Valve Syndrome*,
https://doi.org/10.1007/978-3-031-85589-4_1

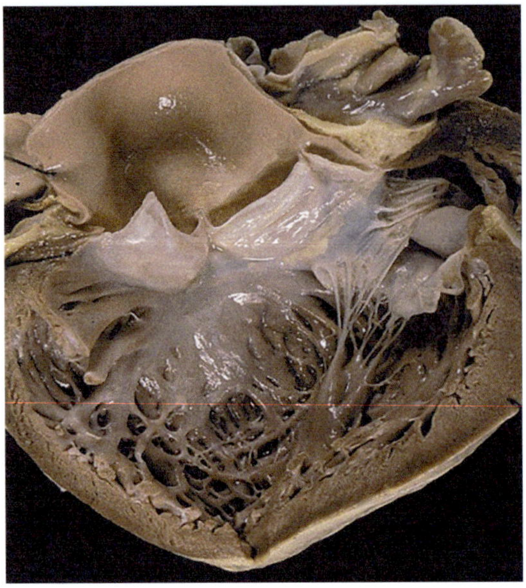

Fig. 1.1 Heart specimen with bicuspid aortic valve (without raphe). (*Reproduced from Thiene, G., Basso, C., Rizzo, S., Della Barbera, M., Valente, M., Bortolotti, U. (eds) Pathology of Cardiac Valve Disease. Springer, Cham.* https://doi.org/10.1007/978-3-031-35498-4_5)

The frequent association of BAV with dilatation of the aorta and aortic arch coarctation was noted by Maude Abbott in 1928 and 1936 [2, 3]. In 1928 Wauchope, with his autopsy experience, proved that BAV is the commonest congenital heart disease [6].

Thereafter several investigators confirmed the propensity of BAV to develop calcific stenosis, distinct from of rheumatic valve disease.

In 1984 Larson and Edwards calculated that BAV had ninefold greater risk of aortic dissection than normal population [10].

In 1972 McKusick, at the John Hopkins Hospital in Baltimore, observed Erdheim's idiopathic cystic medial necrosis in BAV aortic wall, accounting with aneurysmal dilatation [17].

Hemodynamic relation in between BAV and aortopathy, in terms of traumatic flow impact on the ascending aorta, was taken into consideration in 2004 by Robicsek et al. [18] who investigated experimentally the hemodynamic factors with a computerized digital modelling on cryopreserved, non diseased BAVs. Unlike in the normal tricuspid valve, the function of BAV was characterized by asymmetrical flow patterns and turbulence, with excessive folding, extended area of cusp contact and significant stenosis. He postulated that these abnormal flows might cause higher and uneven stress distribution also in the ascending aorta, as to be the cause of aortic dilatation at risk of dissection.

In 2005 Guntheroth and Spiers wondered "Does aortic root dilatation with BAV occur as a primary tissue abnormality or as a relatively benign post stenotic phenomenon?" [19]. They added that the most important issue is "should we necessarily fright an 1–2% of the general population that have BAV with the specter of aortic dissection"?

Clinico-pathologic investigations, carried out also at the University of Padua, confirmed the congenital nature of the syndrome [20–31].

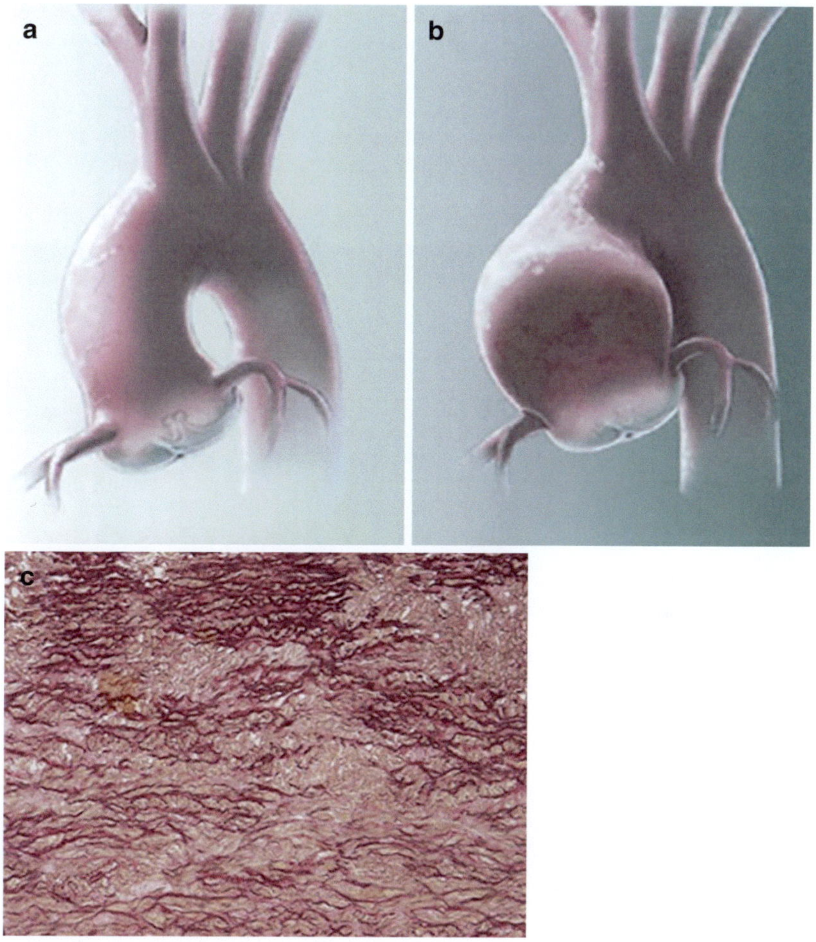

Fig. 1.2 Normal ascending aorta (**a**) vs. dilated ascending aorta in BAV syndrome (**b**). Disrupted elastic lamellae at histology of the aortic wall (**c**), Weigert-Van Gieson stain. (*Reproduced from Thiene, G., Basso, C., & Della Barbera, M. (2021). Pathology of the Aorta and Aorta as Homograft. Journal of Cardiovascular Development and Disease, 8(7), 76.* https://doi.org/10.3390/jcdd8070076)

Table 1.1 Prevalence of bicuspid aortic valve reported in necropsy studies

Author (references)	Year	Study population (n)	Male/ female	BAV prevalence (%)
Osler [4]	1886	800	NA	1.2
Lewis and Grant [5]	1923	215	NA	1.39
Wauchope [6]	1928	9966	NA	0.5
Grant et al. [7]	1928	1350	NA	0.89
Gross [8]	1937	5000	NA	0.56
Roberts [9]	1970	1440	NA	0.9
Larson and Edwards [10]	1984	21,417	NA	1.37
Datta et al. [11]	1988	8800	NA	0.59
Pauperio et al. [12]	1999	2000	3/1	0.65

NA not acquired

Table 1.2 Prevalence of bicuspid aortic valve in vivo echo studies

Author, year	N. patients	N. BAV	Prevalence (%)
Basso et al. Am J Cardiol, 2004 [13]	817	4	0.5
Nistri et al. Am J Cardiol, 2005 [14]	20,946	167	0.8

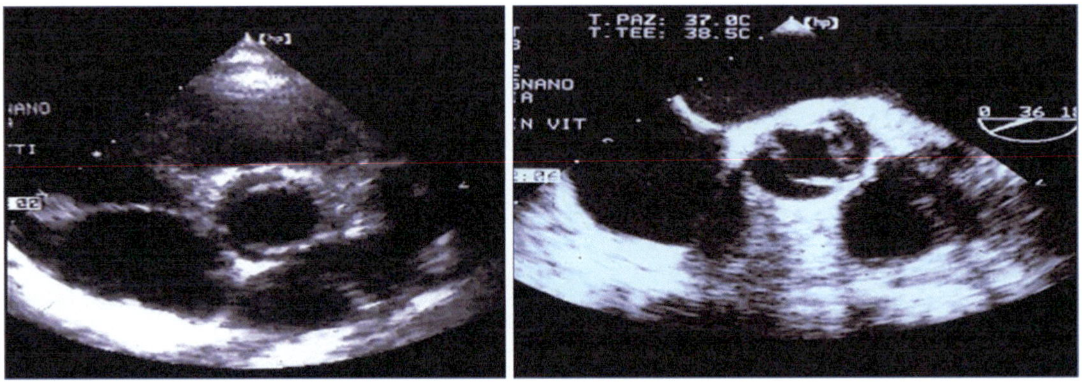

Fig. 1.3 2D echocardiography, basal cross sections: view of BAV

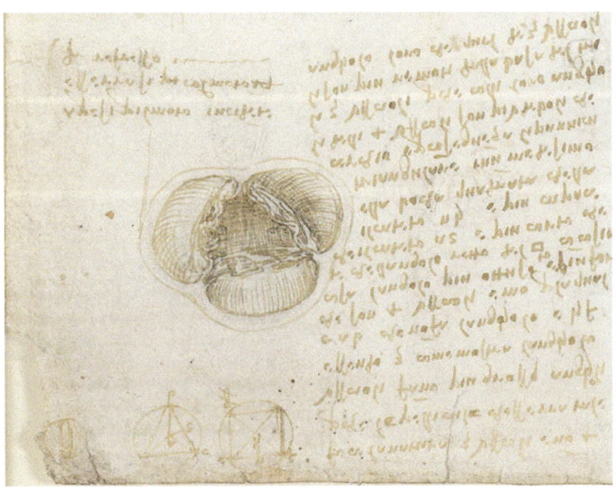

Fig. 1.4 Leonardo da Vinci and his drawings of bicuspid, tricuspid, and quadricuspid aortic valve. From his point of view, the tricuspid valve works better. (*Courtesy of © Royal Collection Enterprises Limited 2025 | Royal Collection Trust*)

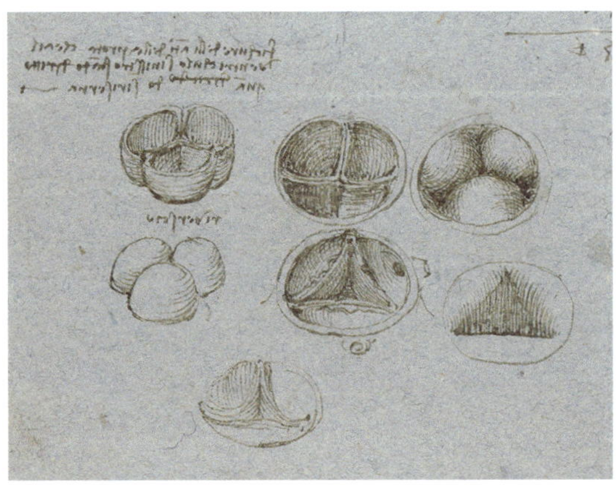

Fig. 1.5 Drawings of Leonardo da Vinci with the normal tricuspid aortic valve apparatus caged in the sinus portion of the ascending aorta, seen from above or below. (*Courtesy of © Royal Collection Enterprises Limited 2025 | Royal Collection Trust*)

References

1. Erdheim J. Medionecrosis aortae idiopathica (cystica). Virchows Arch Pathol Anat. 1929;273:454–79.
2. Abbott Maude E. Coartation of the aorta of adult type; statistical study and historical retrospect of 200 recorded cases with autopsy; of stenosis or obliteration of descending arch in subjects above age of two years. Am Heart J. 1928;3(5):574–618.
3. Abbott Maude E. Atlas of congenital cardiac disease. New York: American Heart Association; 1936. Chart I.
4. Osler W. The bicuspid condition of the aortic valves. Trans Assos Am Physicians. 1886;2:185–92.
5. Lewis T, Grant RT. Observations relating to subacute infective endocarditis. Heart. 1923;10:21–77.
6. Wauchope G. The clinical importance of variations in the number of cusps forming the aortic and pulmonary valves. Q J Med. 1928;21:383–99.
7. Grant RT, Wood GE Jr, Jones TD. Heart valve irregularities in relation to subacute bacterial endocarditis. Heart. 1928;14:247–55.
8. Gross L. So-called congenital bicuspid aortic valve. Arch Pathol. 1937;23:350–62.
9. Roberts WC. The congenitally bicuspid aortic valve. A study of 85 autopsy cases. Am J Cardiol. 1970;26(1):72–83.
10. Larson EW, Edwards WD. Risk factors for aortic dissection: a necropsy study of 161 cases. Am J Cardiol. 1984;53(6):849–55.
11. Datta BN, Bhusnurmath B, Khattri HN, Sapru RP, Bidwai PS, Wahi PL. Anatomically isolated aortic valve disease. Morphologic study of 100 cases at autopsy. Jpn Heart J. 1988;29(5):661–70.
12. Pauperio HM, Azevedo AC, Ferreira CS. The aortic valve with two leaflets—a study in 2,000 autopsies. Cardiol Young. 1999;9(5):488–98.
13. Basso C, Boschello M, Perrone C, Mecenero A, Cera A, Bicego D, et al. An echocardiographic survey of primary school children for bicuspid aortic valve. Am J Cardiol. 2004;93(5):661–3.
14. Nistri S, Basso C, Marzari C, Mormino P, Thiene G. Frequency of bicuspid aortic valve in young male conscripts by echocardiogram. Am J Cardiol. 2005;96(5):718–21.
15. Paget J. On obstructions of the branches of the pulmonary artery. Med Chir Trans. 1844;27:162–494.4.
16. Peacock T. Malformations occuring the later periods of foetal life. On malformations of the human heart. London: Churchill; 1866. p. 133.
17. McKusick VA. Association of congenital bicuspid aortic valve and erdheim's cystic medial necrosis. Lancet. 1972;1(7758):1026–7.
18. Robicsek F, Thubrikar MJ, Cook JW, Fowler B. The congenitally bicuspid aortic valve: how does it function? Why does it fail? Ann Thorac Surg. 2004;77(1):177–85.
19. Guntheroth WG, Spiers PS. Does aortic root dilatation with bicuspid aortic valves occur as a primary tissue abnormality or as a relatively benign poststenotic phenomenon? Am J Cardiol. 2005;95(6):820.
20. Angelini A, Ho SY, Anderson RH, Devine WA, Zuberbuhler JR, Becker AE, et al. The morphology of the normal aortic valve as compared with the aortic valve having two leaflets. J Thorac Cardiovasc Surg. 1989;98(3):362–7.
21. Basso C, Frescura C, Corrado D, Muriago M, Angelini A, Daliento L, et al. Congenital heart disease and sudden death in the young. Hum Pathol. 1995;26(10):1065–72.
22. Clementi M, Notari L, Borghi A, Tenconi R. Familial congenital bicuspid aortic valve: a disorder of uncertain inheritance. Am J Med Genet. 1996;62(4):336–8.
23. Duran AC, Frescura C, Sans-Coma V, Angelini A, Basso C, Thiene G. Bicuspid aortic valves in hearts with other congenital heart disease. J Heart Valve Dis. 1995;4(6):581–90.
24. Giusti S, Cocco P, Thiene G. [The bicuspid aortic valve: a «minor» congenital cardiopathy at risk of serious complications]. G Ital Cardiol. 1991;21(2):189–201.
25. Mancuso D, Basso C, Cardaioli P, Thiene G. Clefted bicuspid aortic valve. Cardiovasc Pathol. 2002;11(4):217–20.
26. Nistri S, Grande-Allen J, Noale M, Basso C, Siviero P, Maggi S, et al. Aortic elasticity and size in bicuspid aortic valve syndrome. Eur Heart J. 2008;29(4):472–9.
27. Nistri S, Sorbo MD, Marin M, Palisi M, Scognamiglio R, Thiene G. Aortic root dilatation in young men with normally functioning bicuspid aortic valves. Heart. 1999;82(1):19–22.
28. Thiene G, Basso C. Pathology and pathogenesis of infective endocarditis in native heart valves. Cardiovasc Pathol. 2006;15(5):256–63.
29. Thiene G, Corrado D, Basso C. Sudden cardiac death in the young and athletes. Text atlas of pathology and clinical correlates. Milano: Springer; 2016.
30. Thiene G, Rizzo S, Basso C. Bicuspid aortic valve: the most frequent and not so benign congenital heart disease. Cardiovasc Pathol. 2024;70:107604.
31. Rizzo S, Basso C, Thiene G. Aorta bicuspide dall'anatomia patologica alla clinica. Cardiol Sci. 2024;12:8–14.

Embryology of Normal and Bicuspid Valves

2

Borja Fernández, María Teresa Soto-Navarrete, Valentín Sans-Coma, and Ana Carmen Durán

2.1 Introduction (Historical Development)

Isolated bicuspid aortic valve (BAV) is well known to be the more common congenital cardiac malformation in adult humans. Its incidence reaches almost 0.4–2% in autopsy populations [1–4]. The tricuspid, normal condition of the aortic valve (TAV) shows three leaflets (cusps), three aortic sinuses and three interleaflets triangles interposed between adjacent sinuses. By contrast, BAVs have two leaflets and their respective two aortic sinuses (Fig. 2.1).

BAV can remain functional, with no clinical signs of disease during a normal life span [5]. However, this valvular anomaly may carry complications such as stenosis by fibrocalcific thickening, progressive incompetence, and infective endocarditis (often becoming regurgitant). Furthermore, it is not uncommon for BAV to appear associated with dilation and dissection of the ascending aorta [2, 6–10]. This is why the congenital condition of the aortic valve and its aortopathy is interpreted as a genetically determined complex disease, often not so benign [11–14]. In addition, BAV is frequently associated with distinctive genetic syndromes such as Turner, DiGeorge, Marfan, and Shone syndromes, which determine the morphology of the anomalous valve [10].

BAVs may show two basic types, anteroposterior or latero-lateral (side-to-side), depending on the spatial relationships of its components [2, 15]. In the BAVs of anteroposterior type (ventrodorsal in murine animal models), the leaflets and sinuses are present at the anterior and posterior position of the aortic root. Moreover, a raphe of varying height may appear as a ridge located at the anterior aortic sinus. The coronary arteries arise from the anterior aortic sinus. In the laterolateral BAVs, which are less common in humans, their components are located at the right and left position of the aortic root. The right and left coronary arteries usually originate from the corresponding aortic sinus (Fig. 2.1).

B. Fernández
Departamento de Biología Animal, Facultad de Ciencias, Universidad de Málaga, Málaga, Spain

Instituto de Investigación Biomédica de Málaga (IBIMA-Plataforma BIONAND), Málaga, Spain

Centro de Investigación Biomédica en Red Enfermedades Cardiovasculares (CIBERCV), Málaga, Spain
e-mail: borjafe@uma.es

M. T. Soto-Navarrete (✉) · V. Sans-Coma · A. C. Durán
Departamento de Biología Animal, Facultad de Ciencias, Universidad de Málaga, Málaga, Spain

Instituto de Investigación Biomédica de Málaga (IBIMA-Plataforma BIONAND), Málaga, Spain
e-mail: mtsoto@uma.es; sanscoma@uma.es; acduran@uma.es

G. Thiene et al. (eds.), *Bicuspid Aortic Valve Syndrome*,
https://doi.org/10.1007/978-3-031-85589-4_2

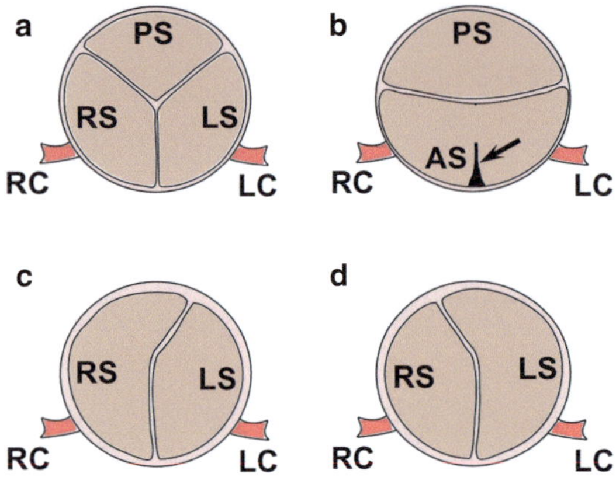

Fig. 2.1 Schematic representations of the aortic valve anatomic types. Cranial views. (**a**) Tricuspid aortic valve (TAV). (**b–d**) Bicuspid aortic valves (BAV). (**b**) Anteroposterior (BAV-A) or right-left (R-L). (**c**) Latero-lateral (BAV-B) or right-non-coronary (R-NC). (**d**) Latero-lateral (BAV-C) or left-non-coronary (L-NC). The arrow points to a raphe. AS, anterior sinus; LC, left coronary artery; LS, left sinus; PS, posterior sinus; RC, right coronary artery; RS, right sinus. (Reproduced from Soto-Navarrete et al. 2020 [48])

One of the most debated aspects of the anatomy of the BAV, especially that with a raphe, has been the acquired versus congenital nature of the anomalous valve. A degenerative (senile) or pathological rheumatic process in a TAV can produce the thickening of the leaflet tissue and varying degrees of commissural fusion between adjacent leaflets, leading to an acquired form of BAV with raphe. Instead, a BAV is considered to be of congenital origin when its anatomical arrangement, including the raphe, had developed as such during embryonic development and not as a result of a pathological process after birth [16]. It should be noted that differentiation between acquired (sometimes misguidedly called "functionally bicuspid") and congenital BAV is important from the clinical viewpoint for ongoing surveillance, medical management and eventual surgery of patients, and research purposes [17].

The first reference about the genesis of a BAV was given by Osler in 1886 [18], based on the autopsy of an 8-month-old human fetus. Osler disclosed his unproven empirically impression that some error associated especially with the development of the affected leaflets was the responsible for the anteroposterior type of BAV.

Then, until the 90s decade, knowledge about the BAV had almost entirely rested on findings in human patients. As a consequence, little direct information was available about its etiology. In humans, the cause and mechanism responsible for the development of a BAV has been a controversial issue during decades [1–3, 19–24].

Between the 60s and 90s of the twentieth century, two main approaches were considered. On one hand, some authors maintained that BAVs initially originate as tricuspid structures and acquired bicuspid morphology by the fusion of two of their leaflets at some point during their pre- or perinatal growth [22, 23, 25]. In this regard, the hypothesis based on the hemodynamic molding held that obstructive vascular defects such as aortic coarctation might decrease blood flow through the valve during development resulting in a failure of the leaflets to separate [22, 25]. Nevertheless, studies based on experimental alteration of developmental haemodynamics failed to explain how an isolated BAV is generated [26, 27]. On the other hand, experimental studies carried out mainly on bird embryos revealed the role of neural crest cells in cardiovascular development [28–33]. This finding prompted a novel approach to the etiology of BAV and also allowed a better understanding of both the isolated and syndromic, non-isolated condition of this anomalous valve [10, 34].

The first evidence of the congenital origin of the BAV was provided by Shaner [25]. He reported the occurrence of the defect in four pig embryos, in which both anteroposterior and latero-lateral types were recognized by histological reconstruction. Afterwards, the identification of suitable animal models of BAV has been a challenging task during the past recent decades. Experimental research on the congenitally BAV started at the University of Málaga thanks to the discovery of Syrian hamster specimens spontaneously affected of this valve defect (Fig. 2.2) [35–37].

The systematic inbreeding of these animals generated the first spontaneous animal model of isolated congenital BAV. In this inbred strain, the mean incidence of BAV has remained around

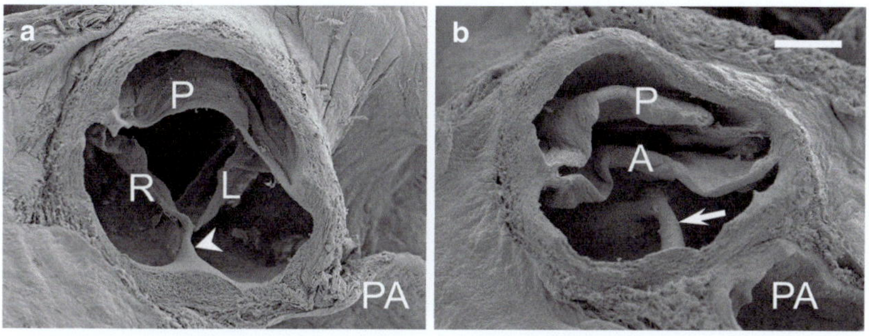

Fig. 2.2 Scanning electron micrographs of aortic valves of hamsters. Cranial views. (**a**) Tricuspid aortic valve. (**b**) Bicuspid aortic valve of anteroposterior type. A, anterior (ventral) leaflet; L, left leaflet; P, posterior (dorsal) leaflet; PA, pulmonary artery; R, right leaflet. The arrowhead indicates the anterior commissure. The arrow points to the raphe. Bar: 250 μm. (Reproduced from Martínez-Vargas et al. 2017 [87])

40% and no statistical difference has been found between the sexes. The study of the animal model demonstrated that congenital anteroposterior (ventro-dorsal) BAVs are expressions of a trait whose variation shows a continuous phenotypic spectrum, ranging from a TAV with no fusion of the ventral (anterior in man) commissure to a BAV devoid of any raphe (Fig. 2.3) [38, 39].

The intermediate phenotypes of the continuum are represented by TAVs with a more or less extensive fusion of the anterior commissure and BAVs with a more or less developed raphe located in the anterior aortic sinus. In TAVs showing fusion of the anterior commissure, the anterior interleaflet triangle existing between the right and left sinuses decreased in size according to the degree of fusion of the adjacent leaflets, which are fused at the most cephalad point. Anteroposterior BAVs lacks an interleaflet triangle, but may have a raphe. Its size can vary from a ridge confined to the wall of the anterior aortic sinus to a raphe that encroaches on the leaflet, but does not extend until the cranial level of the anterior leaflet (Figs. 2.3 and 2.4). Histologically, the raphe is similar to a true commissure (Fig. 2.4). In BAVs with raphe, the distance between the right and left commissures is usually longer anteriorly than posteriorly. In contrast, in BAVs lacking a raphe, the distance between both commissures is similar (Fig. 2.4). The hamster aortic valve anatomical spectrum is analogous to that described around 30 years later in patients, leading to a new nomenclature based on the gradation of the phenotypical continuum ([40]; partial-fusion BAV; fused BAV very asymmetric; fused BAV asymmetric; fused BAV symmetric; fused BAV symmetric no raphe; 2-sinus BAV).

The development of the spontaneous hamster model allowed to address one mechanism for the formation of BAV and its mode of inheritance. Embryological studies revealed that all the phenotypes of the anatomic spectrum develop at an early stage of the valvulogenesis, by a variable degree of fusion of the valvular primordia existing at the developing cardiac outflow tract (OFT) (Fig. 2.5). These evidences confirmed the congenital nature of the valve anomaly and pointed to an abnormal behavior of the migrating neural crest cells as a possible mechanism for the fusion [39]. Accordingly, classics hypotheses concerning the origin of BAV already proposed (i.e., improper initial development of the conotruncal ridges [19]), erroneous septation of the conotruncus [1, 19–21], absence of one of the three valve cushions [19, 20, 25] were definitively discarded. Moreover, it was demonstrated that the aortic valve anatomical spectrum is the result of a complex quantitative inheritance [38], as previously suggested from the analysis of familial clusters of patients [41, 42]. Given that genetically alike Syrian hamsters show all phenotypes that form the continuum of aortic valve morphologies, factors other than genetics may play a role in the development of the BAV [38, 43].

Years after the establishment of the spontaneous Syrian hamster model, BAV was found in an

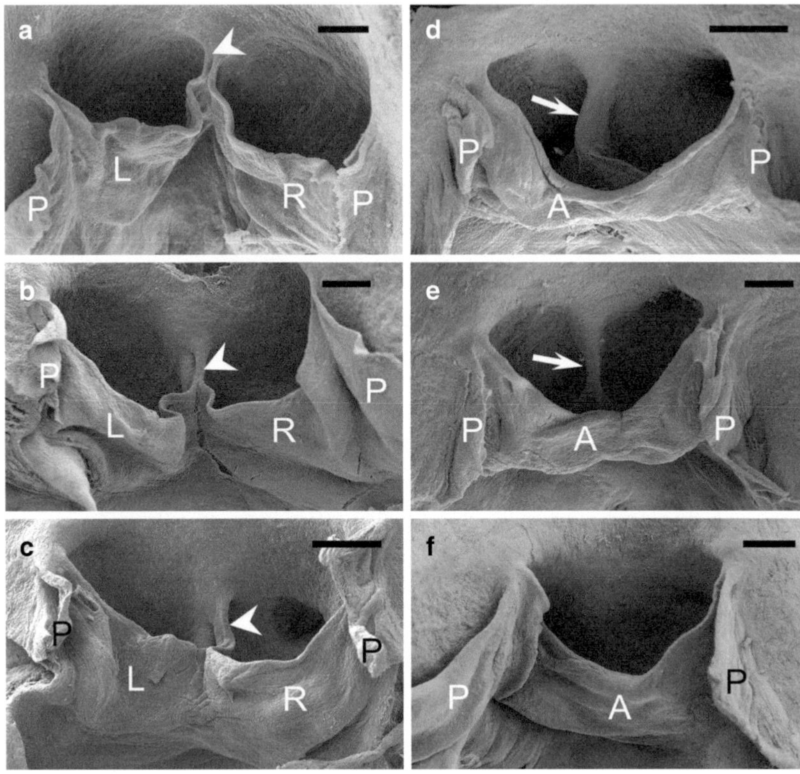

Fig. 2.3 Scanning electron micrographs of tricuspid (TAV) (**a–c**) and bicuspid (BAV) (**e**, **f**) aortic valves of hamsters. Frontal views after dissecting the posterior leaflet. (**a**) TAV without leaflet fusion. (**b**) TAV with a small fusion of the right and left leaflets. (**c**) TAV with an extensive fusion of the leaflets. (**d**) Anteroposterior BAV with a large raphe protruding toward and encroaching on the anterior leaflet. (**e**) Anteroposterior BAV with a small raphe. (**f**) Anteroposterior BAV without a raphe. The arrowheads point to the anterior (ventral) commissure. The arrows indicate the raphe. A, anterior (ventral) leaflet; L, left leaflet; P, posterior (dorsal) leaflet; R, right leaflet. Bars: 200 μm. (Reproduced from Sans-Coma et al. 2012 [43] and Martínez-Vargas et al. 2017 [87])

endothelium nitric oxide synthase knock-out (*Nos3*^{tm1Unc}; eNOS-KO) mouse strain [44], although no details were initially given on the valvular phenotype. The identification of latero-lateral type of BAV with a relatively high incidence in these knockout mice provided the opportunity to have the second animal model of BAV, which has substantially contributed to increase knowledge of its etiology (Fig. 2.6) [45]. In 2009 the comparative study of the Syrian hamster spontaneous model of anteroposterior BAV and the eNOS-KO murine model of latero-lateral BAV empirically revealed that both types of BAVs are different etiological entities, that is, they rely on distinct developmental mechanisms that will be explained below [45]. This finding was a milestone in the investigation of this valvular anomaly and led to the routine inclusion of the aortic valve morphotype in many human clinical and surgical pathological studies ([46]; revised in Michelena et al. [40]). In addition, the studies on the embryology of BAV multiplied during the last decade and later on due to the incorporation of the valvular morphology in the phenotypic characterization of genetically engineered mice, which has generated about 30 rodent strains with high incidence of BAV, mostly syndromic (reviewed in [47, 48]). These strains have allowed to assess etiological and developmental aspects of the disease that

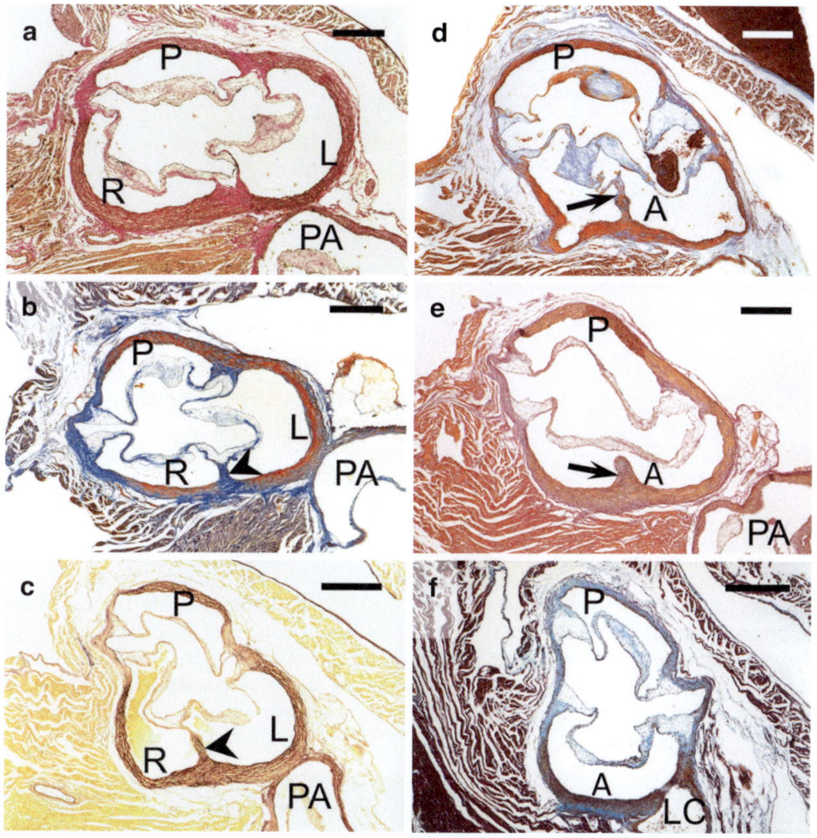

Fig. 2.4 Transverse histological sections of tricuspid (TAV) and bicuspid (BAV) aortic valves of hamsters. (**a**) TAV without leaflet fusion. (**b**) TAV with a small fusion of the right and left leaflets. (**c**) TAV with an extensive fusion of the leaflets. (**d**) Anteroposterior BAV with a large raphe protruding toward and encroaching on the anterior leaflet. (**e**) Anteroposterior BAV with a small raphe. (**f**) Anteroposterior BAV without a raphe. A, anterior (ventral) leaflet; L, left leaflet; LC, left coronary artery; P, posterior (dorsal) leaflet; PA, pulmonary artery; R, right leaflet. Arrowheads indicate fused leaflets. Arrows point to raphes. (**a**) Weigert-van Gieson stain. (**b**, **d**, **e**) Mallory's trichrome stain. (**c**) Orcein-picrofuchsin stain. (**f**) Masson-Goldner stain. Bars: 300 μm (**f** reproduced from Fernández et al. 2009 [45])

resulted in a whole new body of data, revealing new cellular and molecular mechanisms involved in the formation of the diverse types of BAV [48]. However, recent research has demonstrated interspecific and intraspecific differences between humans and animal models regarding the phenotypic expression and heritability of aortic valve malformations, highlighting the importance of these considerations for results extrapolation [47]. Interestingly, studies with patients driven by results from animal models have shown that the different anomalous aortic valves do not share a common natural history and associated pathologies [40].

2.2 Normal Valvulogenesis

The heart is the first organ to form during embryonic development in a process known as cardiogenesis, which has been extensively studied in animal models to extrapolate to humans (reviewed in [49, 50]). The process takes place on embryonic day (E) 7.5–8 in mice and Carnegie Stages (CS) 10–13 in humans [51]. It begins in the lateral anterior splanchnic mesoderm, composed of the left and right cardiogenic plates that merge to form two distinct cell populations: the cardiac crescent or first heart field (FHF) and the second

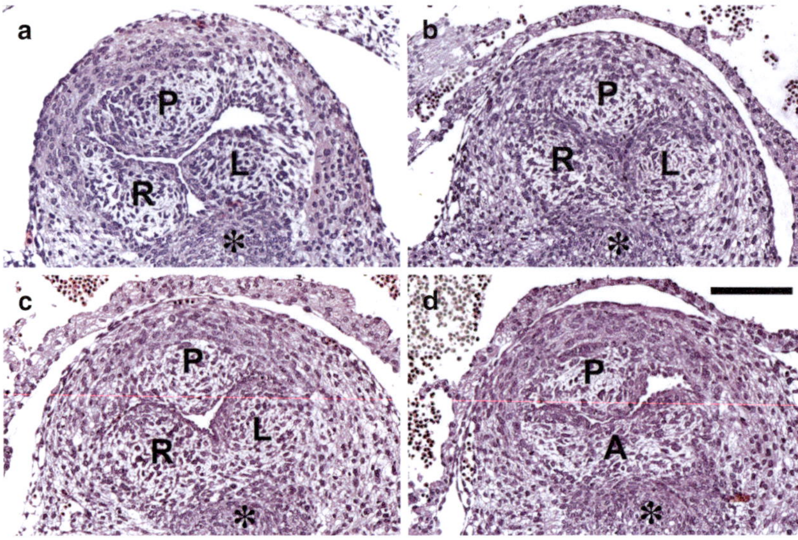

Fig. 2.5 Transverse sections of the developing aortic valve of hamster embryos, just before the excavation of the aortic valve primordia (cushions). The sections correspond to the most cephalad level of the cushions. In panels **a–c**, three cushions, right (R), left (L) and posterior (P), can be identified. In panels **b**, **c**, the right and left cushions are partially fused. In panel **d**, only two cushions, anterior (A) and posterior (P), can be recognized. The anterior cushion results from the complete fusion of the right and left cushions. The asterisks indicate the condensed mesenchyme of the aorticopulmonary septum. Embryonic stage: E12. Haematoxylin-eosin. Bar: 100 µm

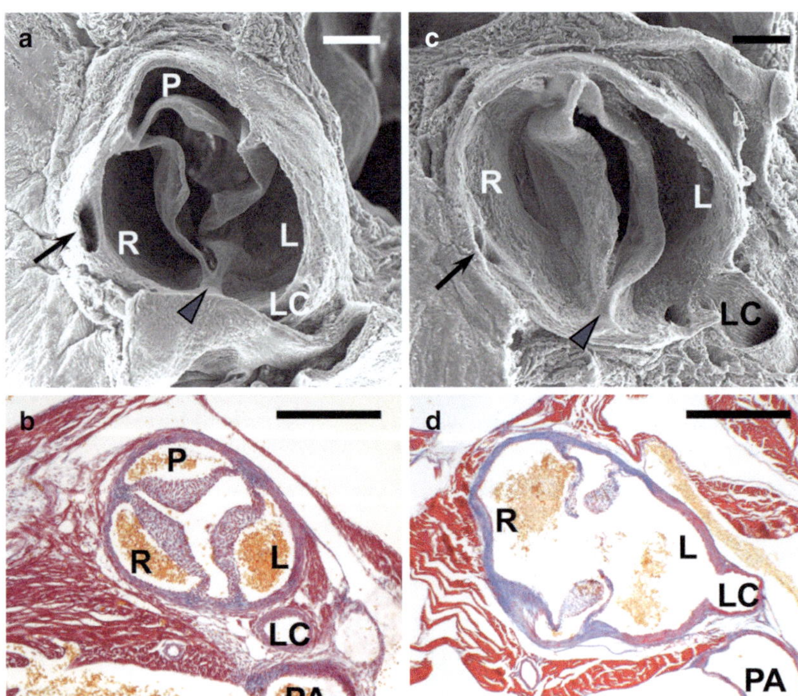

Fig. 2.6 Tricuspid aortic valves (**a**, **b**) and bicuspid aortic valves (BAV) (**c**, **d**) in endothelium nitric oxide synthase knock-out ($eNOS^{-/-}$) mice. (**a**, **c**) Scanning electron micrographs. Cranial views. (**b**, **d**) Transverse sections stained with Mallory trichrome stain. The arrows point to the right coronary artery. The arrowheads indicate the anterior commissure. In BAVs, R indicates the right aortic sinus supporting the fused right and posterior leaflets. L, left aortic sinus; LC, left coronary artery; P, posterior aortic sinus; PA, pulmonary artery; R, right aortic sinus. Bars: 200 µm. (Reproduced from Fernández et al. 2009 [45])

heart field (SHF) [52]. In early stages, the FHF coalesces in the embryonic ventral-midline to give rise to the primary myocardial heart tube [53]. Cells constituting the SHF remain in the cyrcumpharyngeal mesoderm until they are incorporated into the heart tube later in development [53]. After SHF migration and concomitant remodeling by looping and ballooning, this early heart tube becomes an embryonic chambered heart with two atria, two ventricles and the outflow tract. The semilunar valves primordia are formed in the OFT or *conotruncus*, a single cylindrical myocardial tube located in the anterior portion of the heart. It connects distally to the aortic sac, which converges with the aortic arch arteries, and proximally to the right ventricle (Fig. 2.7) [52–55].

At early stages of cardiogenesis, the *conotruncus* is composed of an outer myocardial layer and an inner endocardial layer separated by an acellular extracellular matrix, named cardiac jelly, enriched in hyaluronic acid, glycosaminoglycans, and proteoglycans [56]. The cellularization of the cardiac jelly occurs concomitantly with the process of heart remodeling, in which two distinct cellular mechanisms are involved, migration and endocardial-mesenchymal transition (EMT) [57–61]. The latter mechanism, better known in mice than in humans, occurs between E10-11 and CS10-14, respectively [49–51]. Initially, cells from the endocardial layer lose polarity, detach and migrate into the cardiac jelly, transforming into mesenchymal cells [57, 61]. These mesenchymal cells distribute homogeneously along the OFT in the cardiac jelly, forming two long, helical and facing endocardial cushions called septal and parietal conotruncal ridges (CRs). The parietal CR distributes ventro-laterally in the OFT, while the septal CR is located dorso-medially (Fig. 2.7) [51, 54, 57, 62–64]. In the OFT, two additional small endocardial cushions, known as intercalated cushions (ICs), are formed in the anterior and posterior distal portions of the *conotruncus*, and participate in the development of the semilunar valve primordia [54, 62]. Hereafter, a third embryonic cell population, the cardiac neural crest (CNC) becomes relevant for OFT development.

As said before, cell migration is the second cellular mechanism involved in the cellularization of the OFT cardiac jelly, incorporating CNC derived mesenchymal cells in the forming CRs and ICs [49–51]. CNC cells are ectomesenchymal cells that migrate from the neuroectodermal junction of rhombomers 3–8 at the hindbrain into the heart through the fourth and sixth aortic arch arteries and the aortic sac (Fig. 2.7) [28, 65]. These cells migrate proximally into the CRs in two condensed prongs (Figs. 2.7 and 2.8) [28, 51, 65]. Distally in the aortic sac, the CNC cells form a septum together with the two CRs prongs, showing an inverted U-shape mesenchymal structure, the aorticopulmonary septation complex, responsible for the division of the *conotruncus* in two independent OFTs (Figs. 2.7 and 2.8) [58, 63, 64]. CNC cells contribute to the formation of the thoracic arteries and the semilunar valves together with SHF cells [28, 52, 53, 65].

The OFT septation, which occurs at E11–12.5 in mice and CS14-16 in humans [51], is the consequence of the disto-proximal

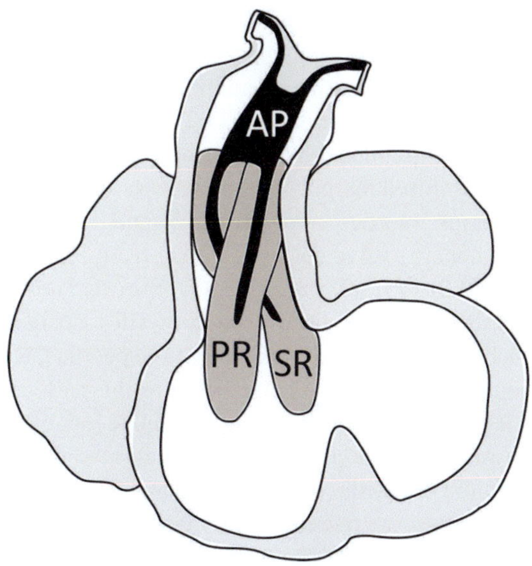

Fig. 2.7 Schematic representation of a developing heart at the stage of outflow tract (OFT) septation. The U-shaped aorticopulmonary septation complex (AP) is formed by cardiac neural crest cells (black) that migrate from the pharyngeal arch arteries along the OFT, invading the two helicoidal and opposing parietal (PR) and septal (SR) conotruncal ridges in two prongs. (Reproduced from Soto-Navarrete et al. 2020 [48])

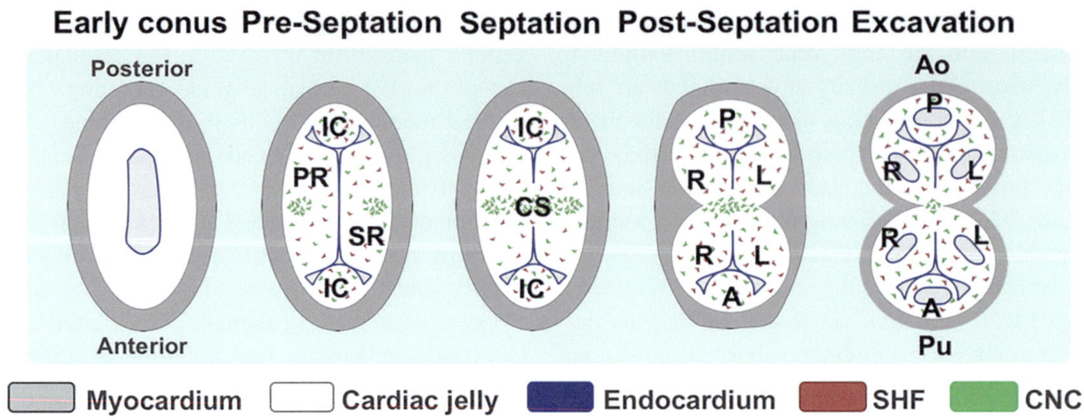

Fig. 2.8 Schematic representation of stages of normal valvulogenesis in transversal view. The early embryonic cardiac outflow tract (OFT) is a cylindrical tube (Early conus) composed of an outer epicardium covering a myocardial layer, and an inner endocardial layer, separated by the acellular cardiac jelly. The cardiac jelly is progressively populated by second heart field (SHF) and cardiac neural crest (CNC) derived mesenchymal cells. Four cushions are formed (Pre-Septation): two intercalated cushions (IC) and two conotruncal ridges (PR, SR). The single OFT is then divided (Septation) by a conotruncal septum (CS) composed of CNC, delineating the six aortic and pulmonary valve cushions (Post-Septation): right (R), left (L), anterior (A) and posterior (P) valve cushions. During Excavation, valve cushions remodel to form the final aortic (Ao) and pulmonary (Pu) valve leaflets and sinuses. (Reproduced from Soto-Navarrete et al. 2020 [48])

migration of the aorticopulmonary septation complex, by which CNC induces the fusion of the central portion of the CRs by EMT (Figs. 2.7 and 2.8) [66, 67], giving rise to the conotruncal septum and the aortic and pulmonary valve cushions (Fig. 2.8) [58, 63, 64, 68]. The right and left aortic and pulmonary valve cushions correspond to the unfused margins of the two CRs, while the posterior (aortic, non-coronary) and anterior (pulmonary) valve cushions derive from the ICs (Fig. 2.8) [45, 62, 68]. The ICs form as lateral extensions of the distal portion of CRs, so that each IC is connected proximally to a specific CR, being the posterior IC connected to the parietal CR and the anterior IC to the septal CR [64, 69].

At this developmental stage, the valve cushions, the primordia of the semilunar valve leaflets and sinuses, are mesenchymal masses covered by endocardium that show an inverted pyramid shape (Fig. 2.9). In subsequent stages (E12.5–18.5 in mouse; CS15-23 in humans), the valve cushions acquire the adult semilunar morphology through a remodeling process called excavation, which includes thinning and elongation of the distal margins of the cushions (Fig. 2.10) [39, 51, 70–72]. Although the maturation of leaflets and sinuses is poorly understood, proliferation, apoptosis, and extracellular matrix secretion are key cellular mechanisms [70, 73–75]. During excavation, valvular interstitial cells are differentiated from two distinct mesenchymal cell lineages, the SHF and the CNC cells, the latter being responsible for regulating the cellular processes necessary for maturation that continue in post-natal stages [65, 71, 72, 74].

2.3 Embryonic Development of BAV

2.3.1 The Early Studies

Most of our knowledge on BAV embryonic development has been obtained from studies in animal models. As discussed above, the anatomy, histology, and embryology of the heart, particularly of the cardiac OFT, are quite similar and comparable among mammals, at least among those of which studies have been conducted, especially rodents, whose research is usually used to extrapolate to humans [38, 45, 47, 51, 76–80].

Fig. 2.9 Scanning electron micrograph of an aortic valve cushion from a hamster embryo at E11.5. The excavation process of the cushion has not begun. The cushion appears as a semiconical swelling with the apex pointing toward the ventricle. The asterisk indicates the site at which the aortic sinus develops. Bar: 25 μm. (Reproduced from Sans-Coma et al. 1996 [39])

The first description of a morphogenetic mechanism leading to BAV formation was performed in the hamster model [39]. During cardiac OFT septation, just after the fusion of the opposing CRs (Fig. 2.8), the right and left aortic valve cushions (the primordia of the right and left aortic valve leaflets and sinuses) appeared fused, forming a single ventral (anterior in humans) cushion, whereas the non-coronary valve cushion was normal (Fig. 2.5d). This fusion of the valve cushions did not appear as a discrete qualitative defect, but rather as a defect subjected to a continuous variation. The anatomical variability was based on the degree of fusion of the adjacent cushions, i.e., the amount of mesenchymal tissue shared by the right and left aortic valve cushions (Fig. 2.5). It ranged from clearly independent to completely fused cushions that formed a big ventral aortic valve cushion. At a later embryonic stage, the process of excavation in partially or totally fused aortic valve cushions leads to the whole spectrum of adult BAV anatomical variations (compare Figs. 2.3, 2.5 and 2.11). This is a relevant but often overlooked conclusion of the study of Sans-Coma et al. [39], which indicates that the different adult anatomical variants of the hamster aortic valve clearly correlate with the degree of severity of the original embryonic defect.

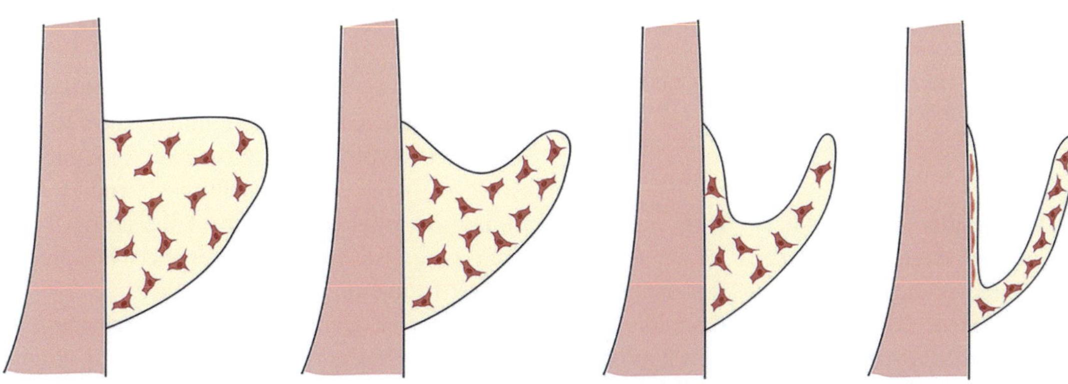

Fig. 2.10 Schematic representation of valve excavation in sagittal view during normal valvulogenesis. Excavation is a process of maturation that involves the transformation of a mesenchymal cushion covered by an endocardial endothelium with an inverted pyramid shape into a valve with an adult semilunar shape. (Reproduced from Soto-Navarrete et al. 2020 [48])

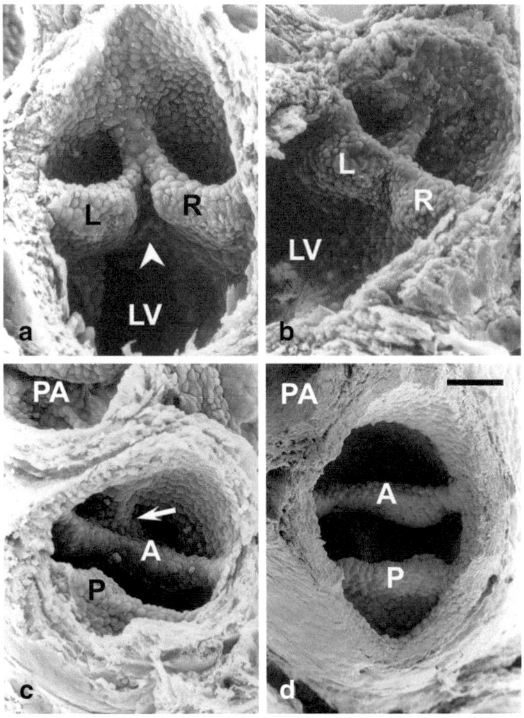

Fig. 2.11 Scanning electron micrographs of aortic valves of hamster embryos. The valve cushions show a marked degree of excavation. (**a**) Anterior aspect of a tricuspid aortic valve (TAV) with partial fusion of the anterior commissure. A small triangle (arrowhead) can be seen between the right and left leaflets. (**b**) Anterior aspect of a TAV with complete fusion of the anterior commissure. No anterior interleaflet triangle is present. (**c**) Cranial view of a bicuspid aortic valve (BAV). The ridge (arrow) located in the anterior aortic sinus outlines a future raphe. (**d**) Cranial view of a BAV with no raphe. A, anterior (ventral) leaflet; L, left leaflet; LV, left ventricle; P, posterior (dorsal) leaflet; PA, pulmonary artery; R, right leaflet. Embryonic stages: E12 to E13. Bar: 50 μm. (Reproduced from Sans-Coma et al. 1996 [39])

The results of this study were important because of three main reasons. First, as mentioned above, they establish the first embryonic mechanism of BAV formation. Secondly, they offer empirical evidence supporting a hypothesis on the formation of commissural fusions and raphes: these may result from the excavation process in more or less fused aortic valve cushions. Detailed inspection of this embryonic mechanism in the hamster model might further help understanding the morphological significance of raphes, the identity of which has been subjected to debate by anatomists, clinicians, and surgeons for decades (see Sect. 2.1). Last but not least, the embryonic and anatomical results in the hamster model establish a polarity of the trait in terms of severity, which can be extrapolated to humans (see Figs. 2.5, 2.11 and 2.3 in order). According to the most up-to-date nomenclature [40], human BAV anatomies can be ordered from least to most severe, in analogy to the hamster model as follows: partial-fusion BAV, fused BAV very asymmetric, fused BAV asymmetric, fused BAV symmetric, fused BAV symmetric no raphe, 2-sinus BAV.

2.3.2 The Breakthrough

The first genetically manipulated mouse models of BAV appeared in the new century. The *Nkx2.5* (*Nkx2–5*^{tm1Rph}) and *eNOS* (*Nos3*^{tm1Unc}) knockout mice and the *Fgf8* conditional knockout mice (*Hoxa3*$^{tm1(cre)Moon}$; *Fgf8*^{tm1Mrc}/*Fgf8*^{tm1Moon}) were shown to develop BAV together with other cardiac and extra-cardiac malformations [44, 81, 82]. BAV in *Fgf8* conditional knockout mice [82] results from defective behavior of CNC cells that colonize the OFT, serving as an experimental confirmation of the involvement of the CNC in BAV etiology. *Nkx2.5* heterozygous mice [81] displayed an eightfold increase in the prevalence of stenotic BAVs, what was strictly genetic background–dependent, being found only in the C57Bl/6 strain. This finding already highlighted the complex nature of BAV etiology, showing the necessary involvement of genetic modulators. However, none of the mentioned studies performed a complete analysis of the aortic valve embryology, so the morphogenetic mechanisms of BAV formation were still missing.

Indeed, the second embryonic mechanism of BAV formation was not described until almost one decade later [45]. In that study, the formation of BAV was compared between embryos from the hamster model and those from eNOS-KO mice. The latter had been found to develop a relatively high incidence of BAV [44], but the study of Fernández et al. [45] showed that BAVs in eNOS-KO mice are of the latero-lateral type, as opposed to the anteroposterior BAVs of the hamster model (Fig. 2.6). In addition, the embryonic mechanism of latero-lateral BAV formation was found to be different to that of anteroposterior BAVs. In eNOS-KO embryos, defective endocardial cushion formation was found during early valvulogenesis. In the early OFT, prior to septation, the posterior IC appeared anatomically fused to the parietal CR, giving rise to three instead of four endocardial cushions (Figs. 2.12 and 2.13). As a result of the septation process, which occurs normally, two instead of three aortic valve cushions formed, namely, one normal left aortic valve cushion, and one big right aortic valve cushion.

Although Sans-Coma et al. [39] already demonstrated that anteroposterior BAVs result from the fusion of the right and left aortic valve cushions, the study of Fernández et al. [45] showed that these cushions do not fuse after normal OFT septation, but they form as coalesced structures from the onset (Figs. 2.14 and2.15). Excessive fusion of CRs during septation leads to the formation of right and left aortic valve cushions with a variable degree of fusion, and an unaffected non-coronary cushion derived from the posterior IC. Therefore, anteroposterior BAVs in hamsters develop as a consequence of abnormal conotruncal septation, whereas latero-lateral BAVs in eNOS-KO mice result from abnormal formation of the endocardial cushions during an earlier developmental stage. It is noteworthy that in both hamster and eNOS-KO embryos, the aortic valve cushions as such do not fuse, but rather develop as fused primordia from their onset, leading to anatomically fused adult valve leaflets and sinuses.

Interestingly, while the variable severity of the hamster embryonic defect results in a continuous valve anatomical variation, as discussed above, BAVs in eNOS-KO mice showed a constant anatomy (symmetrical latero-lateral BAVs without a raphe; Fig. 2.6). As it will be discussed later in this chapter, the two mentioned embryonic defects underlying BAV formation have been described in different mutant mouse models. In all cases, the defect was found to be expressed as a discrete trait, without variable severity, resulting in symmetrical BAVs without a raphe. Indeed, anatomical variability of the BAV in the form of raphes or

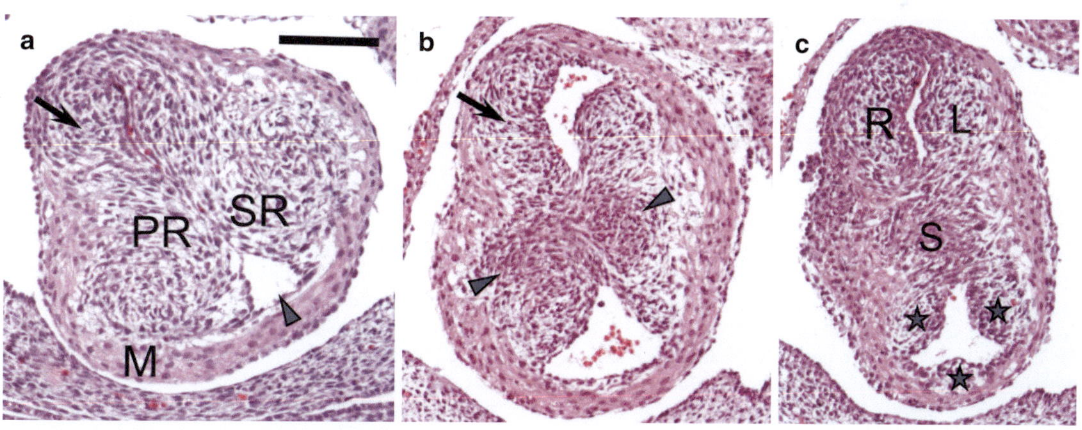

Fig. 2.12 Outflow tract (OFT) septation in mice developing bicuspid aortic valve. Transverse sections of the OFT before (**a**), during (**b**), and after (**c**) fusion of the conotruncal ridges. (**a**) The septal (SR) and parietal (PR) ridges are still independent. The posterior intercalated cushion is fused (arrow) to the PR. The anterior intercalated cushion is not visible because of the orientation of the section. The arrowhead points to its expected location. M, myocardium. (**b**) Fusion of the SR and PR. Only the central portions of the ridges are fused. However, the posterior intercalated cushion is abnormally fused (arrow) to the PR. Arrowheads indicate OFT septum anlagen. (**c**) After fusion of the conotruncal ridges, only two aortic valve cushions, right (R) and left (L), and three pulmonary valve cushions (stars) are present. The OFT septum (S) is normal and divides the OFT into an aortic (upper) and a pulmonary (lower) tract. Embryonic stages: E12 to E12.5. Haematoxylin-eosin. Bar: 100 µm. (Reproduced from Fernández et al. 2009 [45])

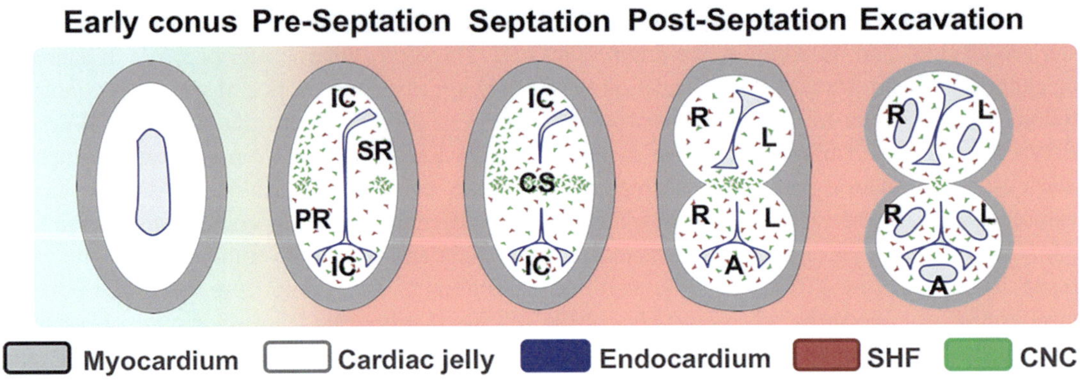

Fig. 2.13 Schematic representation of valvulogenesis with abnormal endocardial cushion formation during pre-septation in transversal view. This abnormality causes the anatomical fusion of the posterior accessory with the right (less frequently the left) aortic cushions, resulting in latero-lateral bicuspid aortic valves. The underlying cellular mechanism consists of alterations in either endocardial-to-mesenchymal transformation and/or cardiac neural crest cell migration (both mechanisms involved in the cellularization of the cardiac jelly). A, anterior valve cushion; CS, conotruncal septum; IC, intercalated valve cushion; L, left valve cushion; PR, parietal conotruncal ridge; R, right valve cushion; SR, septal conotruncal ridge. (Reproduced from Soto-Navarrete et al. 2020 [48])

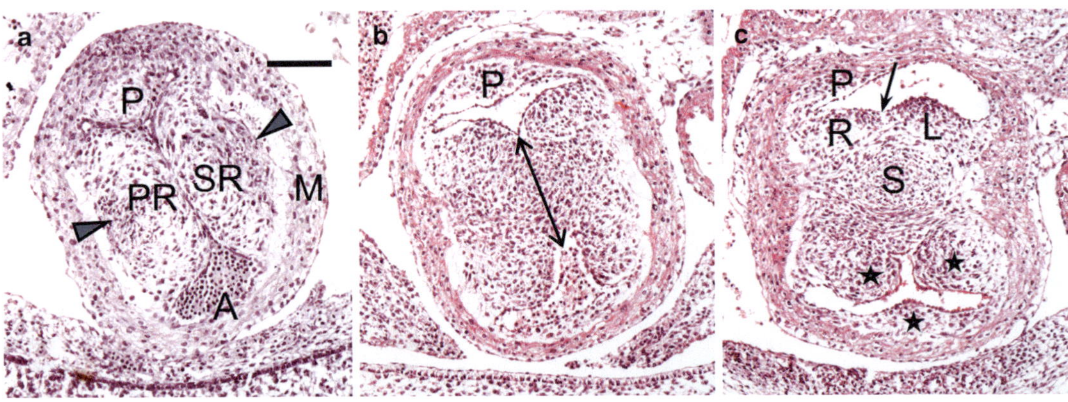

Fig. 2.14 Outflow tract (OFT) septation in hamsters developing bicuspid aortic valve. Transverse sections of the OFT before (**a**), during (**b**), and after (**c**) fusion of the conotruncal ridges. (**a**) The OFT septum anlagen consists of two cellular condensations (arrowheads), located at the center of the septal (SR) and parietal (PR) ridges, close to the myocardium (M). A, anterior intercalated cushion; P, posterior intercalated cushion. (**b**) The fusion of the OFT ridges (double-headed arrow) extends toward the posterior margins of the ridges. P, posterior intercalated cushion. (**c**) The right (R) and left (L) aortic valve cushions are extensively fused (arrow). The layers of cells of the OFT septum (S) are less condensed and less arranged concentrically than in the normal embryos. P, posterior aortic valve cushion. The stars indicate the pulmonary valve cushions. Embryos age: 11–12 days postcoitum. Hematoxylin-eosin. Bar: 100 μm. (Reproduced from Fernández et al. 2009 [45])

fused leaflets has only been detected in a few individuals of three BAV mouse models [83–85]. This fact allowed to propose that the hamster species might be a better model for extrapolation to humans than the mouse model [47].

In hamsters, the most severe cases of BAV formation appear as a normal posterior (non-coronary) cushion and a single big anterior cushion (Fig. 2.5). This anterior cushion has almost the same size as the sum of the right and left valve cushions of a normal embryo, whereas the posterior non-coronary valve cushion has always a normal, small size. However, the most severe form of BAV in the adult shows two leaf-

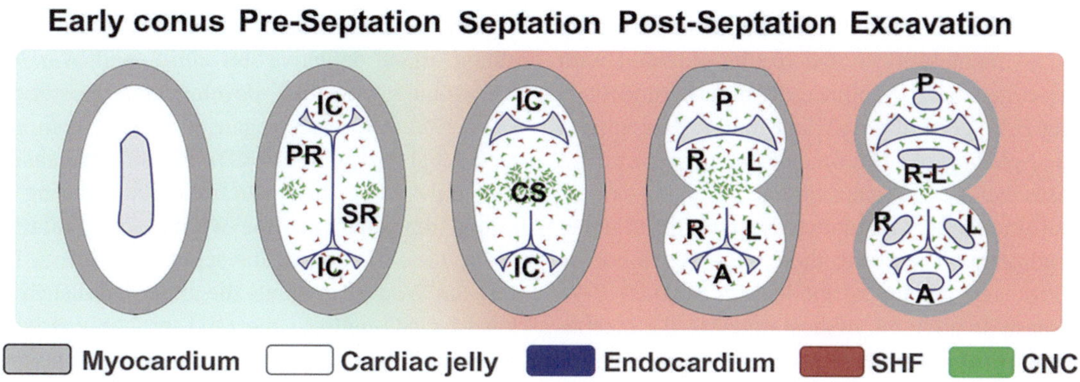

Fig. 2.15 Schematic representation of valvulogenesis with abnormal conotruncal septation in transversal view. The abnormality consists in the excessive fusion of the parietal (PR) and septal (SR) conotruncal ridges (CR) leading to the anatomical fusion of the right (R) and left (L) aortic valve primordia and anteroposterior BAVs. The continuum relies on the extended fusion of the posterior margins of CRs due to alterations of the endocardial-to-mesenchymal transformation process induced by migrating cardiac neural crest cells (more explanations in the text). A, anterior valve cushion; CS, conotruncal septum; IC, intercalated valve cushion; P, posterior valve cushion. (Reproduced from Soto-Navarrete et al. 2020 [48])

lets and sinuses, anterior and posterior, almost equal in size (Fig. 2.2). Something similar occurs in latero-lateral BAVs of eNOS-KO mice. The anatomically fused right and non-coronary cushions show almost the same size as the sum of these two cushions in a normal embryo, and a normal sized left valve cushion (Fig. 2.12). However, latero-lateral BAVs of adult eNOS-KO mice show almost equally sized right and left leaflets and sinuses (Fig. 2.6). Although the growth of leaflets and sinuses during valve cushion excavation has not been investigated in detail, it can be assumed that a differential growth between fused cushions and the normal cushion compensates the final size of leaflets and sinuses. The scarce information on the process of valve cushion excavation hinders deepening into this compensation effect, but blood flow through a developing OFT with asymmetrical valve cushions is probably a key factor.

2.3.3 The Modern Era

The study by Fernández et al. [45] was a breakthrough in our understanding of BAV because it demonstrated that the different BAV types (anteroposterior and latero-lateral) are not mere anatomical variants of the same congenital defect. Instead, each anatomical type of BAV is the result of a distinct etiology. Over the next decade, the number of human studies that considered BAV type as an important clinical factor multiplied, as did studies in genetically manipulated mice that included the aortic valve in the phenotypic screening. As a result, more than 20 new genetically modified mouse models of BAV disease were described (reviewed in Soto-Navarrete et al. [48]). Although these models show a high incidence of BAV associated with specific genetic defects, in most studies the morphology and development of the valve were only one of many characteristics to be assessed in the phenotypic evaluation of the transgenic strain. In the majority of these studies, the primary objective was not to investigate the process of valvulogenesis per se, but rather to examine the functionality of specific molecular pathways and cellular mechanisms or the specific role of the modified genes, without limiting the scope of the study to the cardiovascular system. In their study, Soto-Navarrete et al. [48] not only drew the relevant conclusions from these studies on the embryology of BAV, but also reinterpreted the results of those studies whose main objective excluded cardiogenesis under the prism of valvulogenesis.

Another important consideration regarding the extrapolability of the results obtained with these genetically engineered mouse models is that BAV generally does not manifest as an isolated phenotype. The mutant phenotype in most cases includes complex cardiac and extra-cardiac malformations, sometimes resembling human syndromes. Therefore, these models cannot be considered preclinical models for isolated BAV disease. In addition, in most mutants, the mechanism of BAV formation is clearly a direct consequence of genetic modification. However, in some cases, it cannot be excluded that the development of BAV is secondary to an associated malformation. In any case, these models are invaluable for elucidating the morphogenetic mechanisms and the molecular and cellular pathways involved in the development of BAV.

The involvement of CNC in BAV formation is well established by direct evidence in the hamster [45] and the mouse [69, 86] models, and by indirect evidence in both the hamster model [87, 88] and patients [89]. As previously stated, OFT septation occurs through the fusion of the CRs and the subsequent formation of the conotruncal septum by CNC cells, giving rise to two independent OFTs and the development of the aortic and pulmonary valve cushions (Fig. 2.8). However, when fusion does not strictly involve the center of the CRs, but extends to their posterior margins, the resulting right and left aortic valve cushions appear anatomically fused, leading to anteroposterior BAV (Figs. 2.14 and 2.15). As mentioned above, this morphogenetic defect was first described in the hamster model, where an association was demonstrated between alterations in CNC cells migration and excessive fusion of the CRs [39, 44, 48, 67]. More recently, defective conditional expression of genes regulating CNC migration in mice has provided new evidence on the involvement of this cell lineage in BAV formation [69, 74, 75, 85, 90–92]. The JAG/NOTCH signaling pathway, a known regulator of CNC behavior [93, 94], is a good example of this progress. Activation of JAG1 in CNC cells by NOTCH1 expressed in endocardial cells allows the correct distribution of CNC cells in the embryonic *conotruncus* to form the aorticopulmonary septation complex [92]. Conditional mutant mice defective for *Jag1* and *Notch1* expression in the OFT develop anteroposterior BAVs [75, 92]. MacGrogan et al. [92] demonstrated that reduction of NOTCH1 activity causes BAV only when this reduction occurs during a limited development time window, particularly during the initiation of the septation process. In *Jag1* and *Notch1* mutants, the abnormal distribution of CNC cells in the OFT probably drives excessive fusion of the CRs, leading to anteroposterior BAVs, as it occurs in the hamster model [45, 75, 92, 95].

However, the mechanism by which CNC cells induce the fusion of the CRs in the OFT remains unknown. Indeed, the cellular and molecular mechanisms involved in the normal fusion of CRs were also unknown until recently. Soto Navarrete et al. [67] newly showed evidence that the CRs get fused when the endocardial cells covering the two facing CRs transform to mesenchymal cells through a process of complete EMT that takes place sequentially, from distal to proximal (Fig. 2.16). Moreover, this study, which was conducted in the hamster model, showed that the same process of EMT takes place in the areas of the CRs affected by extra-fusion in BAV developing embryos. It is proposed that the CNC cells that invade the distal OFT induce the process of EMT of the endocardial cells that cover the CRs [45, 67]. Soto-Navarrete et al. [67] suggest that the altered migration of CNC cells along the OFT may entail an extension of the endocardial territory affected by the EMT induction, increasing the number of activated endocardial cells that results in an excessive fusion of the CRs.

CNC migration and EMT are known to be related processes. During early endocardial cushion formation, CNC cells invading the OFT induce EMT, promoting the cellularization of the CRs [66]. Class 3 semaphorin (SEMA3C) and its receptor neuropilin 1 (NRP1) dominate this crosstalk between CNC and endocardial cells, regulating migration of the former and EMT of the latter. In addition, this pathway has been demonstrated to be required for OFT septation, specifically for the migration of CNC cells to the fusion area of CRs [66]. These evidences support

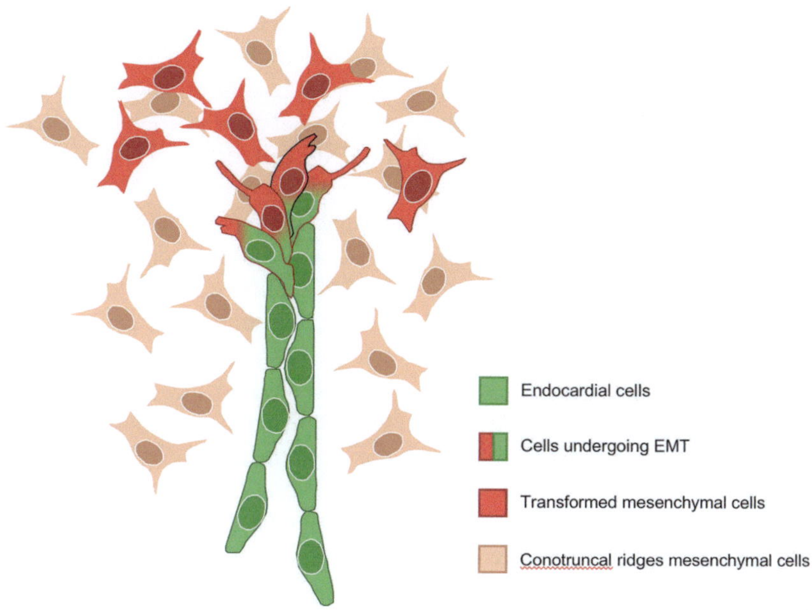

Fig. 2.16 Schematic representation of the endocardial-to-mesenchymal transformation (EMT) process involved in the fusion of the conotruncal ridges during conotruncal septation. EMT occurs in a sequential manner along the distal-proximal axis of the outflow tract: the cells of the caudal endocardium (in green) are quiescent, whereas cranial endocardial cells (in red and green), with a mixed endocardial-mesenchymal morphology, undergo EMT and transform in mesenchymal cells (shown in red). (Reproduced from Soto-Navarrete et al. 2025 [67])

the notion that CNC cells are responsible for the EMT process underlying the fusion of the CRs. Another molecular pathway possibly involved is the already mentioned JAG1-NOTCH1, whose alterations give rise to anteroposterior BAV. It is a crucial regulator of CNC migration but also of EMT and other cellular processes [92, 96]. MacGrogan et al. [92] proposed that the endocardium is likely responsible for the regulation of the CNC migration through the NOTCH pathway, and the alteration of this signaling entails an abnormal migration that leads to defective fusion of the CRs.

In eNOS-KO mice, latero-lateral BAVs form by the abnormal formation of the OFT endocardial cushions (Figs. 2.12 and 2.13). Given the known involvement of eNOS in the EMT process that cellularizes the endocardial cushions [97], it was hypothesized that absence of this enzyme may over-activate the EMT process, resulting in the formation of an aberrant CR anatomically fused with the posterior IC [45]. Indeed, absence of eNOS has been shown to induce atrioventricular valve defects through a defective EMT process [98]. eNOS regulation of EMT is dependent on the NOTCH transcription factor activation [99]. Alterations of the NOTCH pathway in transgenic mice cause EMT defects resulting in abnormal endocardial cushion formation and BAV [83, 100–102]. Moreover, mutant mice defective in the NOTCH regulator BRG1 develop BAV through disruption of the EMT process in the endocardial cushions, which is partially compensated by an increase in CNC mesenchymal cell proliferation [83]. In addition, hemizygote loss of the EMT inductor NPR2 in transgenic mice leads to low penetrant BAV, what may occur following both, NOTCH dependent and independent mechanisms [102]. It has been proposed that a slight reduction of NOTCH1 dosage in the cardiac OFT can alter EMT induction, predisposing to a variety of malformations, including BAV [103]. Other reports [104, 105] have identified alterations of the *Notch* signaling pathway associated with BAV development in different mutant mouse strains, but the cellular and morphogenetic mechanisms involved were not investigated.

All these data point to a dysregulation of the EMT process, in which NOTCH1 seems to be a key factor, as the embryonic origin of latero-lateral BAVs. Remarkably, excessive EMT also appears to involve extra-fusion of the CRs during anteroposterior BAV formation [67]. Therefore, both latero-lateral and anteroposterior BAVs may develop as a consequence of the same defective cellular mechanism (EMT) occurring at two different developmental stages, namely endocardial cushion formation and conotruncal septation, respectively.

The idea that EMT alterations constitute a central mechanism in BAV morphogenesis is especially attractive in the context of BAV aortopathy, one of the main associated pathologies in BAV patients. The mechanism of thoracic aortic dilatation is a subject of intense research due to its clinical importance, and the lack of therapeutic strategies besides surgery [40, 106, 107]. Aortic smooth muscle cells are especially affected, as apoptosis and decellularization are common histopathological features in aneurismatic aortas of BAV patients [108–110]. A recent report has shown activation of genes involved in the embryological EMT cellular pathway in adult smooth muscle cells of the ascending aorta of this patient population [111]. These findings are particularly relevant in the context of the genetic hypothesis of BAV aortopathy. Although altered hemodynamics in the ascending aorta and aortic root caused by the abnormal valve anatomy certainly play a key role in the arterial pathogenesis, genetic (i.e., congenital) factors shared with the mechanism of BAV morphogenesis are currently considered key elements in the etiology of BAV aortopathy [89, 112–114]. The embryonic development of the aortic valve and the ascending aorta are closely linked, since CNC cells are involved in both processes [28, 52, 53, 65]. So that, a defect in the regulation of EMT by CNC cells may lead to embryonic extra-fusion of aortic valve primordia and alterations in smooth muscle cells derived from CNC in the adult aorta [66, 89, 115, 116].

Excessive fusion of the CRs during anteroposterior BAV formation seems to be induced by abnormally migrating CNC cells, so it can be hypothesized that this cell lineage is also responsible for the abnormal formation of endocardial cushions leading to latero-lateral BAVs. Indeed, Peterson et al. [69] found an altered distribution of CNC cells in the *conotruncus* of eNOS-KO mice, suggesting that eNOS expressed by endocardial cells has an active role in the guidance of CNC and SHF mesenchymal cells through the OFT. Therefore, abnormal migration of CNC through the embryonic cardiac OFT may constitute the primary defect for both types of BAV formation, altering the EMT induction activity of these cells on the endocardium during early (i.e., during endocardial cushion formation to form a latero-lateral BAV; Fig. 2.13), or late (i.e., during OFT septation to result in anteroposterior BAV; Fig. 2.15) valvulogenesis. This connection between the mechanisms of EMT and CNC migration allows to explain how alterations of these two embryonic mechanisms can result in the same abnormal valve morphology.

However, abnormal EMT is not the only cellular mechanism that has been adduced as the cause of abnormal endocardial cushion formation leading to latero-lateral BAVs. Disruption of the *Rock* gene specifically in the CNC of mouse embryos leads to a variety of semilunar valve defects, including BAV [91]. This protein is required for appropriate cell-cell adhesion and communication among CNC cells invading the cardiac OFT [117]. Defective ROCK expression during OFT development causes altered aggregation of CNC cells in the *conotruncus*, what may lead to mispositioning of the endocardial cushions [91]. As explained in the previous chapter, each IC is formed as an extension of a specific CR [69], being the posterior IC connected proximally with the parietal CR. Therefore, an abnormal aggregation of CNC cells in the posterior portion of the cardiac OFT could result in anatomical fusion specifically of the right (parietal) CR with the posterior IC (Fig. 2.13). This defect would lead specifically to BAV type B (R-NC fusion). Interestingly, this BAV morphotype is by far much more frequent in humans and animal models than BAV type C (L-NC fusion) [40, 47].

PBX is another transcription factor that regulates CNC behavior [118]. Combined mutants for different murine *Pbx* genes give rise to a spectrum of cardiac OFT malformations that includes BAV [90]. These defects have been shown to be caused by alterations of the distribution of CNC, but whether these alterations cause defective formation of endocardial cushions or abnormal septation remains to be elucidated.

Besides abnormal formation of the CR and the corresponding IC, an additional mechanism of latero-lateral BAV formation has been proposed. Deletion of *Alk2* specifically in SHF-derived cells of transgenic mice causes a significant reduction of cell proliferation in mesenchymal cells of all the valvular cushions, leading in some cases to the absence of the posterior IC cushion [84] (Fig. 2.17). As a consequence, only one right and one left aortic valve cushions form, giving rise to a latero-lateral BAV. In most affected embryos, defective *Alk2* expression causes hypoplasia instead of loss of the posterior IC, forming a TAV with a very small non-coronary leaflet and sinus. In this model, BAVs seem not to result from the coalescence of the parietal CR and the posterior IC, but from the underdevelopment of the latter. However, in the study of Thomas et al. [84], the higher affection of the posterior IC compared to other OFT cushions is not justified.

The morphogenetic defects leading to BAV described until now occur early in valvulogenesis. However, several conditional mutant mouse models have shown that a BAV may be the consequence of defects during late valvulogenesis, particularly during the excavation of the semilunar valves (Fig. 2.18) [74, 75, 85, 119–121]. Interestingly, excavation defects in these models result in latero-lateral and anteroposterior BAVs that show a variety of intermediate morphologies (Fig. 2.18), as opposed to the other mutant mouse models whose BAVs show a single morphology, i.e., symmetrical BAV without raphes or commissural fusions (Fig. 2.6) [47]. Although the excavation of the semilunar valves is a poorly understood embryonic process of maturation, it is known that CNC and SHF-derived mesenchymal cell proliferation, apoptosis, differentiation, and extracellular matrix secretion are the main cellular processes involved [70, 73–75]. In addition, several studies indicate that CNC-derived mesenchymal cells regulate some of the cellular mechanisms involved in cushion remodeling [65, 71, 72, 74].

KROX20 deficient mice develop latero-lateral BAVs with a variety of intermediate phenotypes [85]. In *Krox20* mutant embryos, Odelin et al. [85] identified defective maturation of aortic valve cushions after normal endocardial cushion formation and normal conotruncal septation. Absence of KROX20 caused alterations in the number and distribution of CNC cells in the aortic valve cushions, especially at the hinges, the anchorage

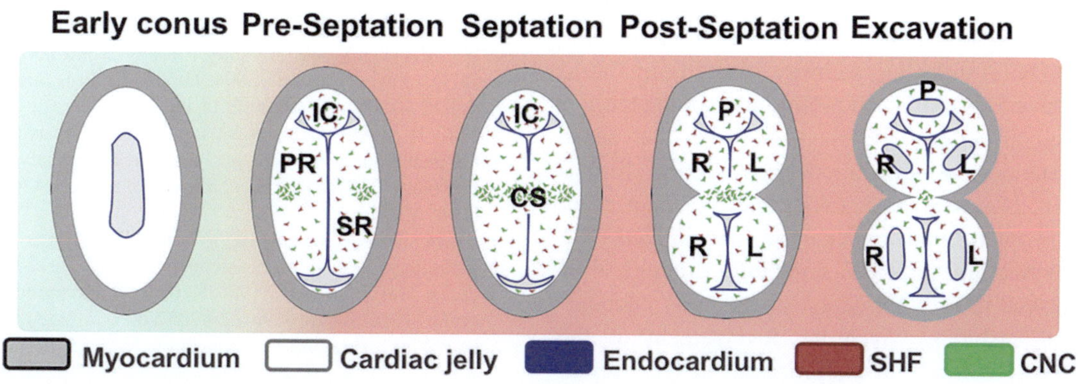

Fig. 2.17 Schematic representation of valvulogenesis with abnormal intercalated cushion formation in transversal view. The alteration of cells derived from the secondary heart field (SHF), specifically in the posterior intercalated cushion (P), may give rise to its absence, resulting in a latero-lateral bicuspid aortic valve. CS, contruncal septum; IC, intercalated valve cushion; L, left valve cushion; PR, parietal conotruncal ridge; R, right valve cushion; SR, septal conotruncal ridge. (Reproduced from Soto-Navarrete et al. 2020 [48])

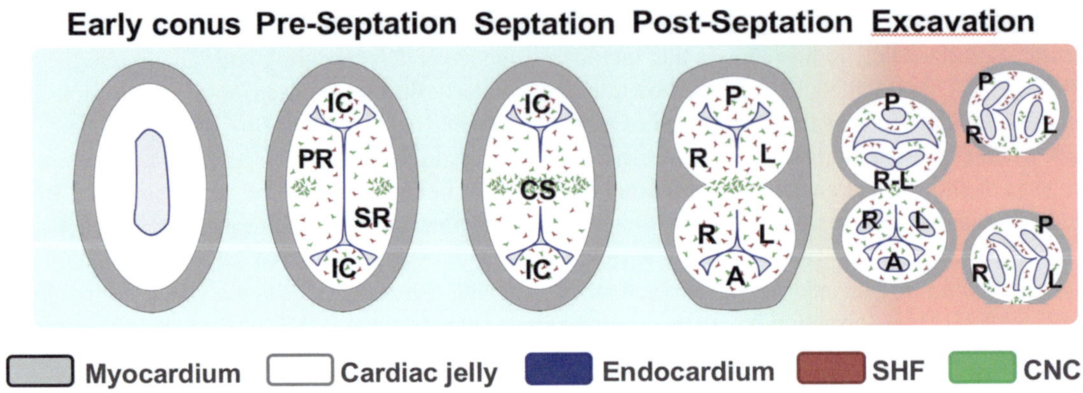

Fig. 2.18 Schematic representation of valvulogenesis with abnormal excavation in transversal view. The alteration of excavation usually affects the leaflet hinges, leading to the anatomical fusion of adjacent valve leaflets and both latero-lateral or antero-posterior BAVs, depending on the affected valve cushions, with a variety of intermediate phenotypes. It occurs due to incorrect interactions between cardiac neural crest (CNC) and second heart field (SHF) mesenchymal cells, which induce defective cell proliferation and/or matrix secretion. A, anterior valve cushion; CS, conotruncal septum; IC, intercalated valve cushion; L, left valve cushion; P, posterior valve cushion; PR, parietal conotruncal ridge; R, right valve cushion; SR, septal conotruncal ridge. (Reproduced from Soto-Navarrete et al. 2020 [48])

regions of the developing leaflets to the wall of the OFT. This was associated with a defective production of extracellular matrix in these particular areas during the excavation period, giving rise to partial anatomical fusions of the developing leaflets consistent with intermediate phenotypes, including raphes, of latero-lateral BAVs.

A mechanism of BAV formation similar to that proposed by Odelin et al. [85] in KROX20 deficient mice might operate in *Pax3* [74] and *Notch1* [75] conditional mutant mice. In *Pax3* mutants, the genetic manipulation affects the NOTCH1 signaling pathway, resulting in alterations of apoptosis and proliferation of the mesenchymal cells in the valve cushions that cause abnormal cushion remodeling and BAV formation. However, in *Notch1* mutants, Wang et al. [75] did not study in detail the effect of NOTCH1 deletion during OFT septation, although they showed normal formation of the early endocardial cushions. Given the involvement of the *Notch1* pathway in the processes of OFT septation [75, 92], excessive fusion of the CRs leading to anteroposterior BAV, as in the hamster model, cannot be excluded [75, 95].

References

1. Wauchope GM. The clinical importance of variations in the number of cusps forming the aortic and pulmonary valves. Q J Med. 1928;21:383–99.
2. Roberts WC. The congenitally bicuspid aortic valve. A study of 85 autopsy cases. Am J Cardiol. 1970;26:72–83. https://doi.org/10.1016/0002-9149(70)90761-7.
3. Roberts WC. Living with a congenitally bicuspid aortic valve. Am J Cardiol. 1989;64:1408–9. https://doi.org/10.1016/0002-9149(89)90598-5.
4. Basso C, Boschello M, Perrone C, Mecenero A, Cera A, Bicego D, et al. An echocardiographic survey of primary school children for bicuspid aortic valve. Am J Cardiol. 2004;93:661–3. https://doi.org/10.1016/j.amjcard.2003.11.031.
5. Mills P, Leech G, Davies M, Leatham A. The natural history of a non-stenotic bicuspid aortic valve. Br Heart J. 1978;40:951–7. https://doi.org/10.1136/hrt.40.9.951.
6. Subramanian R, Olson LJ, Edwards WD. Surgical pathology of pure aortic stenosis: a study of 374 cases. Mayo Clin Proc. 1984;59:683–90. https://doi.org/10.1016/S0025-6196(12)62057-6.
7. Durán AC, Frescura C, Sans-Coma V, Angelini A, Basso C, Thiene G. Bicuspid aortic valves in hearts with other congenital heart disease. J Heart Valve Dis. 1995;4:581–90.
8. Fedak PWM, Verma S, David TE, Leask RL, Weisel RD, Butany J. Clinical and pathophysiological

implications of a bicuspid aortic valve. Circulation. 2002;106:900–4. https://doi.org/10.1161/01.CIR.0000027905.26586.E8.
9. Siu SC, Silversides CK. Bicuspid aortic valve disease. J Am Coll Cardiol. 2010;55:2789–800. https://www.jacc.org/doi/full/10.1016/j.jacc.2009.12.068.
10. Niaz T, Poterucha JT, Olson TM, Johnson JN, Craviari C, Nienaber T, et al. Characteristic morphologies of the bicuspid aortic valve in patients with genetic syndromes. J Am Soc Echocardiograph. 2018;31:194–200. https://doi.org/10.1016/j.echo.2017.10.008.
11. Nistri S, Sorbo MD, Marin M, Palisi M, Scognamiglio R, Thiene G. Aortic root dilatation in young men with normally functioning bicuspid aortic valves. Heart. 1999;82:19–22. https://doi.org/10.1136/hrt.82.1.19.
12. Michelena HI, Khanna AD, Mahoney D, Margaryan E, Topilsky Y, Suri RM, et al. Incidence of aortic complications in patients with bicuspid aortic valves. JAMA Netw Open. 2011;306:1104–12. https://doi.org/10.1001/jama.2011.1286.
13. Braveman AC, Roman MJ. Bicuspid aortic valve in Marfan syndrome: the root of the problem. Circ Cardiovasc Imaging. 2019;12(3):e008860. https://doi.org/10.1161/CIRCIMAGING.119.00886.
14. Thiene G, Rizzo S, Basso C. Bicuspid aortic valve: the most frequent and not so benign congenital heart disease. Cardiovasc Pathol. 2024;70:107604. https://doi.org/10.1016/j.carpath.2024.107604.
15. Sabet HY, Edwards WD, Tazelaar HD, Daly RC. Congenitally bicuspid aortic valves: a surgical pathology study of 542 cases (1991 through 1996) and a literature review of 2,715 additional cases. Mayo Clin Proc. 1999;74:14–26. https://doi.org/10.4065/74.1.14.
16. Thubrikar M. The aortic valve. 1st ed. Florida: CRC Press; 1990.
17. Michelena HI, Maleszewski JJ, Della Corte A, Otto CM, Schäfers HJ. The international bicuspid aortic valve consensus statement: enduring the test of time. Ann Thorac Surg. 2023;2023(116):1347–8. https://www.annalsthoracicsurgery.org/article/S0003-4975(23)00850-0/abstract.
18. Osler W. The bicuspid condition of the aortic valves. Trans Assoc Am Physicians. 1886;1:185–92.
19. De Vries WM. Ueber Abweichungen in der Zahl der Semilunarklappen. Beitr Pathol Anat. 1918;64:39–54.
20. Simonds JP. Congenital malformations of the aortic and pulmonary valves. Am J Med Sci. 1923;166:585–95.
21. Koletsky S. Congenital bicuspid pulmonary valves. Arch Pathol. 1941;31:338–53.
22. Moore GM, Hutchins GM, Brito JC, Kang H. Congenital malformations of the semilunar valves. Hum Pathol. 1980;11:367–72. https://doi.org/10.1016/S0046-8177(80)80033-5.
23. Angelini A, Ho SY, Anderson RH, Devine WA, Zuberbuhler JR, Becker AE, et al. The morphology of the normal aortic valve as compared with the aortic valve having two leaflets. J Thorac Cardiovasc Surg. 1989;98:362–7. https://doi.org/10.1016/S0022-5223(19)34382-X.
24. Losenno KL, Goodman RL, Chu MW. Bicuspid aortic valve disease and ascending aortic aneurysms: gaps in knowledge. Cardiol Res Pract. 2012;2012:145202. https://doi.org/10.1155/2012/145202.
25. Shaner RF. Abnormal pulmonary and aortic semilunar valves in embryos. Anat Rec. 1963;143:5–13. https://doi.org/10.1002/ar.1091470103.
26. Colvee E, Hurle JM. Malformations of the semilunar valves produced in chick embryos by mechanical interference with cardiogenesis. Anat Embryol. 1983;168:59–71. https://doi.org/10.1007/BF00305399.
27. Pérez de Miguelsanz J, Puerta AJ, Peña AL, Prados JC, Viejo F. Alteración de los arcos aórticos en embriones de pollo como consecuencia de factores hemodinámicos modificados. Rev Esp Cardiol. 1989;42:394–8.
28. Kirby ML, Gale TF, Stewart DE. Neural crest cells contribute to normal aorticopulmonary septation. Science. 1983;220:1059–61. https://doi.org/10.1126/science.6844926.
29. Bockman D, Redmond M, Waldo K, Davis H, Kirby ML. Effect of neural crest ablation on development of the heart and arch arteries in the chick. Am J Anat. 1987;180:433–43. https://doi.org/10.1002/aja.1001800403.
30. Nishibatake M, Kirby ML, Van Mierop LHS. Pathogenesis of persistent truncus arteriosus and dextroposed aorta in the chick embryo after neural crest ablation. Circulation. 1987;75:225–64. https://doi.org/10.1161/01.CIR.75.1.255.
31. Sumida H, Akimoto N, Nakamura H. Distribution of the neural crest cells in the heart of birds: a three dimensional analysis. Anat Embryol. 1989;180:29–35. https://doi.org/10.1007/BF00321897.
32. Takamura K, Okishima T, Ohdo S, Hayakawa K. Association of cephalic neural crest cells with cardiovascular development, particularly that of the semilunar valves. Anat Embryol. 1990;182:263–72. https://doi.org/10.1007/BF00185519.
33. Kirby ML. Cellular and molecular contributions of the neural crest to cardiovascular development. Trends Cardiovasc Med. 1993;3:18–23. https://doi.org/10.1016/1050-1738(93)90023-Y.
34. Kappetein AP, Gittenberger-de Groot AC, Zwinderman AH, Rohmer J, Poelmann RE, Huysmans HA. The neural crest as a possible pathogenetic factor in coarctation of the aorta and bicuspid aortic valve. J Thorac Cardiovasc Surg. 1991;102:830–6. https://doi.org/10.1016/S0022-5223(19)33931-5.
35. Arqué JM, Cruz V, Rosado LM, Sans-Coma V. Congenital anomalies of coronary arteries in rodents. Am J Cardiol. 1986;57:498–9. https://doi.org/10.1016/0002-9149(86)90789-7.

36. Durán AC, Arqué JM, Cardo M, Sans-Coma V. Descripción morfológica de una nueva anomalía arterial coronaria en el hamster sirio *Mesocricetus auratus* (Waterhouse, 1839); interpretación genética y filogenética. Misc Zool. 1988;12:329–38.
37. Sans-Coma V, Arqué JM, Durán AC, Cardo M, Fernández B. Coronary artery anomalies and bicuspid aortic valves in the Syrian hamster. Basic Res Cardiol. 1991;86:148–53. https://doi.org/10.1007/BF02190547.
38. Sans-Coma V, Cardo M, Durán AC, Franco D, Fernández B, Arqué JM. Evidence for a quantitative genetic influence on the formation of aortic valves with two leaflets in the Syrian hamster. Cardiol Young. 1993;3:132–40. https://doi.org/10.1017/S1047951100001396.
39. Sans-Coma V, Fernández B, Durán AC, Thiene G, Arqué JM, Muñoz-Chápuli R, et al. Fusion of valve cushions as a key factor in the formation of congenital bicuspid aortic valves in Syrian hamsters. Anat Rec. 1996;244:490–8. https://doi.org/10.1002/(SICI)1097-0185(199604)244:4<490::AID--AR7>3.0.CO;2-Z.
40. Michelena HI, Della Corte A, Evangelista A, Maleszewski JJ, Edwards WD, Roman MJ, et al. International consensus statement on nomenclature and classification of the congenital bicuspid aortic valve and its aortopathy, for clinical, surgical, interventional and research purposes. Eur J Cardiothorac Surg. 2021;60:448–76. https://doi.org/10.1093/ejcts/ezab038.
41. Emanuel R, Withers R, O'Brien K, Ross P, Feizi O. Congenitally bicuspid aortic valves. Clinicogenetic study of 41 families. Br Heart J. 1978;40:1402–7. https://doi.org/10.1136/hrt.40.12.1402.
42. Godden DJ, Sandhu PS, Kerr F. Stenosed bicuspid aortic valves in twins. Eur Heart J. 1987;8:316–8. https://doi.org/10.1093/oxfordjournals.eurheartj.a062276.
43. Sans-Coma V, Fernández MC, Fernández B, Durán AC, Anderson RH, Arqué JM. Genetically alike Syrian hamsters display both bifoliate and trifoliate aortic valves. J Anat. 2012;220:92–101. https://doi.org/10.1111/j.1469-7580.2011.01440.x.
44. Lee C, Zhao YD, Courtman DW, Stewart DJ. Abnormal aortic valve development in mice lacking endothelial nitric oxide synthase. Circulation. 2000;101:2345–8. https://doi.org/10.1161/01.CIR.101.20.2345.
45. Fernández B, Durán AC, Fernández-Gallego T, Fernández MC, Such M, Arqué JM, et al. Bicuspid aortic valves with different spatial orientations of the leaflets are distinct etiological entities. J Am Coll Cardiol. 2009;54:2312–8. https://doi.org/10.1016/j.jacc.2009.07.044.
46. Evangelista A, Gallego P, Calvo-Iglesias F, Bermejo J, Robledo-Carmona J, Sánchez V, et al. Anatomical and clinical predictors of valve dysfunction and aortic dilation in bicuspid aortic valve disease. Heart. 2018;104:566–73. https://doi.org/10.1136/heartjnl-2017-311560.
47. Fernández B, Soto-Navarrete MT, López-García A, López-Unzu MA, Durán AC, Fernández MC. Bicuspid aortic valve in 2 model species and review of the literature. Vet Pathol. 2020;57:321–31. https://doi.org/10.1177/0300985819900018.
48. Soto-Navarrete MT, López-Unzu MÁ, Durán AC, Fernández B. Embryonic development of bicuspid aortic valves. Prog Cardiovasc Dis. 2020;63:407–18. https://doi.org/10.1016/j.pcad.2020.06.008.
49. O'Donnell A, Yutzey KE. Mechanisms of heart valve development and disease. Development. 2020;147:dev183020. https://doi.org/10.1242/dev.183020.
50. Henderson DJ, Eley L, Chaudhry B. New concepts in the development and malformation of the arterial valves. J Cardiovasc Dev Dis. 2020;7:38. https://doi.org/10.3390/jcdd7040038.
51. Henderson DJ, Eley L, Turner JE, Chaudhry B. Development of the human arterial valves: understanding bicuspid aortic valve. Front Cardiovasc Med. 2022;8:802930. https://doi.org/10.3389/fcvm.2021.802930.
52. Meilhac SM, Buckingham ME. The deployment of cell lineages that form the mammalian heart. Nat Rev Cardiol. 2018;15:705–24. https://doi.org/10.1038/s41569-018-0086-9.
53. Kelly RG, Buckingham ME, Moorman AF. Heart fields and cardiac morphogenesis. Cold Spring Harb Perspect Med. 2014;4:a015750. https://doi.org/10.1101/cshperspect.a015750.
54. Gittenberger-de Groot AC, Bartelings MM, DeRuiter MC, Poelmann RE. Basics of cardiac development for the understanding of congenital heart malformations. Pediatr Res. 2005;57:169–76. https://doi.org/10.1203/01.PDR.0000148710.69159.61.
55. Rana MS, Sizarov A, Chistoffels V, Moorman AFM. Development of the human aortic arch system captured in an interactive three-dimensional reference model. Am J Med Genet A. 2014;164:1372–83. https://doi.org/10.1002/ajmg.a.35881.
56. Davis CL. The cardiac jelly of the chick embryo. Anat Rec. 1924;27:201–2.
57. Markwald RR, Fitzharris TP, Manasek FJ. Structural development of endocardial cushions. Am J Anat. 1977;148:85–119. https://doi.org/10.1002/aja.1001480108.
58. Waldo KL, Miyagawa-Tomita S, Kumiski D, Kirby ML. Cardiac neural crest cells provide new insight into septation of the outflow tract: aortic sac to ventricular septal closure. Dev Biol. 1998;196:129–44. https://doi.org/10.1006/dbio.1998.8860.
59. Waldo KL, Lo CW, Kirby ML. Connexin 43 expression reflects neural crest patterns during cardiovascular development. Dev Biol. 1999a;208:307–23. https://doi.org/10.1006/dbio.1999.9219.
60. Waldo K, Zdanowicz M, Burch J, Kumiski DH, Stadt HA, Godt RE, et al. A novel role for cardiac

neural crest in heart development. J Clin Invest. 1999b;103:1499–507. https://doi.org/10.1172/JCI6501.
61. Monaghan MG, Linneweh M, Liebscher S, Van Handel B, Layland SL, Schenke-Layland K. Endocardial-to-mesenchymal transformation and mesenchymal cell colonization at the onset of human cardiac valve development. Development. 2016;143:473–82. https://doi.org/10.1242/dev.133843.
62. Anderson RH, Webb S, Brown NA, Lamers W, Moorman A. Development of the heart: (3) formation of the ventricular outflow tracts, arterial valves, and intrapericardial arterial trunks. Heart. 2003;89:1110–8. https://doi.org/10.1136/heart.89.9.1110.
63. Anderson RH, Chaudhry B, Mohun TJ, Bamforth SD, Hoyland D, Phillips HM, et al. Normal and abnormal development of the intrapericardial arterial trunks in humans and mice. Cardiovasc Res. 2012;95:108–15. https://doi.org/10.1093/cvr/cvs147.
64. Spicer DE, Bridgeman JM, Brown NA, Mohun TJ, Anderson RH. The anatomy and development of the cardiac valves. Cardiol Young. 2014;24:1008–22. https://doi.org/10.1017/S1047951114001942.
65. Schussler O, Gharibeh L, Mootoosamy P, Murith N, Tien V, Rougemont AL, et al. Cardiac neural crest cells: their rhombomeric specification, migration, and association with heart and great vessel anomalies. Cell Mol Neurobiol. 2021;41:403–29. https://doi.org/10.1007/s10571-020-00863-w.
66. Plein A, Calmont A, Fantin A, Denti L, Anderson NA, Scambler PJ, et al. Neural crest–derived SEMA3C activates endothelial NRP1 for cardiac outflow tract septation. J Clin Invest. 2015;125:2661–76. https://doi.org/10.1172/JCI79668.
67. Soto-Navarrete MT, Pozo-Vilumbrales B, López-Unzu MA, Martin-Chaves L, Durán AC, Fernández B. Endocardial-to-mesenchymal transition underlies cardiac outflow tract septation and bicuspid aortic valve formation in the Syrian hamster model. Sci Rep. 2025;15:8583. https://doi.org/10.1038/s41598-025-91454-6
68. Kramer TC. The partitioning of the truncus and conus and the formation of the membranous portion of the interventricular septum in the human heart. Am J Anat. 1942;71:343–70. https://doi.org/10.1002/aja.1000710303.
69. Peterson JC, Chughtai M, Wisse LJ, Gittenberger-de Groot AC, Feng Q, Goumans MJT, et al. Bicuspid aortic valve formation: Nos3 mutation leads to abnormal lineage patterning of neural crest cells and the second heart field. Dis Mod Mech. 2018;11:mm034637. https://doi.org/10.1242/dmm.034637.
70. Hurlé JM, Colvee E, Blanco AM. Development of mouse semilunar valves. Anat Embryol. 1980;160:83–91. https://doi.org/10.1007/BF00315651.
71. Aikawa E, Whittaker P, Farber M, Mendelson K, Padera RF, Aikawa M, et al. Human semilunar cardiac valve remodeling by activated cells from fetus to adult: implications for postnatal adaptation, pathology, and tissue engineering. Circulation. 2006;113:1344–52. https://doi.org/10.1161/CIRCULATIONAHA.105.591768.
72. Hinton RB, Yutzey KE. Heart valve structure and function in development and disease. Annu Rev Physiol. 2011;73:29–46. https://doi.org/10.1146/annurev-physiol-012110-142145.
73. Hurlé JM, Colvee E. Changes in the endothelial morphology of the developing semilunar heart valves. Anat Embryol. 1983;1983(167):67–83. https://doi.org/10.1007/BF00304601.
74. Jain R, Engleka KA, Rentschler SL, Manderfield LJ, Li L, Yuan L, et al. Cardiac neural crest orchestrates remodeling and functional maturation of mouse semilunar valves. J Clin Invest. 2011;121:422–30. https://doi.org/10.1172/JCI44244.
75. Wang Y, Wu B, Farrar E, Lui W, Lu P, Zhang D, et al. Notch-Tnf signalling is required for development and homeostasis of arterial valves. Eur Heart J. 2017;38:675–86. https://doi.org/10.1093/eurheartj/ehv520.
76. Fernández MC, Durán AC, Real R, López D, Fernández B, de Andrés AV, et al. Coronary artery anomalies and aortic valve morphology in the Syrian hamster. Lab Anim. 2000;34:145–54. https://doi.org/10.1258/002367700780457545.
77. Fernández B, Durán AC, Fernández MC, Fernández-Gallego T, Icardo JM, Sans-Coma V. The coronary arteries of the C57BL/6 mouse strains: implications for comparison with mutant models. J Anat. 2008;212:12–8. https://doi.org/10.1111/j.1469-7580.2007.00838.x.
78. Wang X, Astrof S. Neural crest cell-autonomous roles of fibronectin in cardiovascular development. Development. 2016;143:88–100. https://doi.org/10.1242/dev.125286.
79. Sans-Coma V, Pozo-Vilumbrales B, Fernández MC, López-Unzu MA, Soto-Navarrete MT, et al. The medical versus zoological concept of outflow tract valves of the vertebrate heart. J Cardiovasc Dev Dis. 2022;9:318. https://doi.org/10.3390/jcdd9100318.
80. Gill E, Bamforth SD. Molecular pathways and animal models of semilunar valve and aortic arch anomalies. In: Rickert-Sperling S, Kelly RG, Haas N, editors. Congenital heart diseases: the broken heart. Advances in experimental medicine and biology. Cham: Springer; 2024. p. 777–96. https://doi.org/10.1007/978-3-031-44087-8_46.
81. Biben C, Weber R, Kesteven S, Stanley E, McDonald L, Elliott DA, et al. Cardiac septal and valvular dysmorphogenesis in mice heterozygous for mutations in the homeobox gene Nkx2-5. Circ Res. 2000;87:888–95. https://doi.org/10.1161/01.RES.87.10.888.
82. Macatee TL, Hammond BP, Arenkiel BR, Francis L, Frank DU, Moon AM. Ablation of specific

expression domains reveals discrete functions of ectoderm- and endoderm-derived FGF8 during cardiovascular and pharyngeal development. Development. 2003;130:6361–74. https://doi.org/10.1242/dev.00850.
83. Akerberg BN, Sarangam ML, Stankunas K. Endocardial Brg1 disruption illustrates the developmental origins of semilunar valve disease. Dev Biol. 2015;407:158–72. https://doi.org/10.1016/j.ydbio.2015.06.015.
84. Thomas PS, Sridurongrit S, Ruiz-Lozano P, Kaartinen V. Deficient signaling via Alk2 (Acvr1) leads to bicuspid aortic valve development. PLoS One. 2012;7(4):e35539. https://doi.org/10.1371/journal.pone.0035539.
85. Odelin G, Faure E, Coulpier F, Di Bonito M, Bajolle F, Studer M, et al. Krox20 defines a subpopulation of cardiac neural crest cells contributing to arterial valves and bicuspid aortic valve. Development. 2018;145:dev151944. https://doi.org/10.1242/dev.151944.
86. Makki N, Capecchi MR. Cardiovascular defects in a mouse model of HOXA1 syndrome. Hum Mol Genet. 2012;21:26–31. https://doi.org/10.1093/hmg/ddr434.
87. Martínez-Vargas J, Ventura J, Machuca A, Muñoz-Muñoz F, Fernández MC, Soto-Navarrete MT, et al. Cardiac, mandibular and thymic phenotypical association indicates that cranial neural crest underlies bicuspid aortic valve formation in hamsters. PLoS One. 2017;12:e0183556. https://doi.org/10.1371/journal.pone.0183556.
88. Soto-Navarrete MT, Pozo-Vilumbrales B, López-Unzu MA, Rueda-Martínez C, Fernández MC, Durán AC, et al. Experimental evidence of the genetic hypothesis on the etiology of bicuspid aortic valve aortopathy in the hamster model. Front Cardiovasc Med. 2022;9:928362. https://doi.org/10.3389/fcvm.2022.928362.
89. Jiao J, Xiong W, Wang L, Yang J, Qiu P, Hirai H, et al. Differentiation defect in neural crest-derived smooth muscle cells in patients with aortopathy associated with bicuspid aortic valves. EBioMedicine. 2016;10:282–90. https://doi.org/10.1016/j.ebiom.2016.06.045.
90. Stankunas K, Shang C, Twu KY, Kao SC, Jenkins NA, Copeland NG, et al. Pbx/Meis deficiencies demonstrate multigenetic origins of congenital heart disease. Circ Res. 2008;103:702–9. https://doi.org/10.1161/CIRCRESAHA.108.175489.
91. Phillips HM, Mahendran P, Singh E, Anderson RH, Chaudhry B, Henderson DJ. Neural crest cells are required for correct positioning of the developing outflow cushions and pattern the arterial valve leaflets. Cardiovasc Res. 2013;99:452–60. https://doi.org/10.1093/cvr/cvt132.
92. MacGrogan D, D'Amato G, Travisano S, Martínez-Poveda B, Luxán G, del Monte-Nieto G, et al. Sequential ligand-dependent notch signaling activation regulates valve primordium formation and morphogenesis. Circ Res. 2016;118:1480–97. https://doi.org/10.1161/CIRCRESAHA.115.308077.
93. High FA, Zhang M, Proweller A, Tu L, Parmacek MS, Pear WS, et al. An essential role for Notch in neural crest during cardiovascular development and smooth muscle differentiation. J Clin Invest. 2007;117:353–63. https://doi.org/10.1172/JCI30070.
94. High FA, Jain R, Stoller JZ, Antonucci NB, Lu MM, Loomes KM, et al. Murine Jagged1/Notch signaling in the second heart field orchestrates Fgf8 expression and tissue-tissue interactions during outflow tract development. J Clin Invest. 2009;2009(119):1986–96. https://doi.org/10.1172/JCI38922.
95. Fernández B, Durán AC. Comment on: Wang et al. Notch-Tnf signalling is required for development and homeostasis of arterial valves. Eur Heart J. 2016;38:675–86. https://academic.oup.com/eurheartj/article/38/9/675/3056887#usercomments.
96. Torregrosa-Carrión R, Luna-Zurita L, García-Marques F, D'Amato G, Piñeiro-Sabarís R, Bonzon-Kulichenko E, et al. NOTCH activation promotes valve formation by regulating the endocardial secretome. Mol Cell Proteomics. 2019;18:1782–95. https://doi.org/10.1074/mcp.RA119.001492.
97. O'Riordan E, Mendelev N, Patschan S, Patschan D, Eskander J, Cohen-Gould L, et al. Chronic NOS inhibition activates endothelial-mesenchymal transformation. Am J Physiol Heart Circ Physiol. 2007;292:H285–94. https://doi.org/10.1152/ajpheart.00560.2006.
98. Liu Y, Feng Q. NOing the heart: role of nitric oxide synthase-3 in heart development. Differentiation. 2012;84:54–61. https://doi.org/10.1016/j.diff.2012.04.004.
99. Chang AC, Fu Y, Garside VC, Niessen K, Chang L, Fuller M, et al. Notch initiates the endothelial-to-mesenchymal transition in the atrioventricular canal through autocrine activation of soluble guanylyl cyclase. Dev Cell. 2011;21:288–300. https://doi.org/10.1016/j.devcel.2011.06.022.
100. Timmerman LA, Grego-Bessa J, Raya A, Bertrán E, Pérez-Pomares JM, Díez J, et al. Notch promotes epithelial-mesenchymal transition during cardiac development and oncogenic transformation. Genes Dev. 2004;2004(18):99–115. https://doi.org/10.1101/gad.276304.
101. Laforest B, Andelfinger G, Nemer M. Loss of Gata5 in mice leads to bicuspid aortic valve. J Clin Invest. 2011;121:2876–87. https://doi.org/10.1172/JCI44555.
102. Blaser MC, Wei K, Adams RLE, Zhou YQ, Caruso LL, Mirzaei Z, et al. Deficiency of natriuretic peptide receptor 2 promotes bicuspid aortic valves, aortic valve disease, left ventricular dysfunction,

and ascending aortic dilatations in mice. Circ Res. 2018;122:405–16. https://doi.org/10.1161/CIRCRESAHA.117.311194.

103. McBride KL, Riley MF, Zender GA, Fitzgerald-Butt SM, Towbin JA, Belmont JW, et al. NOTCH1 mutations in individuals with left ventricular outflow tract malformations reduce ligand-induced signaling. Hum Mol Genet. 2008;17(18):2886–93. https://doi.org/10.1093/hmg/ddn187.
104. Mommersteeg MTM, Yeh ML, Parnavelas JG, Andrews WD. Disrupted Slit-Robo signalling results in membranous ventricular septum defects and bicuspid aortic valves. Cardiovasc Res. 2015;106:55–66. https://doi.org/10.1093/cvr/cvv040.
105. Siguero-Álvarez M, Salguero-Jiménez A, Grego-Bessa J, de la Barrera J, MacGrogan D, Prados B, et al. A human hereditary cardiomyopathy shares a genetic substrate with bicuspid aortic valve. Circulation. 2023;147:47–65. https://doi.org/10.1161/CIRCULATIONAHA.121.058767.
106. Messner B, Bernhard D. Bicuspid aortic valve-associated aortopathy: where do we stand? J Mol Cell Cardiol. 2019;133:76–85. https://doi.org/10.1016/j.yjmcc.2019.05.023.
107. Bravo-Jaimes K, Prakash SK. Genetics in bicuspid aortic valve disease: where are we? Prog Cardiovasc Dis. 2020;63:398–406. https://doi.org/10.1016/j.pcad.2020.06.005.
108. Carlson RG, Lillehei CW, Edwards JE. Cystic medial necrosis of the ascending aorta in relation to age and hypertension. Am J Cardiol. 1970;25:411–5. https://doi.org/10.1016/0002-9149(70)90006-8.
109. Tadros TM, Klein MD, Shapira OM. Ascending aortic dilatation associated with bicuspid aortic valve: pathophysiology, molecular biology, and clinical implications. Circulation. 2009;119:880–90. https://doi.org/10.1161/CIRCULATIONAHA.108.795401.
110. Halushka MK, Angelini A, Bartoloni G, Basso C, Batoroeva L, Bruneval P, et al. Consensus statement on surgical pathology of the aorta from the Society for Cardiovascular Pathology and the Association for European Cardiovascular Pathology: II. Noninflammatory degenerative diseases—nomenclature and diagnostic criteria. Cardiovasc Pathol. 2016;25:247–57. https://doi.org/10.1016/j.carpath.2016.03.002.
111. Freiholtz D, Bergman O, Lång K, Poujade FA, Paloschi V, Granath C, et al. Bicuspid aortic valve aortopathy is characterized by embryonic epithelial to mesenchymal transition and endothelial instability. J Mol Med. 2023;101:801–11. https://doi.org/10.1007/s00109-023-02316-5.
112. Hinton RB. Bicuspid aortic valve and thoracic aortic aneurysm: three patient populations, two disease phenotypes, and one shared genotype. Cardiol Res Pract. 2012;2012:926975. https://doi.org/10.1155/2012/926975.
113. Padang R, Bannon PG, Jeremy R, Richmond DR, Semsarian C, Vallely M, et al. The genetic and molecular basis of bicuspid aortic valve associated thoracic aortopathy: a link to phenotype heterogeneity. Ann Cardiothorac Surg. 2013;2:83–91. https://doi.org/10.3978/j.issn.2225-319X.2012.11.17.
114. Prakash SK, Bossé Y, Muehlschlegel JD. A roadmap to investigate the genetic basis of bicuspid aortic valve and its complications: insights from the international BAVCon (Bicuspid Aortic Valve Consortium). J Am Coll Cardiol. 2014;64:832–9. https://doi.org/10.1016/j.jacc.2014.04.073.
115. Cheung C, Bernardo AS, Trotter MW, Pedersen RA, Sinha S. Generation of human vascular smooth muscle subtypes provides insight into embryological origin-dependent disease susceptibility. Nat Biotechnol. 2012;30:165–73. https://doi.org/10.1038/nbt.2107.
116. Pfaltzgraff ER, Shelton EL, Galindo CL, Nelms BL, Hooper CW, Poole SD, et al. Embryonic domains of the aorta derived from diverse origins exhibit distinct properties that converge into a common phenotype in the adult. J Mol Cell Cardiol. 2014;69:88–96. https://doi.org/10.1016/j.yjmcc.2014.01.016.
117. Satoh K, Fukumoto Y, Shimokawa H. Rho-kinase: important new therapeutic target in cardiovascular diseases. Am J Physiol Heart Circ Physiol. 2011;301:H287–96. https://doi.org/10.1152/ajpheart.00327.2011.
118. Moens CB, Selleri L. Hox cofactors in vertebrate development. Dev Biol. 2006;291:193–206. https://doi.org/10.1016/j.ydbio.2005.10.032.
119. Dupuis LE, Osinska H, Weinstein MB, Hinton RB, Kern CB. Insufficient versican cleavage and Smad2 phosphorylation results in bicuspid aortic and pulmonary valves. J Mol Cell Cardiol. 2013;60:50–9. https://doi.org/10.1016/j.yjmcc.2013.03.010.
120. Toomer KA, Fulmer D, Guo L, Drohan A, Peterson N, Swanson P, et al. A role for primary cilia in aortic valve development and disease. Dev Dyn. 2017;246:625–34. https://doi.org/10.1002/dvdy.24524.
121. Gharibeh L, Komati H, Bossé Y, Boodhwani M, Heydarpour M, Fortier M, et al. GATA6 regulates aortic valve remodeling, and its haploinsufficiency leads to right-left type bicuspid aortic valve. Circulation. 2018;138:1025–38. https://doi.org/10.1161/CIRCULATIONAHA.117.029506.

Congenital Heart Disease Associated with Bicuspid Aortic Valve

3

Carla Frescura, Ana Duran, and Gaetano Thiene

Among the congenital malformations affecting the semilunar valves, the bicuspid aortic valve (BAV) is the most common anomaly [1–5]. In the youngs, BAV is asymptomatic but its natural history is complicated by valvular stenosis due to dystrophic calcification, by dilatation of sinusal aorta, valvular incompetence, infective endocarditis, and aortophaty at risk of dissection with sudden death [6, 7].

In the majority of cases, BAV is present as isolated malformation but in a few cases, it may be associated to other congenital cardiac malformations [8].

Rarely BAV may be observed in hearts with anomalous atrial situs or anomalous atrioventricular (AV) or ventriculo arterial (VA) connections. In our Anatomical Collection of Congenital Heart Disease (CHD), consisting in 1860 specimens, 98 BAVs were associated to other congenital malformations. BAV was associated to anomalies of the AV connection in 11 hearts (absent left AV connection in 7, double inlet left ventricle in 4) (Figs. 3.1 and 3.2) and to anomalies of VA connection in 1 (complete transposition of the great arteries) (Table 3.1).

Most frequently, BAV associated with other congenital cardiac malformations is found in hearts with normal atrial situs and normal connections of the cardiac segments (Table 3.1).

Obstruction of the aortic arch and ventricular septal defect are the most frequent associated malformations (Figs. 3.3, 3.4, 3.5, 3.6 and 3.7) (Table 3.1). BAV may be present also in hearts with AV septal defect (Fig. 3.8), congenital aortic valve stenosis (Fig. 3.9), hypoplastic left ventricle, and tetralogy of Fallot (Fig. 3.10) (Table 3.1).

Aortic arch coarctation may be present with intact ventricular septum (Fig. 3.3) or with ventricular septal defect (Figs. 3.4 and 3.5). In the presence of ventricular septal defect, interruption of the aortic arch may be found [9–11] (Fig. 3.6).

Aortic coarctation was present in all cases with anomalous AV or VA connections, aortic valve stenosis and hypoplastic left ventricle (Table 3.1).

Obstructive anomalies of the aortic arch represent the most frequent congenital malformation associated to bicuspid aortic valve (70% of cases in our Anatomical Collection) [9, 10].

The ventricular septal defect in hearts with bicuspid aortic valve may be perimembranous or muscular or multiple (Fig. 3.7). The

C. Frescura (✉) · G. Thiene
Cardiovascular Pathology Unit, Department of Cardiac, Thoracic and Vascular Sciences and Public Health, School of Medicine, University of Padua, Padua, Italy
e-mail: carla.frescura@tin.it; gaetano.thiene@unipd.it

A. Duran
Departamento de Biología Animal, Facultad de Ciencias, Universidad de Málaga, Málaga, Spain

Instituto de Investigación Biomédica de Málaga (IBIMA-Plataforma BIONAND), Málaga, Spain
e-mail: acduran@uma.es

G. Thiene et al. (eds.), *Bicuspid Aortic Valve Syndrome*,
https://doi.org/10.1007/978-3-031-85589-4_3

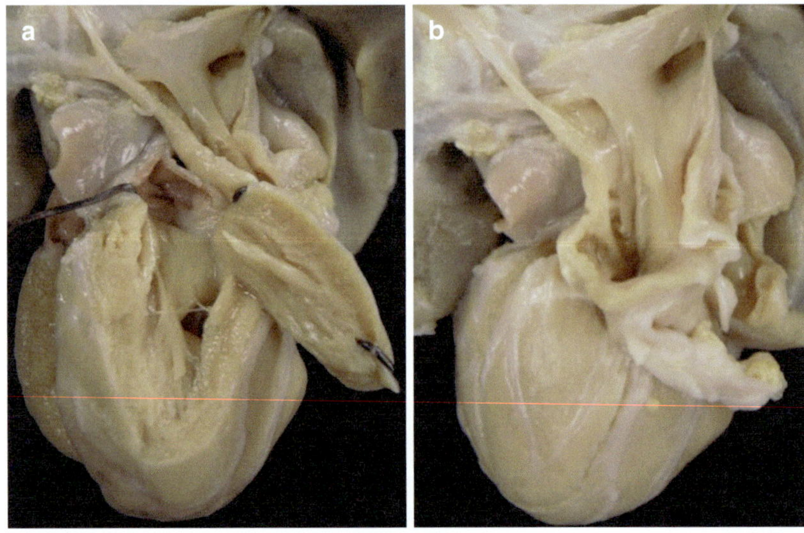

Fig. 3.1 Bicuspid aortic valve and mitral atresia. (**a**) View of the left ventricle and aorta: the left ventricle is hypoplastic and a muscular ventricular defect is present in the trabecular portion of the interventricular septum. The aorta takes origin from the left ventricle and a bicuspid aortic valve is present. (**b**) The left atrium is extremely hypoplastic and the mitral orifice is absent (mitral atresia). From Frescura C., Thiene G. Cardiopatie congenite. Testo Atlante, © 2025 Piccin Nuova Libraria S.p.A

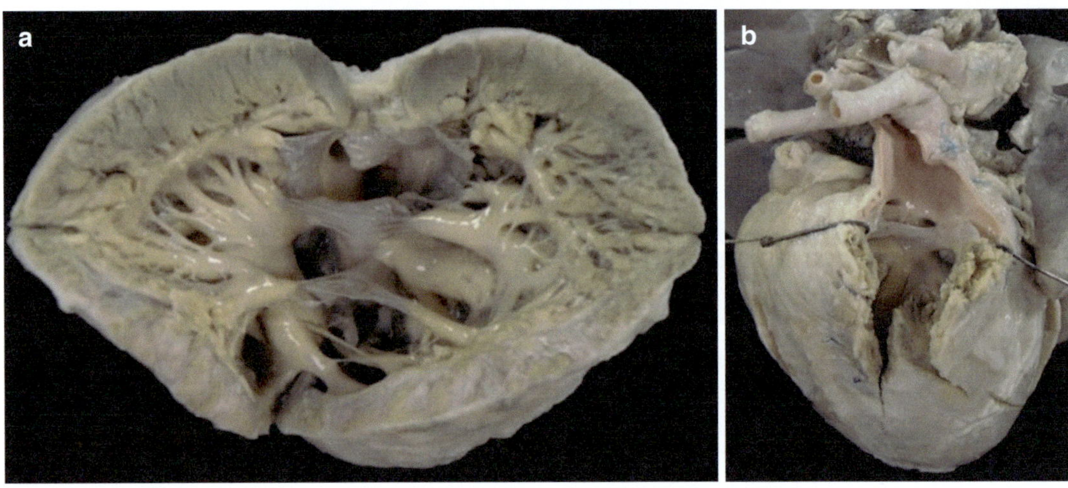

Fig. 3.2 Bicuspid aortic valve and double inlet left ventricle. (**a**) View from the apex of the dominant ventricle: both the atrioventricular valves drain into a ventricle of left morphology (double inlet left ventricle). (**b**) The aorta takes origin from a hypoplastic right ventricle (ventriculo arterial discordance). The aortic valve is bicuspid

Table 3.1 Bicuspid aortic valve in hearts with normal and abnormal sequence of the cardiac cavities (98 hearts)

Cardiac malformations		No cases	With aortic arch obstruction
Anomalous atrial situs		–	–
Anomalous connections	Anomalous atrioventricular connections (absent/double inlet)	11	11
	Anomalous ventriculo arterial connections (discordant)	1	1
Normal connections	Isolated ventricular septal defect	21	–
	Ventricular septal defect with aortic arch coarctation/interruption	32	32
	Isolated aortic coarctation	11	11
	Atrioventricular septal defect	5	3
	Infantile aortic valve stenosis	5	5
	Hypoplastic left ventricle	5	5
	Tetralogy of Fallot	3	–
	Others	4	–
Total		**98**	**68 (70%)**

Data from the Anatomical Collection of Congenital Heart Disease of the University of Padua (1860 hearts)

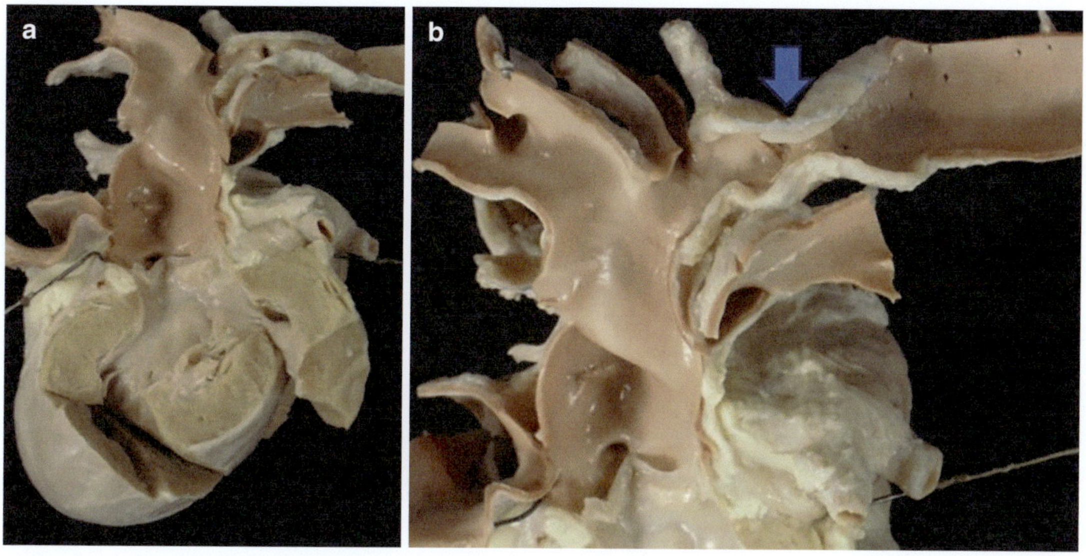

Fig. 3.3 Bicuspid aortic valve with aortic coarctation and intact septum. (**a**) Bicuspid aortic valve in a specimen with intact ventricular septum and aortic arch coarctation (arrow). Note the hypertrophy of the left ventricular free wall. (**b**) Close up of the same specimen showing the aortic coarctation. Note the dilatation of the descending aorta after the arch obstruction. From Frescura C., Thiene G. Cardiopatie congenite. Testo Atlante, © 2025 Piccin Nuova Libraria S.p.A

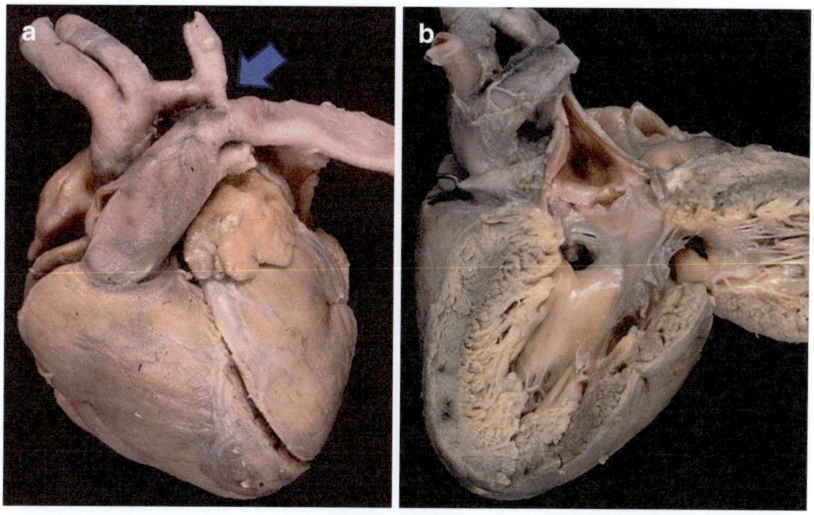

Fig. 3.4 Bicuspid aortic valve with aortic coarctation and ventricular septal defect. (**a**) External view of the heart showing a tubular hypoplasia of the aortic arch between the left carotid and left subclavian arteries. A preductal aortic coarctation is also present (arrow). (**b**) Left ventricular view showing a bicuspid aortic valve with a perimembranous ventricular septal defect

perimembranous defect is most frequently found, both in cases with isolated interventricular defect and in cases with associated aortic coarctation [11] (Table 3.2).

BAV may be also part of a polyvalvular disease [12]. In hearts with associated congenital cardiac anomalies, BAV may show an antero-posterior or side-by-side right-left orientation of the leaflets, with a predominant right-left orientation, as compared to hearts with isolated bicuspid valve [13].

Mitral pathology (cleft, parachute, dysplasia) and obstruction of the left outflow tract (fibrous stenosis, persistence of antero-lateral muscle band) may be found in hearts with bicuspid aortic valve (Table 3.3).

Trisomy 18 may be present in cases of BAV with associated bicuspid pulmonary valve and De George syndrome in cases with interruption of the aortic arch.

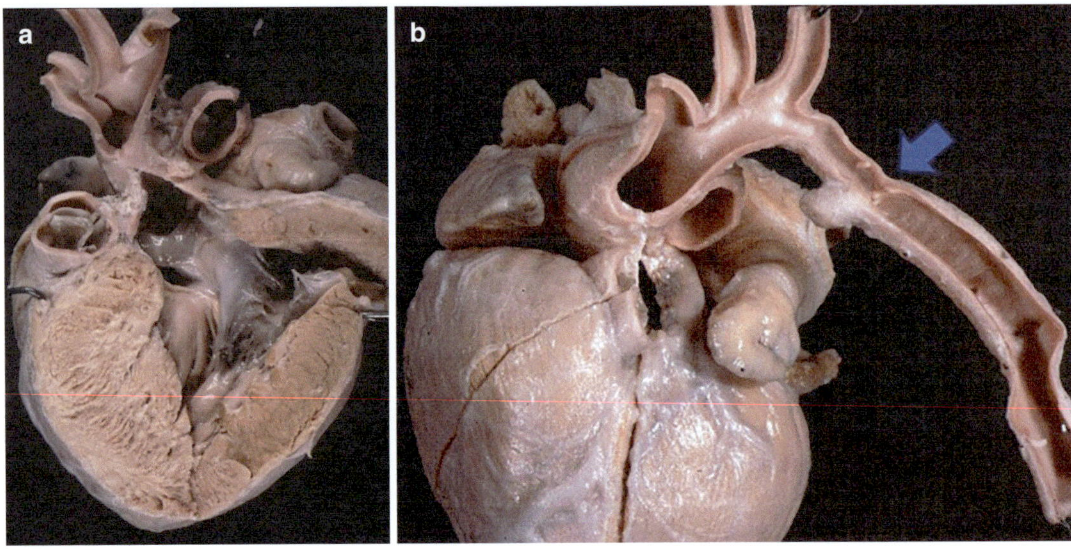

Fig. 3.5 Bicuspid aortic valve with aortic coarctation and ventricular septal defect. (**a**) View of the left ventricular outflow: a bicuspid aortic valve and a perimembranous ventricular septal defect are present. Note the severe hypertrophy of the left ventricle. (**b**) External view of the heart: the arrow indicates the aortic coarctation just in front of the ductus arteriosus

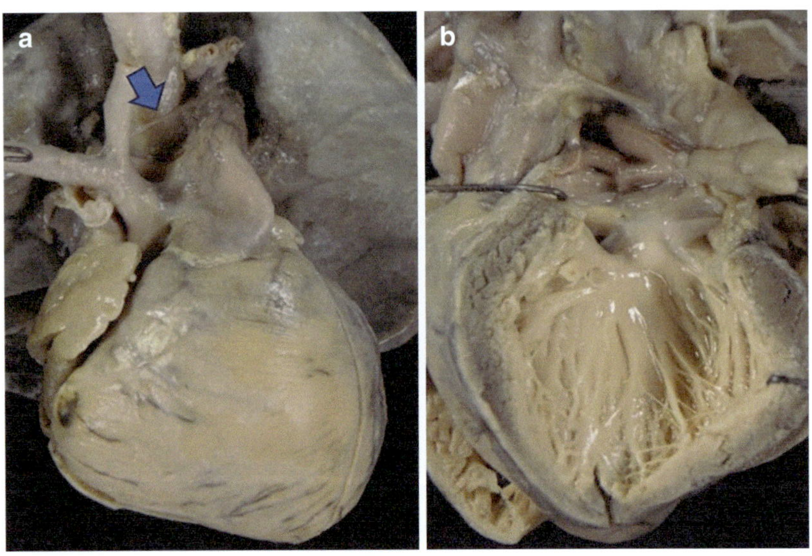

Fig. 3.6 Bicuspid aortic valve with interrupted aortic arch and ventricular septal defect. (**a**) External view of the heart: the segment of the aortic arch located between the left carotid and left subclavian arteries is reduced to a fibrous strand (arrow) (interruption of the aortic arch). The descending aorta is supplied by a patent ductus arteriosus. (**b**) Left ventricular view: a bicuspid aortic valve and a perimembranous ventricular septal defect are present. From Frescura C., Thiene G. Cardiopatie congenite. Testo Atlante, © 2025 Piccin Nuova Libraria S.p.A

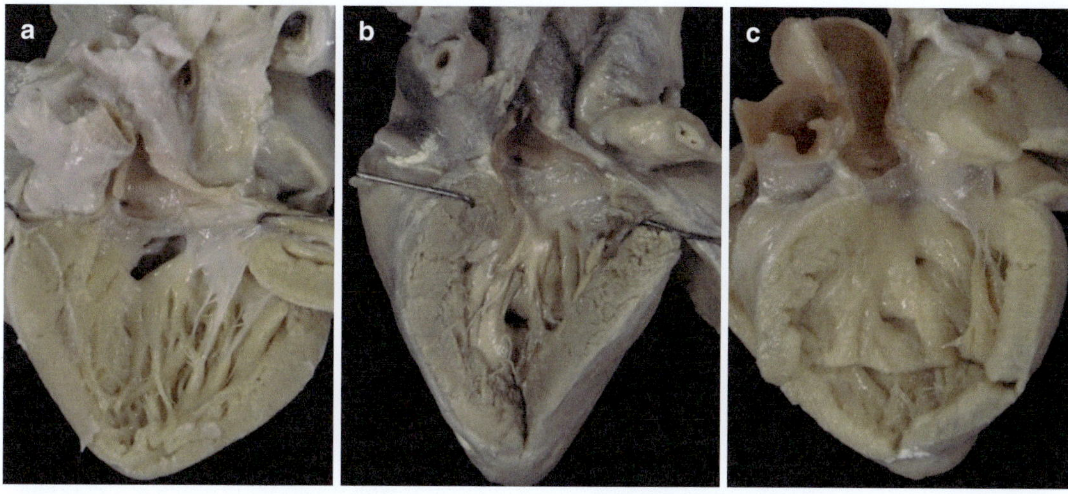

Fig. 3.7 Bicuspid aortic valve: type of ventricular septal defect. View from the left ventricle in different cardiac specimens with bicuspid aortic valve: the associated ventricular defect is perimembranous (**a**), muscular (**b**) or multiple (**c**)

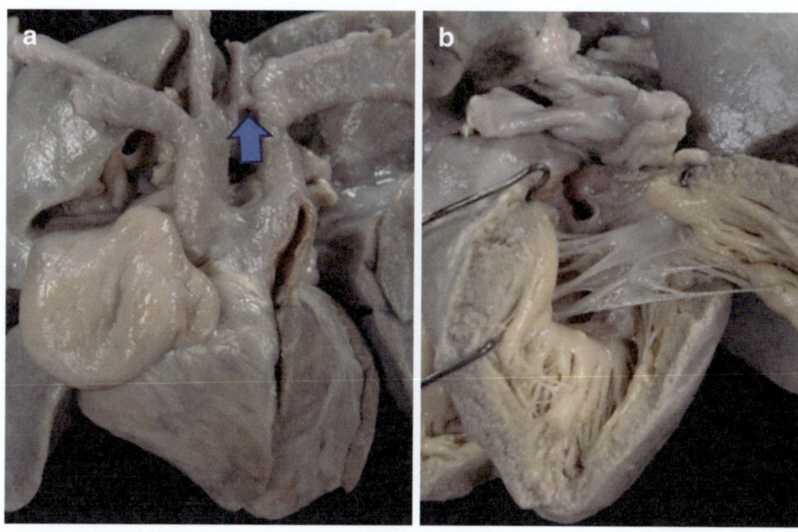

Fig. 3.8 Bicuspid aortic valve with atrioventricular septal defect and aortic coarctation. (**a**) External view of the heart showing hypoplasia of the aortic arch and aortic coarctation (arrow). (**b**) View of the left ventricular outflow: a bicuspid aortic valve is present in association with a complete, type C atrioventricular septal defect

The etiology of BAV is still unclear and genetic [14–16] and hemodynamic factors [10] are considered possible causes of aortic valve maldevelopment. Also, an anomalous migration of cells from the neural crest was advanced to be responsible of the contemporary occurrence of anomalies of aortic valve, aortic arch, and tunica media of the ascending aorta (aortopathy) [17].

Coexistence of bicuspid aortic and pulmonary valves is a rare occurrence [18, 19] (Fig. 3.11) and is usually observed in patients with chromosomal anomalies, especially Trisomy 18 (Table 3.4). The bicuspid morphology of both the semilunar valves seems to be an independent occurrence.

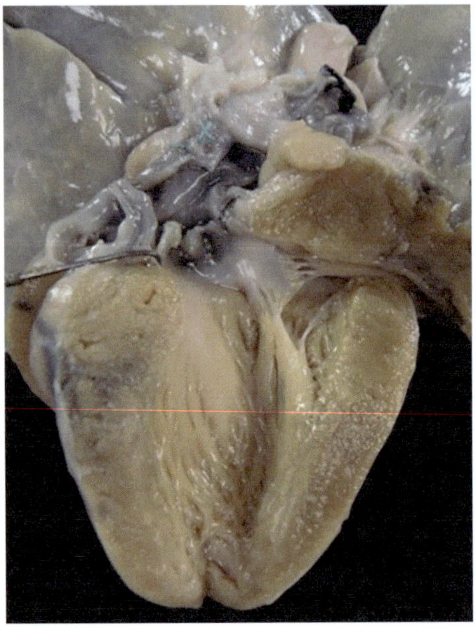

Fig. 3.9 Bicuspid and stenotic aortic valve. View of the left ventricular outflow tract: the bicuspid aortic valve is stenotic with two dysplastic leaflets with myxoid nodules. Note the hypertrophy of the left ventricular free walls. From Frescura C., Thiene G. Cardiopatie congenite. Testo Atlante, © 2025 Piccin Nuova Libraria S.p.A

Variable coronary anatomy although not specific has been reported in patients with BAV [20].

The team on congenital heart disease in Leiden, led by the late Adriana Gittenberger-De Groot, studied coronary anatomy in 186 patients with BAV (79% men, mean age 51 ± 14 years) by CT and selective coronary angiography.

Both patients with type 1B BAVs and patients with BAV and aortic coarctation had a left dominant coronary artery systems. Moreover, patients with type 1B BAVs were more at risk of developing significant atherosclerotic coronary artery disease.

Moreover, type 1B BAVs more often had separate ostia, which can be of clinical importance while performing selective coronary angiography and using coronary artery perfusion for myocardial protection during surgical aortic valve replacement.

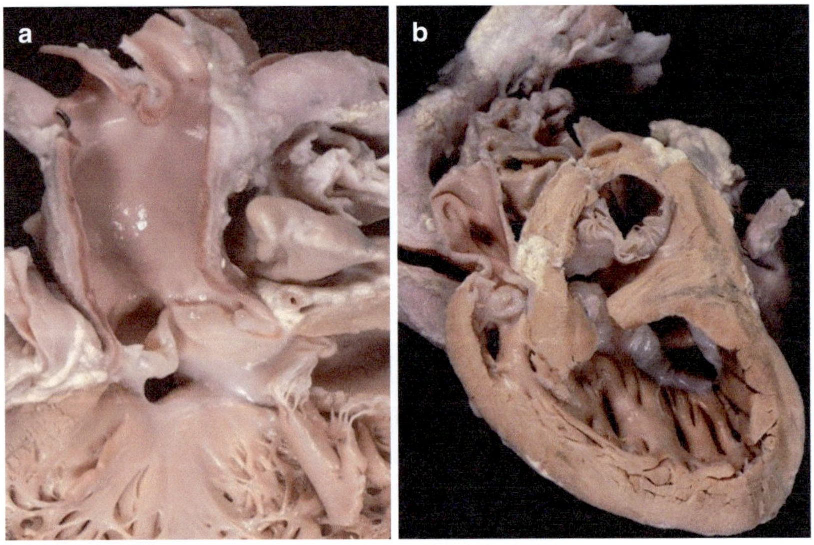

Fig. 3.10 Bicuspid aortic valve with tetralogy of Fallot. (**a**) View of left ventricle and aorta: a bicuspid and dysplastic aortic valve and a perimembranous ventricular septal defect are present. The aorta overrides the ventricular septum. (**b**) Echocardiographic section of the same specimen showing the subpulmonary stenosis due to deviation of the infundibular septum and the pulmonary valve stenosis, typical features of Tetralogy of Fallot

Table 3.2 Type of ventricular septal defect associated with bicuspid aortic valve

Type of ventricular septal defect	Isolated ventricular septal defect 21 cases	Ventricular septal defect with aortic coarctation/interruption 32 cases
Perimembranous	17	23
Muscular	–	5
Multiple	4	4

Data from the Anatomical Collection of Congenital Heart Disease of the University of Padua (1860 hearts)

Table 3.3 Anomalies associated to bicuspid aortic valve: ventricular septal defect and aortic coarctation

Type of anomaly	Isolated ventricular septal defect	Ventricular septal defect with aortic coarctation/interruption	Isolated aortic coarctation
Mitral pathology (cleft, parachute, dysplasia)	3	7	3
Left ventricular tract obstruction	13	10	–
Trisomy 18	5	–	–
De George syndrome	–	2	–

Data from the Registry of Congenital Heart Disease of the University of Padua (1860 hearts)

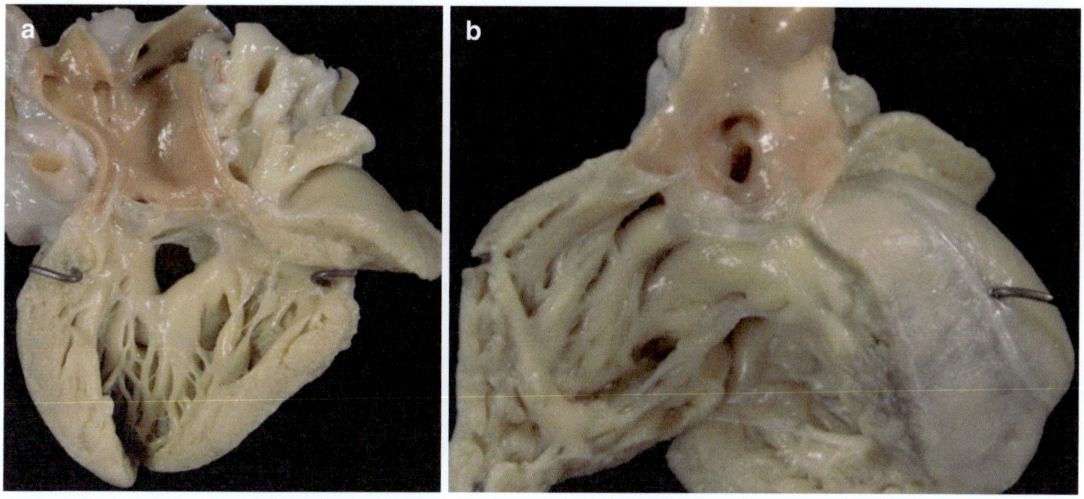

Fig. 3.11 Bicuspid aortic and pulmonary valve. (**a**) View of the left ventricle and aorta showing a bicuspid aortic valve and a large perimembranous ventricular septal defect. (**b**) View of the right ventricular outflow: a bicuspid pulmonary valve is present

Table 3.4 Coexistence of bicuspid aortic and pulmonary valves (11 hearts)

Cardiac malformations	Bicuspid aortic and pulmonary valve
Isolated ventricular septal defect	8
Ventricular septal defect with aortic arch coarctation/interruption	2
Tetralogy of Fallot	1
Trisomy 18	5

Data from the Anatomical Collection of Congenital Heart Disease of the University of Padua (1860 hearts)

References

1. Angelini A, Ho SY, Anderson RH, Devine WA, Zuberbuhler JR, Becker AE. The morphology of the normal aortic valve as compared with the aortic valve having two leaflets. J Thorac Cardiovasc Surg. 1989;98:362–7.
2. Fernandez SM, Sanders SP, Khairy P, Jenkins KJ, Gauvreau K, Lang P, Simonds H, Colan SD. Morphology of bicuspid aortic valve in children and adolescents. J Am Coll Cardiol. 2004;44:1648–51.
3. Nistri S, Basso C, Thiene G. Bicuspid aortic valve. G Ital Cardiol. 2012;13:25–37.
4. Roberts WC. The congenitally bicuspid aortic valve. A study of 85 autopsy cases. Am J Cardiol. 1970;26:72–83.
5. Roberts WC. Valvular, subvalvular and supravalvular stenosis: morphologic features. Cardiovasc Clin. 1973;5:79–126.
6. Basso C, Frescura C, Corrado D, Muriago M, Angelini A, Daliento L, Thiene G. Congenital heart disease and sudden death in the young. Hum Pathol. 1995;26:1065–72.
7. Giusti S, Cocco P, Thiene G. The bicuspid aortic valve: a "minor" congenital cardiopathy at risk of serious complications. G Ital Cardiol. 1991;21:189–201.
8. Duran AC, Frescura C, Sans-Coma V, Angelini A, Basso C, Thiene G. Bicuspid aortic valves in hearts with other congenital heart disease. J Heart Valve Dis. 1995;4:581–90.
9. Becker AE, Becker MJ, Edwards JE. Anomalies associated with coarctation of aorta: particular reference to infancy. Circulation. 1970;41:1067–75.
10. Moore GW, Hutchins GM. Association of interrupted aortic arch with malformations producing reduced blood flow to the fourth aortic arches. Am J Cardiol. 1978;42:467–72.
11. Moene RJ, Gittenbergerde Groot AC, Oppenheimer-Dekker A, Bartelings M. Anatomic characteristics of ventricular septal defect associated with coarctation of the aorta. Am J Cardiol. 1987;59:952–5.
12. Bartram U, Bartelings MM, Kramer HH, Gittenberger de Groot AC. Congenital polyvalvular disease: a review. Pediatr Cardiol. 2001;22:93–101.
13. Fernández B, Durán AC, Fernández-Gallego T, Fernandez MC, Such M, Arquè JM, Sans-Coma V. Bicuspid aortic valves with different spatial orientations of the leaflets are distinct etiological entities. J Am Coll Cardiol. 2009;54:2312–8.
14. Sans-Coma V, Fernández B, Durán AC, Thiene G, Arqué JM, Muñoz-Chápuli R, Cardo M. Fusion of valve cushions as a key factor in the formation of congenital bicuspid aortic valves in Syrian hamsters. Anat Rec. 1996;244:490–8.
15. Sans-Coma V, Cardo M, Durán AC, Franco D, Fernandez B, Arque JM. Evidence for a quantitative genetic influence on the formation of aortic valves with two leaflets in the Syrian hamster. Cardiol Young. 1993;3:132–40.
16. Clementi M, Notari L, Borghi A, Tenconi R. Familial congenital bicuspid aortic valve: a disorder of uncertain inheritance. Am J Med Genet. 1996;62:336–8.
17. Kappetein AP, Gittenberger de Groot AC, Zwinderman AH, Rohmer J, Poelmann RE, Huysmans HA. The neural crest as a possible pathogenetic factor in coarctation of the aorta and bicuspid aortic valve. J Thorac Cardiovasc Surg. 1991;102:830–6.
18. Kemaloğlu Öz T, Karadeniz FÖ, Gundlapalli H, Erer B, Sharma RK, Ahmed M, Nanda NC, Yıldırım A, Orhan G, Öz A, Eren M. Coexisting bicuspid aortic and pulmonary valves with normally related great vessels diagnosed by live/real time three-dimensional transesophageal echocardiography. Echocardiography. 2014;31:218–21.
19. Koenraadt WMC, Bartelings M, Gittenberger-de Groot AC, Bökenkamp R, DeRuiter MC, Schalij MJ, Jongbloed MRM. Pulmonary valve morphology in patients with bicuspid aortic valves. Pediatr Cardiol. 2018;39:690–4.
20. Koenraadt WMC, Tokmaji G, DeRuiter MC, Vliegen HW, Scholte AJ, Siebelink HM, Gittenberger de Groot AC, de Graaf MA, Wolterbeek R, Mulder BJ, Bouma BJ, Schalij MJ, Jongbloed MR. Coronary anatomy in children with bicuspid aortic valves and associated congenital heart disease. Heart. 2016;15(102):943–9.

4 The Normal Anatomy of the Aortic Valve Apparatus

Stefania Rizzo, Monica De Gaspari, Cristina Basso, and Gaetano Thiene

The aortic valve apparatus is located into the sinusal ascending aorta, separate from the tubular portion by the sino-tubular junction.

The normal aortic valve consists of three swallow's nest cusps, separate by interleaflet triangles and commissures. The latter are the highest attachment of the cusps at the sino-tubular junction (Fig. 4.1) [1–4].

Normally the cusps are three: one posterior in continuity with the anterior leaflet of the mitral valve (Fig. 4.2), two anterior, right and left, lying over myocardium (the right upon the ventricular septum and the left over the anterolateral band of the free wall basal left ventricle) (Fig. 4.3).

The interleaflet triangle between the posterior and right cusps overlaps the membranous septum, under which the His bundle runs [5] (Fig. 4.4).

The aortic valve apparatus is located posteriorly and on the right of the pulmonary one, wedged between the mitral and tricuspid valves (Fig. 4.5).

The aortic valve is low, so that the infundibular septum separates the subpolmonary right ventricular outflow from the sinusal part of the aorta. The anatomy of the aortic valve and its relationship with the pulmonary outflow is explarable by short axis parasternal 2D echo (Fig. 4.6).

There are three different borders ("anuli") between the left ventricle and the aorta (Fig. 4.7): (1) the basal ring, the lowest attachment of the cusps; (2) The anatomic ring, crown-shaped, corresponding to semilunar cusps attachement to the aortic wall; (3) The commissural ring, corresponding to the sino-tubular junction, which represents the border of the sinusal aortic root into the tubular ascending aorta.

The coronary arteries, right and left, originate from the anterior aortic sinuses, just under the sinotubular junction (Fig. 4.8). During systole the cusps may obstruct the ostia. Fortunately, the blood stream to the myocardium occurs during diastole, when the cusps get not to close the aortic orifice [6].

The role of the Valsalva sinuses is to relieve abnormal stress on the cusps. Leonardo was the first to consider the aortic wall an intrinsic component of the aortic valve apparatus (Fig. 4.9). Moreover, he perceived that, when the heart contracts, the blood leaves violently the left ventricle creating a vortex against the sinus aortic wall and the same cusps, so that the aortic orifice begins to close already during systole, as confirmed by cardiac magnetic resonance (Fig. 4.10). Recently Katamyana et al. confirmed that Valsalva sinuses facilitate smooth closure of the aortic cusps (Fig. 4.11).

S. Rizzo (✉) · M. De Gaspari · C. Basso · G. Thiene
Cardiovascular Pathology Unit, Department of Cardiac, Thoracic and Vascular Sciences and Public Health, School of Medicine, University of Padua, Padua, Italy
e-mail: s.rizzo@unipd.it; monica.degaspari@unipd.it; cristina.basso@unipd.it; gaetano.thiene@unipd.it

G. Thiene et al. (eds.), *Bicuspid Aortic Valve Syndrome*,
https://doi.org/10.1007/978-3-031-85589-4_4

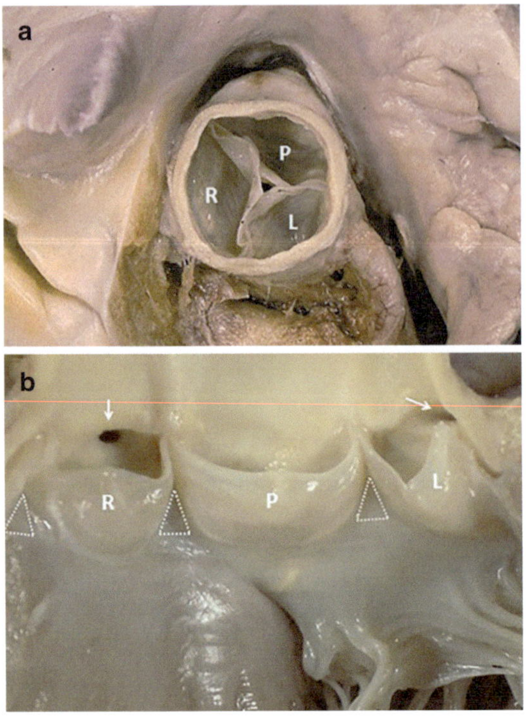

Fig. 4.1 The aortic valve apparatus. (**a**) View from above of the cusps within the sinusal aorta; P = Posterior, non coronary; L = left anterior, coronary; R = right anterior, coronary. (**b**) Open aortic valve apparatus. Arrows indicate coronary ostia at the sino-tubular junction and triangles the interleaflet trigones. Note the discontinuity of the cusps at the commissures

As far as valve cusps histology, they are made of three layers: (1) the "ventricularis," facing the left ventricular cavity and consisting mostly of thin elastic fibers; (2) the "fibrosa," facing the aortic wall and consisting mostly of collagen fibers; (3) the "spongiosa," the central layer, consisting mainly of mucoid extracellular matrix (Fig. 4.12). The fibrosa is corrugated/crimpted during systole and flattened in diastole to enlarge the size of the cusps while closing the orifice (Fig. 4.13). The number of elastic lamellae is higher in the acending aorta (about 56) and less in the abdominal aorta (about 28) [7].

With regard to the other fundamental part of the aortic valve apparatus, namely the aortic wall, it consists of tunics intima, media, and adventitia. The intima is lined by endhotelium covering a thin myointimal stratum. The tunica media is the thickest and consists of elastic lamellar units (Fig. 4.14), discovered by Wolinsky and Glacov in 1967 (Fig. 4.15). Each lamellar unit consists of elastic fibers encompassing smooth muscle cells and a tiny extracellular matrix like a sandwich (Fig. 4.16). Overall the so many smooth muscle cells should be considered a parenchyma rather than a mesenchyma of the organ aorta. The adventitia is rich of vessels which may penetrate

Fig. 4.2 Fibrous continuity between the posterior, non coronary aortic cusp, and the anterior mitral leaflet. (**a**) Gross view; (**b**) histology, Azan stain

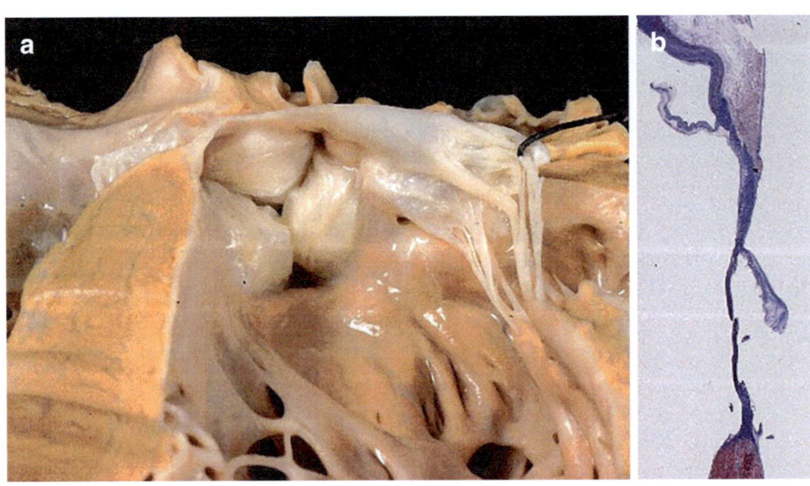

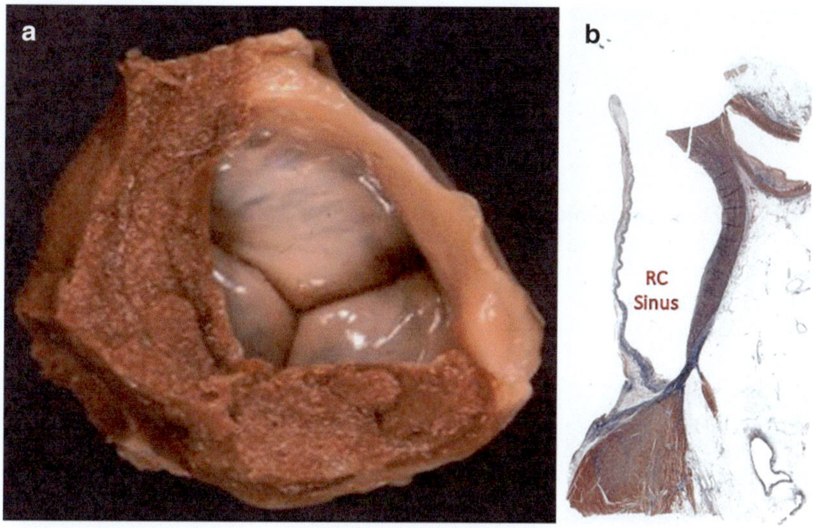

Fig. 4.3 (**a**) Left and right aortic cusps lie over myocardium. (**b**) The right lies over the ventricular septum. Azan stain

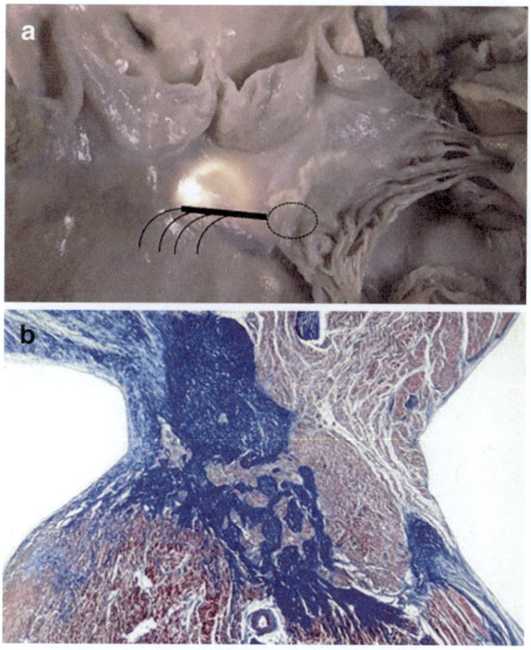

Fig. 4.4 (**a**) The His bundle runs underneath the mebranous septum, just below the interleaflet triangle between the right and posterior aortic cusps. (**b**) Histology of the penetrating bundle, Azan stain

the tunica media (vasa vasorum) and may be the source of intramural haematoma or wall dissection with rupture of the aorta.

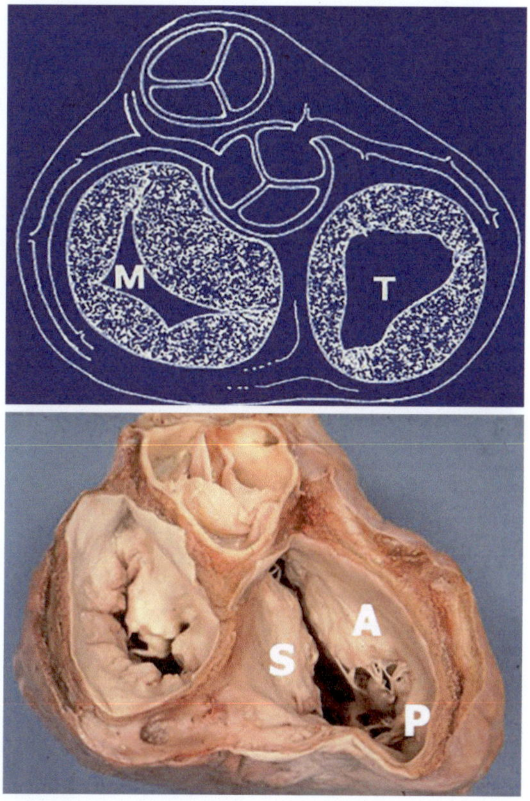

Fig. 4.5 The aortic valve is wedged in between the mitral and tricuspid valves. M = mitral valve; T = Tricuspid valve. A, S and P indicate respectively the anterior, septal and posterior leaflets of the tricuspid valve

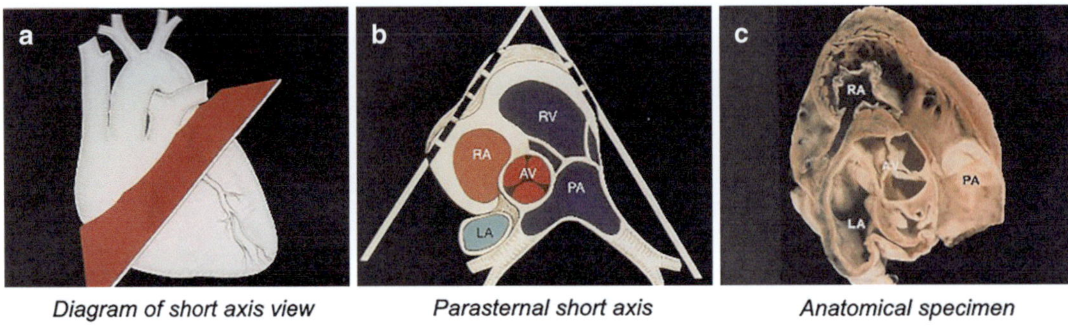

Fig. 4.6 Basal parasternal short axis 2D cardiac echo (**a**, **b**), compared to a normal heart specimen (**c**). Note that the infundibular septum separates the subpulmonary right ventricular outflow tract from the right aortic sinus (**c**)

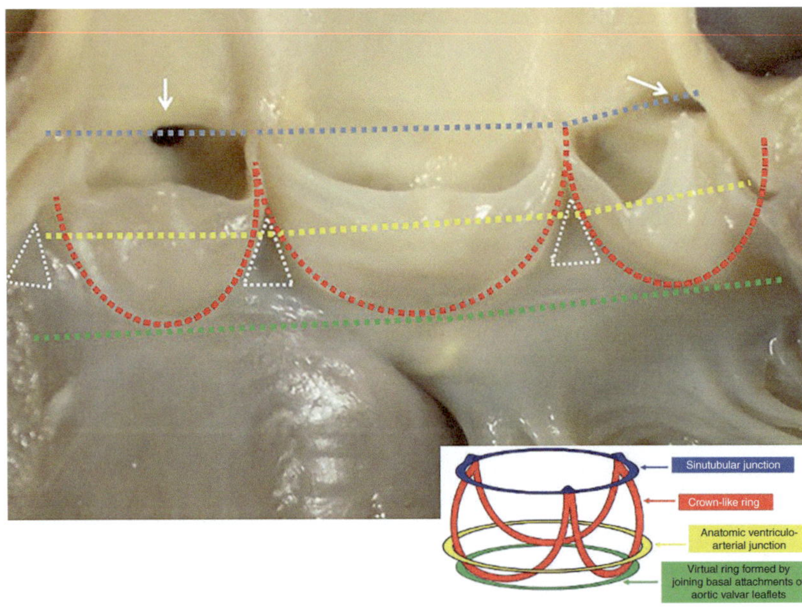

Fig. 4.7 The various rings separating the left ventricle from the aorta

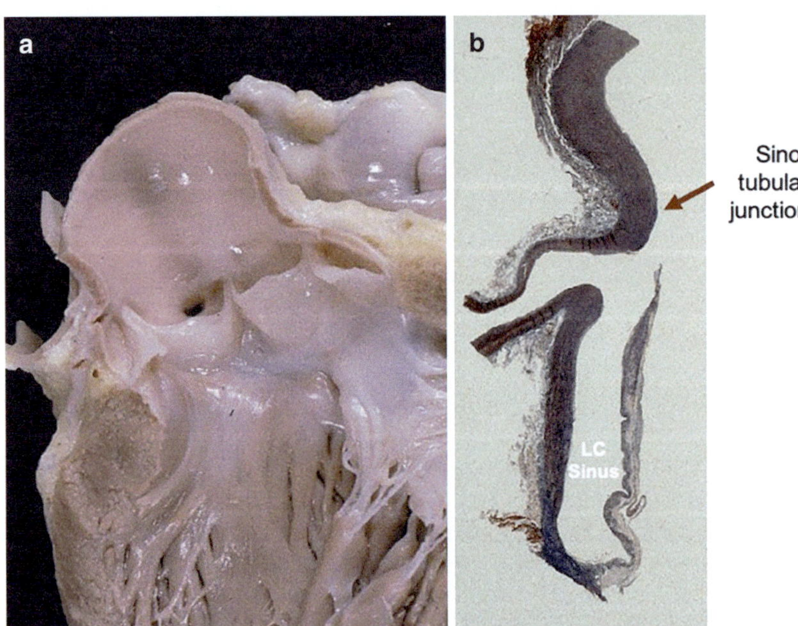

Fig. 4.8 The ostium of the coronary arteries (here the right one) is located just below the sinotubular junction (**a**, **b**). During systole the cusps may interfere with ostia patency. However the blood flow to myocardium occurs during the diastole, when the cusps get near to close the aortic valve orifice

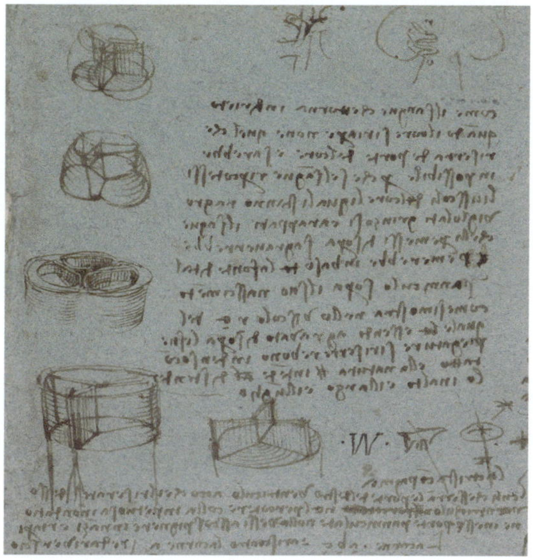

Fig. 4.9 Leonardo considered the aorta an intrinsic part of the aortic valve apparatus. (*Courtesy of © Royal Collection Enterprises Limited 2025 | Royal Collection Trust*)

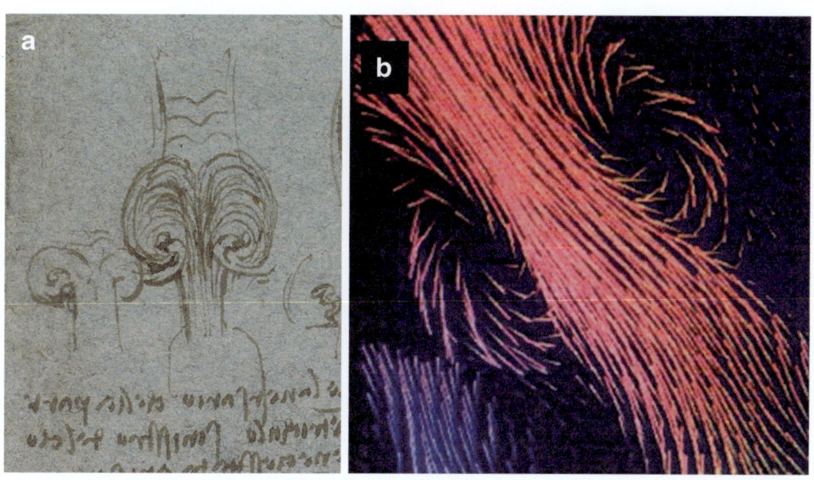

Fig. 4.10 Leonardo postulated that closing of the aortic orifice starts during systole (**a**, *Courtesy of © Royal Collection Enterprises Limited 2025 | Royal Collection Trust*), a hypothesis that was confirmed by cardiac magnetic resonance (**b**, *Courtesy of Prof. M. Perazzolo Marra, from the Archives of Department of Cardiac, Thoracic and Vascular Sciences and Public Health, University of Padua*)

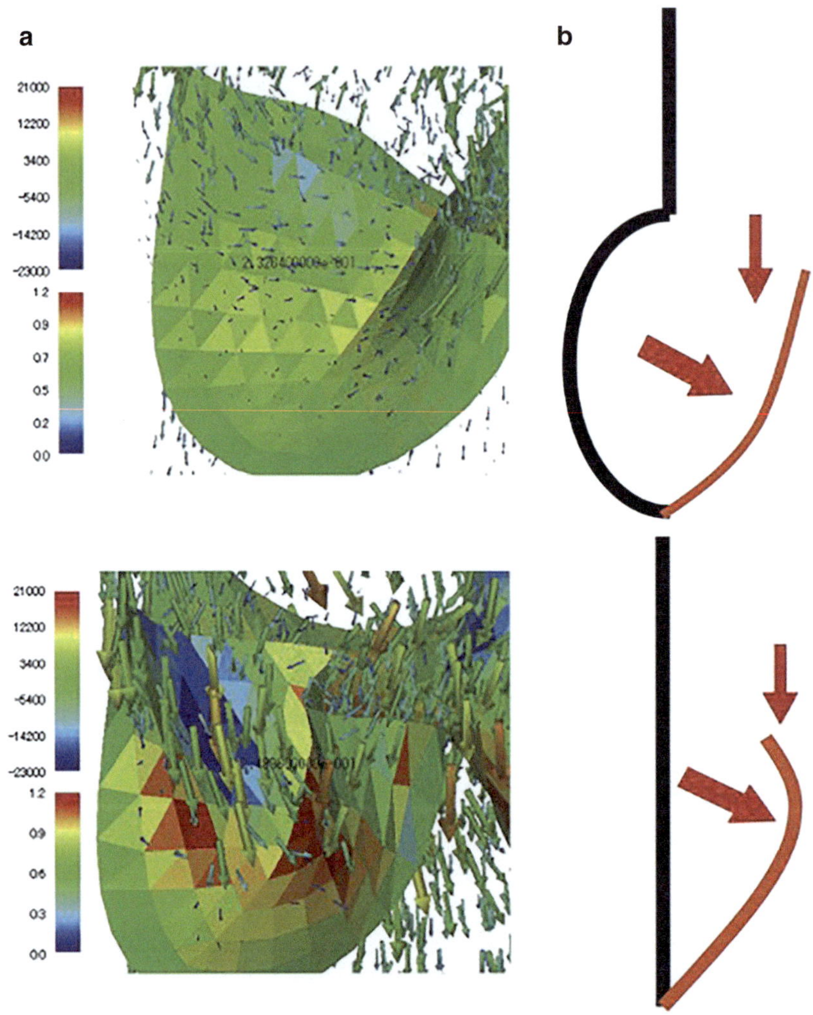

Fig. 4.11 Katamaya et al. confirmed that the role of Valsalva sinuses is to relieve abnormal stress on the valve aortic cusps and to facilitate smooth closure (panel **a** and **b**)

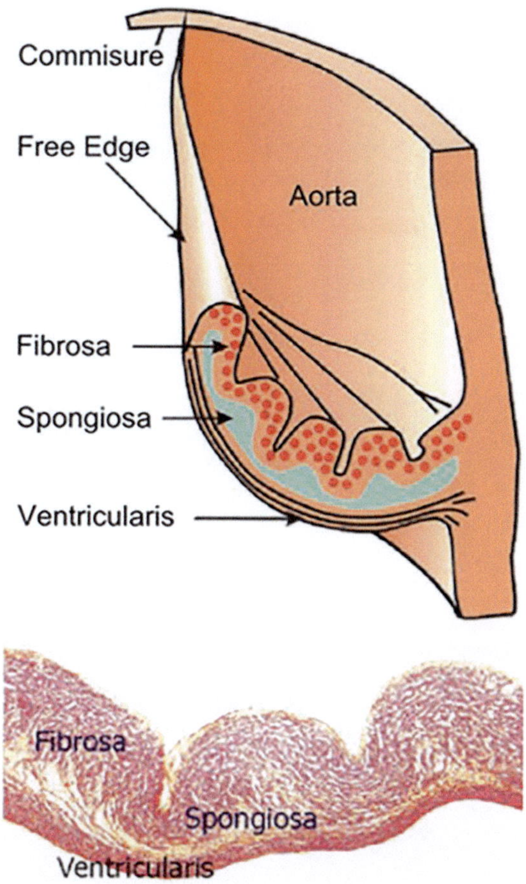

Fig. 4.12 Schematic representation of histological layers of the aortic valve cusps. Note the thin elastic ventricularis, the thick fibrosa rich of collagen fibers and the spongiosa in between, with mucoid extracellular matrix. (Reproduced with permission from Doehring TC, Kahelin M, Vesely I. Mesostructures of the Aortic Valve. J Heart Valve Dis. 2005 Sep;14(5):679–86)

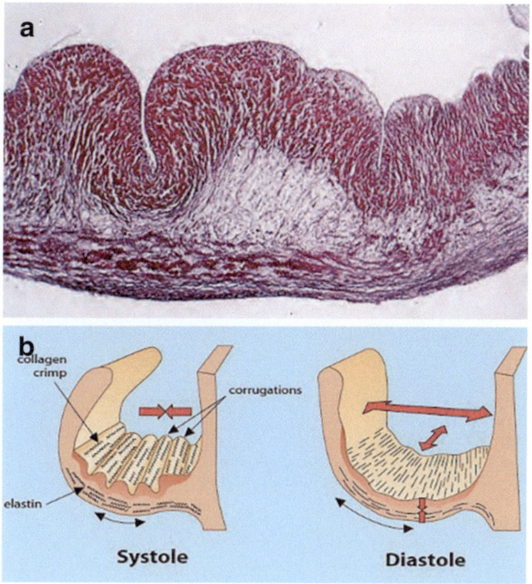

Fig. 4.13 During systole the fibrosa is corrugated and cripted, whereas during diastole is flattened to enlarge the size of the cusps for the orifice closure (panel **a** and **b**)

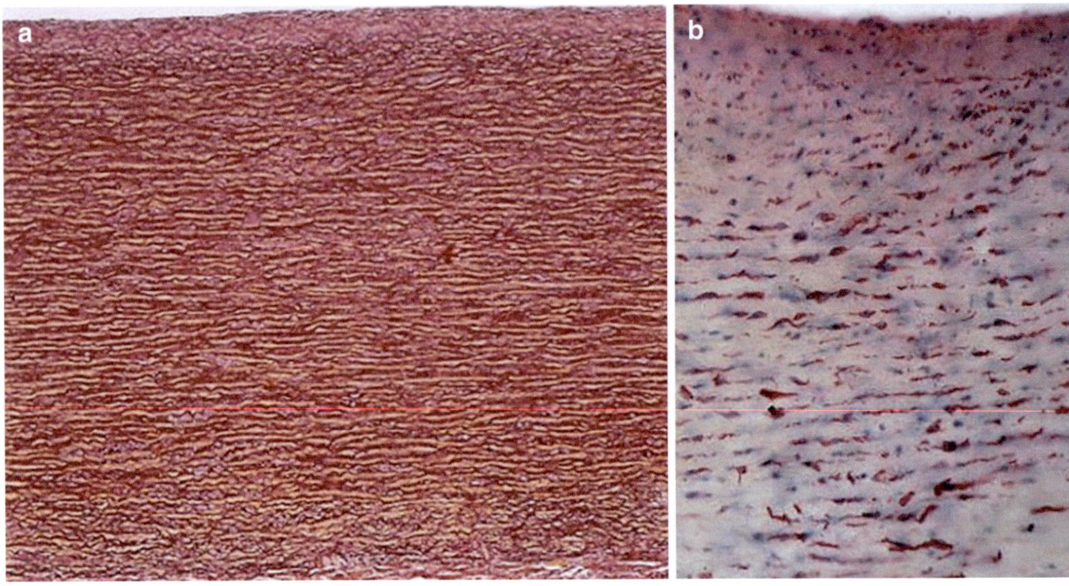

Fig. 4.14 The tunica media is the thickest aortic layer and consists of musculoelastic lamellar units. Weigert-Van Gieson stain (**a**) and smooth muscle cells actin immunohistochemestry (**b**)

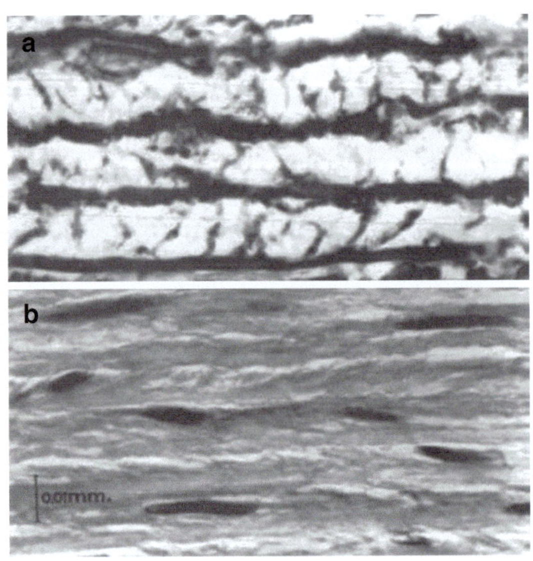

c

A Lamellar Unit of Aortic Medial Structure and Function in Mammals

By **Harvey Wolinsky, M.D., and Seymour Glagov, M.D.**

Fig. 4.15 The original illustration (**a**, **b**) of the paper published by Wolinsky and Glagov (**c**) in 1967, who advanced the concept of lamellar units of elastic fibers and smooth muscle cells in the tunica media of the aorta [8]

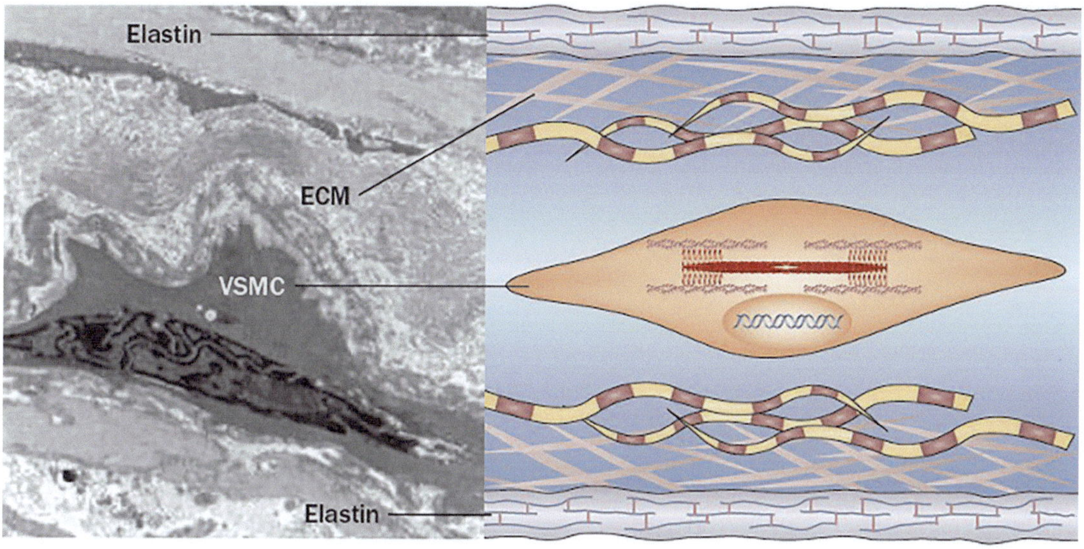

Fig. 4.16 The lamellar unit consists of elastin fibers encompassing vascular smooth muscle cells (vsmc) and extracellular matrix (ecm), like a sandwich [9]

References

1. Anderson RH, Becker AE. Cardiac anatomy. An integrated text and color atlas. Edinburgh: Grover Medical Publishing, Churchill Livingstone; 1986.
2. Anderson RH, Devine WA, Ho SY, et al. The myth of the aortic annulus: the anatomy of the subaortic outflow tract. Ann Thorac Surg. 1991;52:640–6.
3. Anderson RH, Yen HS. The anatomy of the heart. Miami Lakes: Cordis International SA Training & Development; 1995.
4. Anderson RH. Anatomy: clinical anatomy of the aortic root. Heart. 2000;84:670–3.
5. Basso C, Ho Y, Rizzo S, Thiene G. Anatomic and histopathologic characteristics of the conductive tissues of the heart. In: Gussak I, Antzelevitch C, editors. Electrical diseases of the heart. London: Springer; 2013. p. 1–25. ISBN: 9781447148814.
6. Cavalcanti JS, de Melo NC, deVasconcelos RS. Morphometric and topographic study of coronary ostia. Arq Bras Cardiol. 2003;81:359–62.
7. Doehring TC, Kahelin M, Vesely I. Mesostructures of the aortic valve. J Heart Valve Dis. 2005;14(5):679–86.
8. Wolinsky H, Glagov S. A lamellar unit of aortic medial structure and function in mammals. Circ Res. 1967;20(1):99–111.
9. El-Hamamsy I, Yacoub MH. Cellular and molecular mechanisms of thoracic aortic aneurysms. Nat Rev Cardiol. 2009;6(12):771–86.

Morphology Spectrum of Bicuspid Aortic Valve and Atrio-Ventricular Conduction System

5

Chiara Fraccaro, Cristina Basso, and Gaetano Thiene

5.1 Anatomy of the Bicuspid Aortic Valve and Proposed Classifications

The bicuspid aortic valve (BAV) is the most common congenital heart defect, characterized by the presence of two, rather than the normal three, aortic valve leaflets. The leaflets or cusps are typically asymmetric. Unlike the tricuspid aortic valve (TAV), which has three equally sized cusps and three interleaflet triangles and commissures, the BAV generally has two cusps (one larger cusp often termed the conjoined cusp, and another smaller one) and two interleaflet triangles. The larger cusp usually results from the fusion of two cusps, commonly the right (R) and left (L) coronary cusps. A fibrous and/or calcified ridge called *raphe* and resulting from incomplete separation of the fused cusps may be present. Thus, the raphe can be considered as an "aborted" commissure during embryonic development, underneath which there is no interleaflet triangle (Fig. 5.1) [1].

However, three basic anatomical types of BAV are distinguished depending on the anatomical orientation of their constituents. R-L BAVs, as described above, are the result of the anatomical embriogenic fusion of the R and L valve leaflets. They exhibit an anteroposterior orientation of the leaflets, and both coronary arteries typically originate from the common anterior sinus. When the anatomical fusion involves one anterior coronary leaflet and the posterior non-coronary (NC) leaflet, the final configurations are referred to as R-NC and L-NC BAVs, respectively. These variants display a latero-lateral orientation of the leaflets, with each coronary artery normally arising from its respective cusp. Rarely, the leaflets are symmetrical or there is no raphe ("pure" or "true" bicuspid valve). In general, in BAV phenotypes with fusion resulting in two asymmetrical aortic cusps, there are three identifiable sinuses of Valsalva. On the other hand, a symmetrical BAV without raphe has only two identifiable sinuses of Valsalva. Of note, an oval (American football shape) systolic opening of BAV is evident, as opposed to the triangular opening of a tricuspid aortic valve.

According to different published series, the most prevalent BAV phenotype overall is the

C. Fraccaro (✉)
Cardiology Clinic, Department of Cardiac, Thoracic, Vascular Sciences and Public Health, University of Padua, Padua, Italy
e-mail: chiara.fraccaro@unipd.it

C. Basso
Cardiovascular Pathology Unit, Department of Cardiac, Thoracic, Vascular Sciences and Public Health, University of Padua, Padua, Italy
e-mail: cristina.basso@unipd.it

G. Thiene
School of Medicine, University of Padua, Padua, Italy
e-mail: gaetano.thiene@unipd.it

G. Thiene et al. (eds.), *Bicuspid Aortic Valve Syndrome*,
https://doi.org/10.1007/978-3-031-85589-4_5

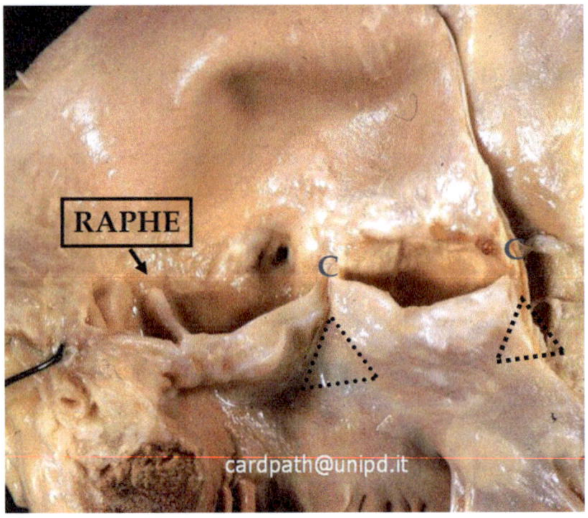

Fig. 5.1 BAV anatomy. BAV with a ventral raphe and no interleaflet triangle underneath in a 57 years-old male patient. (*From Gaetano Thiene, Stefania Rizzo, Cristina Basso, Bicuspid aortic valve: The most frequent and not so benign congenital heart disease. Cardiovasc Pathol. 2024 May-Jun:70:107604. doi: 10.1016/j.carpath.2024.107604. Epub 2024 Jan 20. PMID: 38253300 DOI: 10.1016/j.carpath.2024.107604*)

LR-fusion pattern, followed by the R-NC fusion. However, the morphological characteristics of BAV such as the fusion patterns of the cusps and the presence of a raphe can vary geographically. For example, RL cusp fusion is the most common pattern in Western populations, whereas other fusion types may be more prevalent in different regions [2].

Several classification systems and different nomenclature have been proposed for BAV, focusing on cusp morphology, the presence of a raphe, and the orientation of the cusps, using multiple numbers and letters for the different phenotypes. These classifications are alternatively based on pathological findings or imaging-derived findings (echocardiography, CT scan or cardiac magnetic resonance). Table 5.1 summarizes all the different BAV nomenclature proposed in literature [10]. Among all, the most frequently used classifications across the last decades are described in more detail here below.

Table 5.1 Different proposed BAV classification and relative nomenclature over the last few decades (*Modified with permission from Michelena HI et al. International consensus statement on nomenclature and classification of the congenital bicuspid aortic valve and its aortopathy, for clinical, surgical, interventional and research purposes. Eur J Cardiothorac Surg. 2021;60(3):448–76*)

Year	Authors and references	Study methodology	Patients (*n*)	Nomenclature	Additional considerations
1970	Roberts [3]	Pathology	85	• Anterior–posterior cusps • Right–left cusps Presence of raphe	Discussed differentiating congenital BAV versus acquired
1983	Brandenburg et al. [4]	Echocardiography	115	Clock-face nomenclature: • Commissures at 4–10 o'clock with raphe at 2 o'clock (R-L) • Commissures at 1–6 o'clock with raphe at 10 o'clock (RN) • Commissures at 3–9 o'clock without raphe (L-N)	Noted different sizes of the resulting 2 functional cusps
1989	Angelini et al. [5]	Pathology	64	• Anterior–posterior cusps • Right–left cusps Presence of raphe	Noted presence of 2 (true BAV) versus 3 sinuses
1999	Sabet et al. [6]	Pathology	534	• RL • RN • LN Presence of raphe	Noted symmetry of cusps: equal, unequal, thirds

(continued)

Table 5.1 (continued)

Year	Authors and references	Study methodology	Patients (*n*)	Nomenclature	Additional considerations
2007	Sievers and Schmidtke [7]	Pathology	304	• Type 0 (no raphe): anteroposterior or lateral cusps (true BAV) • Type 1 (1 raphe): R-L, RN, L-N • Type 2 (2 raphes): L-R, RN	Noted type 2 morphology associated with more aortic aneurysms
2008	Schaefer et al. [8]	Echocardiography	186	• Type 1: RL • Type 2: RN • Type 3: LN Presence of raphe Aorta: • Type N: normal shape • Type E: sinus effacement • Type A: ascending aorta dilatations	Noted type 1 BAV was associated with type N aorta with dilated root Noted type 2 BAV associated with type A aorta
2013	Kang et al. [9]	Computed tomography	167	Anteroposterior orientation: • Type 1: R-L with raphe • Type; 2: R-L without raphe Right–left orientation: • Type 3: RN with raphe • Type 4: L-N with raphe • Type 5: symmetrical cusps with 1 coronary artery originating from each cusp Aorta: • Type 0: normal • Type 1: dilated root • Type 2: dilated ascending aorta • Type 3: diffuse involvement of the ascending aorta and arch	Noted AS and type 3 aorta more commonly in right–left orientation and AR and type N aorta more commonly in anteroposterior orientation
2014	Michelena et al. [10]	Echocardiography	Multiple studies	BAVCon nomenclature: • Type 1: R-L • Type 2: RN • Type 3: L-N Presence of raphe	Noted symmetry of cusps and presence of 2 (true BAV) or 3 sinuses Noted predominant ascending aorta dilatation in all BAV and the existence of 'root phenotype'
2016	Jilaihawi et al. [11]	Computed tomography	130	• Tricommissural: functional or acquired bicuspidity of a trileaflet valve • Bicommissural with raphe • Bicommissural without raphe	Noted no association between nomenclature and TAVR complications
2017	Sun et al. [12]	Echocardiography	681	Dichotomous nomenclature: • R-L • Mixed: (RN or L-N)	Noted mixed phenotype was associated with AS and surgery of the aorta Good interobserver variability of phenotypes

(continued)

Table 5.1 (continued)

Year	Authors and references	Study methodology	Patients (*n*)	Nomenclature	Additional considerations
2017	Murphy et al. [13]	Cardiac magnetic resonance	386	Clock-face nomenclature: • Type 0: partial fusion/eccentric leaflet? • Type 1: RN, RL, LN partial fusion/eccentric leaflet? • Type 2: RL and RN, RL and LN, RN and LN partial fusion/eccentric leaflet?	Noted partial fusion and/or eccentric leaflet

AR: aortic regurgitation; AS: aortic stenosis; BAV: bicuspid aortic valve; BAVCon: bicuspid aortic valve consortium; LN: left non-coronary fusion; RL: right–left fusion; RN: right non-coronary fusion; TAVR: transcatheter aortic valve replacement

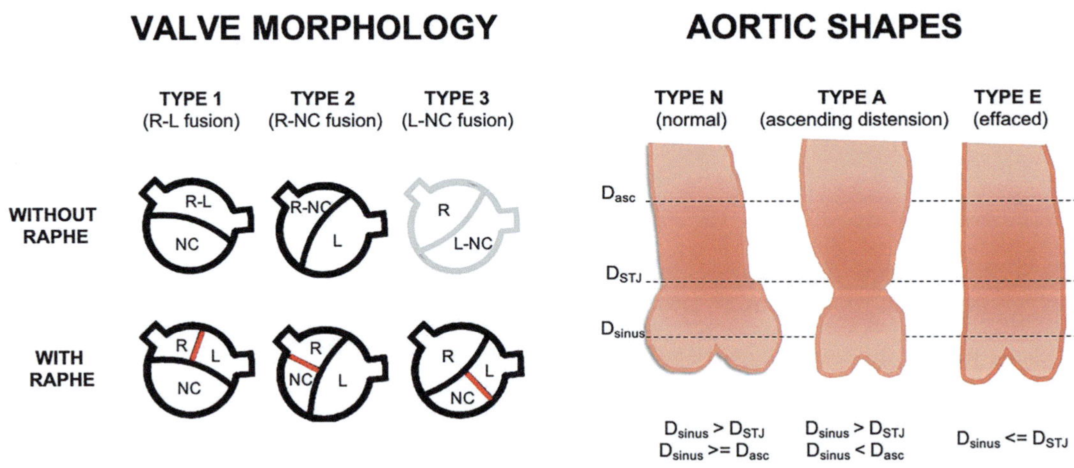

Fig. 5.2 Schäfers' Classification of BAV morphology and aortic shapes. On the left panel, a schematic of BAV phenotypes as seen from a parasternal short axis view on echocardiography. Type 1 has a "fusion" between right and left coronary cusp and this was the most frequent configuration; in type 2 the "fusion" is between right and non-coronary cusp; in type 3, it is between left and non-coronary cusp. Top row, without a raphe; bottom row with raphe (in red). Type 3 with without a raphe was in grey color because this configuration was not seen in the study group. On the right panel, there is a schematic of three aortic shapes. D_{sinus}, diameter at the sinus of Valsalva; D_{STJ}, diameter at the sinotubular junction; D_{asc}, diameter in the mid ascending aorta. (*Modified from Kusner et al. The Natural History of Bicuspid Aortic Valve Disease. Structural Heart. 2023;7(2):100119* [14])

Schäfers' Classification [8]**:** This proposed classification is based on the review of the transthoracic echocardiograms of 191 adult patients with BAV. Three morphologies are identified: type 1 (RL fusion), type 2 (R-NC fusion), and type 3 (L-NC fusion). The presence or not of the raphe is noted. Moreover, three aortic shapes are defined: normal (N), sinus effacement (E), and ascending dilatation (A) (Fig. 5.2).

Sievers and Schmidtke Classification [7]**:** These authors proposed a pathology-based detailed classification system from 304 surgical specimens, that has been widely adopted. This system classifies BAV based on the presence and number of raphes and cusp orienta-

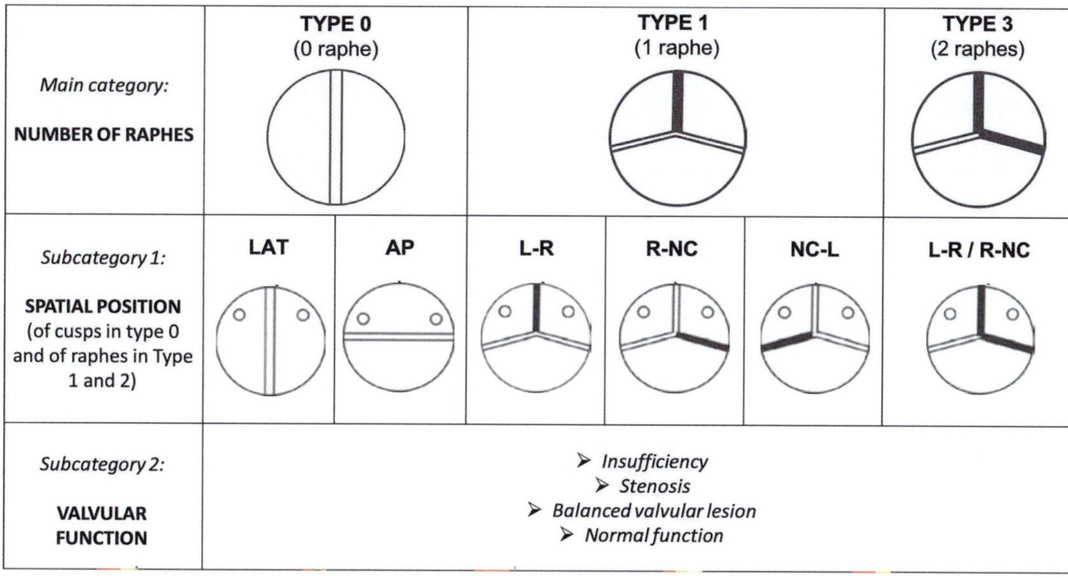

Fig. 5.3 Sievers and Schmidtke Classification of BAV phenotypes. Prominent lines in schematic drawings represent a raphe. The codification for the main category is Type (Type 0 for value with no raphe; Type 1 for valve with one raphe; Type 2 for valve with two raphes); for the first subcategory the special position: ap, lat, L-R, R-N, N-L, and L-R/R-N; and for the second category the valvular function: Insufficiency (I), stenosis (S), balanced valvular lesion (B), and normal function (No). Thus, the classification system may be presented as a codification in three blocks: Type, spatial position, and valvular function (e.g., BAV Type 1, L-R, S, meaning a bicuspid aortic valve with one raphe between the left and right coronary sinus, and with valvular stenosis, which was the most frequent specimen). (*Modified from H-H. Sievers and C. Schmidtke, J Thorac Cardiovasc Surg. 2007 May;133(5):1226–33*)

tion, identifying three types: type 0 (no raphe), type 1 (one raphe), and type 2 (two raphes). It also considers the spatial orientation of the fused cusps (anteroposterior or right-left) as well as functional status of the valve (Fig. 5.3). However, the number of sinuses is not considered (type 0 does not differentiate between a fused BAV with no raphe and a 2-sinus BAV). Also, the symmetry of cusps (important for surgical repair planning) is not included into the classification. Moreover, the recognition of partial fusion (forme fruste) and aortopathy phenotypes is missing.

International Bicuspid Aortic Valve Consortium (BAVCon) classification by Michelena et al. [15]**:** This classification refines the Schäfers' system, emphasizing the clinical implications of the different morphologies. This is an echocardiography-based classifications which highlights the significance of cusp orientation (anteroposterior or right-left) and the presence or absence of raphes. The diagnosis is based on parasternal long- and short-axis imaging of the aortic valve. The schematics presented represent the parasternal short-axis echocardiographic view. Bicuspid valves are classified as type 1 (RL coronary cusp fusion), type 2 (R-NC cusp fusion), and type 3 (L-NC cusp fusion) (Fig. 5.4).

CT-Based Classification by Jilaihawi et al. [11]**:** This classification utilizes multidetector computed tomography (MDCT) to assess the aortic valve morphology in BAV patients undergoing transcatheter aortic valve implantation (TAVI) procedures. The novelty of this classification is the clinical correlation of BAV phenotype with outcome and complications after TAVI. It identifies different subtypes based on cusp symmetry and the presence of a raphe. The authors characterize patients with a bicuspid anatomy into three categories:

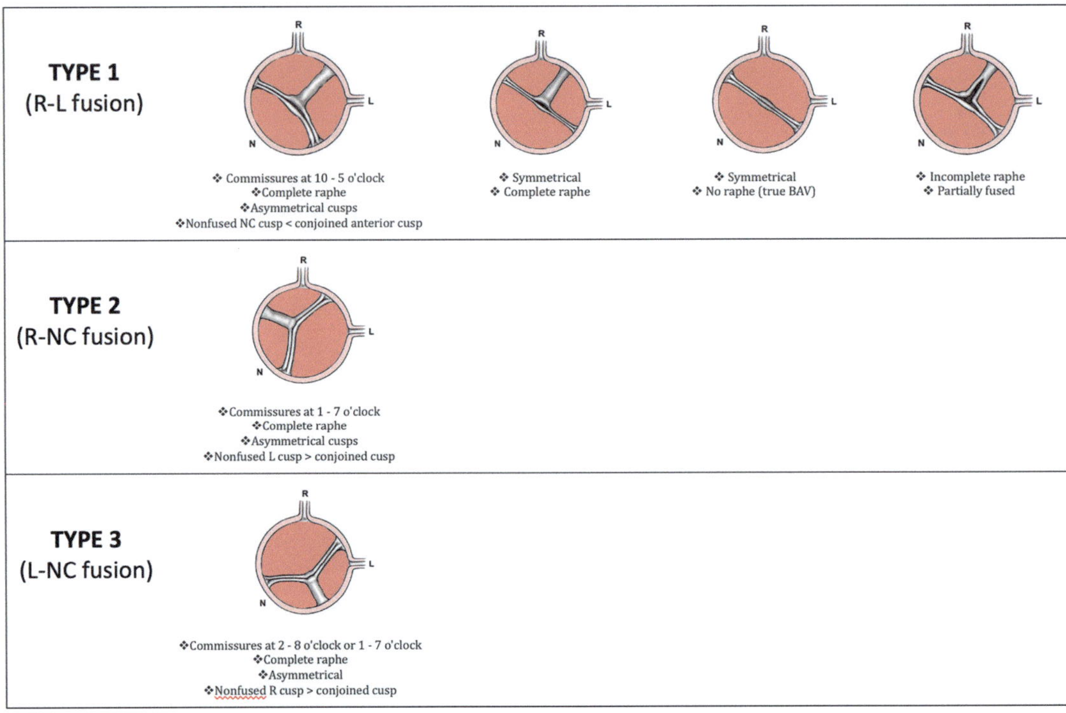

Fig. 5.4 International Bicuspid Aortic Valve Consortium (BAVCon) classification. Schematic of BAV phenotypes as seen by transthoracic echocardiogram. The diagnosis is based on parasternal long- and short-axis imaging of the aortic valve. The schematics presented represent the parasternal short-axis echocardiographic view. Bicuspid valves are classified as type 1 (right-left coronary cusp fusion), type 2 (right-noncoronary cusp fusion), and type 3 (left-noncoronary cusp fusion)

"tricommissural," "bicommissural raphe-type," and "bicommissural non-raphe-type" (Fig. 5.5). This method offers enhanced precision in defining the anatomical features of BAV compared to traditional echocardiographic techniques. This classification was designed as a simplified representation relevant to TAVI to take into account the interface of prosthesis and the aortic-valvular complex, at both the basal leaflet plane (presence or absence of a raphe) and at the commissural level (presence of two or three commissures). A raphe, particularly if calcified, may influence transcaheter heart valve (THV) expansion and apposition at the annular level. Moreover, the presence or absence of a third commissure may also influence THV apposition. The authors found that for patients with bicommissural but not tricommissural BAV, the increased intracommissural distance was associated with increased PVL. They also found a trend towards an increased incidence of new PPM in patients with fused BAV with LR cusp fusion who received a balloon expandable THV.

BAV may not be diagnosed until later in life when complications such as aortic stenosis or regurgitation become symptomatic [8]. However, pediatric patients with BAV might present with different clinical profiles, including early-onset aortopathy. Knowing the BAV phenotype is important to predict the natural history and the risk of complications, including premature aortic stenosis, regurgitation, and ascending aortic aneurysms. In fact, each BAV subtype carries

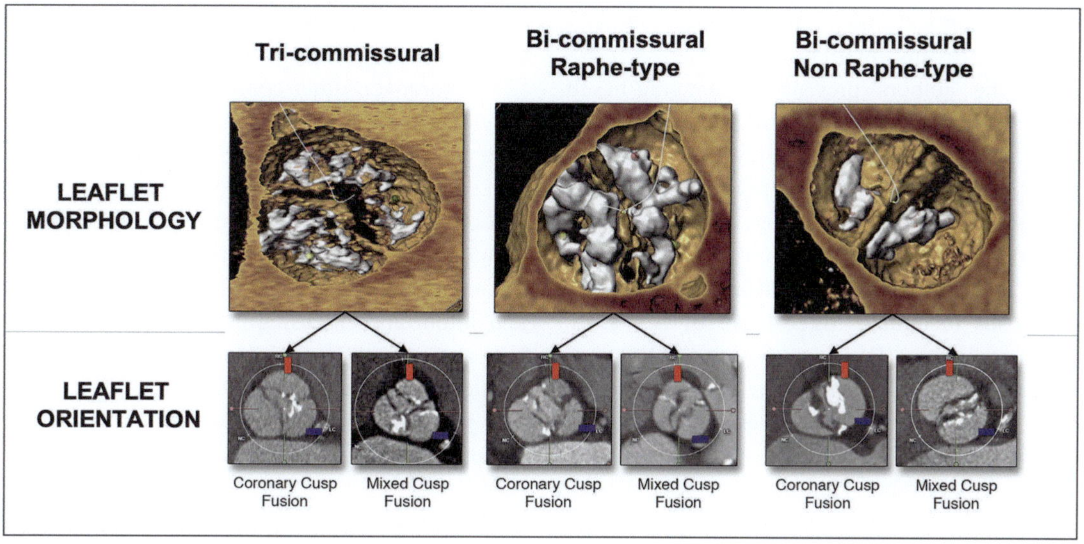

Fig. 5.5 CT-based classification of BAV phenotypes by Jilaihawi et al. Leaflet morphology is classified on the basis of number of commissures (two or three) and, in the presence of two commissures, the presence or absence of a raphe. This classification yields tricommissural, bicommissural raphe type, and bicommissural non-raphe types. Leaflet orientation is classified on the basis of cusp fusion, which is either coronary cusp fusion or mixed non-coronary–coronary cusp fusion. Take off of the right coronary artery is indicated by the red line; take off of the left coronary artery is indicated by the blue line. (*Modified from H. Jilaihawi et al. J Am Coll Cardiol Img. 2016;9:1145–58*)

specific risks. For instance, patients with RL fusion are more prone to ascending aortic aneurysms, whereas those with L-NC fusion may have a higher risk of aortic stenosis. Understanding these epidemiological and morphological variations is crucial for tailoring management strategies and surgical planning to the specific needs of different populations, thereby improving patient outcomes on a global scale [2]. However, the significant heterogeneity of the classifications previously reported risks being of limited clinical utility and may not effectively guide aortic valve repair, TAVI procedures, or predict outcomes. One of the most debated topics is the optimal classification system for BAV. With multiple classifications proposed, ranging from simple morphological classifications like Sievers' system to more detailed imaging-based classifications like Jilaihawi's CT-based system, there is no consensus on which approach best predicts clinical outcomes or guides treatment decisions [7, 11]. The lack of standardization complicates comparative studies and may lead to inconsistencies in clinical management.

To try to overcome these limitations and homogenize the nomenclature of BAV, an *International consensus statement on nomenclature and classification of the congenital bicuspid aortic valve and its aortopathy, for clinical, surgical, interventional and research purposes* has been published in 2021, with the endorsement of the Heart Valve Society (HVS), European Association of Cardiovascular Imaging (EACVI), Society of Thoracic Surgeons (STS), American Association for Thoracic Surgery (AATS), Society for Cardiovascular Magnetic Resonance (SCMR), Society of Cardiovascular Computed Tomography (SCCT), North American Society for Cardiovascular Imaging (NASCI), and the International Bicuspid Aortic Valve Consortium (BAVCon). This **INTERNATIONAL CONSENSUS STATEMENT** was draft with the intention to be universally used by clinicians (both pediatric and adult), echocardiography sonographers and physicians, cardiovascular advanced-imaging specialists, interventional cardiologists, cardiovascular surgeons, pathologists, geneticists, and researchers. It is designed for clinical, surgical, interventional, and research purposes. In this international consensus statement, the authors present an imaging-based, descriptive, simple but-comprehensive and language-intuitive nomenclature and classification system that is based on the English language and not on numbers or letters and is based on important and available anatomical, clinical, surgical, and pathological scientific data [10]. This consensus recognizes three types of BAVs: (1) The fused type (RL cusp fusion, R-NC cusp fusion and L-NC cusp fusion phenotypes); (2) The 2-sinus type (latero-lateral and antero-posterior phenotypes); and (3) The partial-fusion (forme fruste) type. The presence of raphe and the symmetry of the fused type phenotypes are described, since these are critical aspects in particular for surgical repair planning (Fig. 5.6). While the commissural angle of the non-fused cusps is usually symmetric in 2-sinus BAV, in the fused type phenotypes, it can be symmetrical (when the angle is close to 180°) or with different degree of asymmetry (Fig. 5.7).

The International Consensus also recognizes three types of BAV-associated aortopathy: (1) The ascending phenotype; (2) The root phenotype; and (3) Extended phenotypes. In the ascending phenotype, the preferential dilatation is at the level of the tubular ascending aorta; the root phenotype involves preferential dilatation of the root; finally, the extended phenotype shows dilatation of more than one segment among the root, the ascending aorta, and the arch (Fig. 5.8).

5.2 The Coronary Arteries in BAV

Certain common variations in coronary artery patterns frequently occur in BAV, alongside other anatomical anomalies. Notably, the presence of a left-dominant coronary artery system appears particularly prevalent in BAV patients, with rates ranging from 25% to 75%, significantly higher than the 10–12% observed in the general population with normal aortic valves [16–20]. The association of congenital BAV and occurrence of a single coronary artery is a rare described variant [21]. Moreover, other common variants in coronary distribution include short left main coronary artery (LMCA) length and separate left anterior descending and circumflex ostia [16, 17]. Some authors found that BAVs with LR fusion without a raphe show more left dominant coronary artery system and separate ostia, which could be of clinical importance in using coronary artery perfusion as a means of myocardial protection during surgical aortic valve replacement [22]. In addition, this BAV phenotypes also seem more at risk of developing significant coronary artery disease [22].

As previously described, the origins of the coronary arteries can vary according to BAV morphology. In R-L BAVs, both coronary arteries typically originate from the common anterior sinus. In R-NC and L-NC BAVs, each coronary artery normally arises from its respective cusp [10]. In BAV anatomy location of coronary ostia can be higher than usual, closer to the sinotubular junction or more superior from the aorta [20], and/or highly eccentric and closer to the non-coronary commissure than in TAV (Figs. 5.9 and 5.10).

All these anomalies not only can complicate surgical procedures such as aortic valve replacement or repair, but have also important implications for TAVI. For example, the location of the coronary ostia very close to the commissures may lead to an increased risk of coronary obstruction by the native leaflets tilted up by the THV frame, mostly when leaflets are long and bulky. Moreover, the presence and location of the raphe matter in case of TAVI. For instance, a calcific raphe located between the noncoronary and right coronary cusp might favor an asymmetrical displacement of the THV and leaflets towards the left coronary artery ostium, with an increased risk of obstruction. Differently, when the raphe is between the coronary cusps, the THV asymmetrical (ante-raphe) expansion may create a free space in front of the coronary ostia, which might

	BAV TYPES		
	FUSED BAV	**2-SINUS BAV**	**PARTIAL-FUSION BAV (forme fruste)**
PREVALENCE	90-95%	5-7%	unknown
AORTIC SINUSES	3	2	3
CUSPS	2 (Usually different size/shape with asymmetric or symmetric non-fused commissural angle)	2 (Roughly same size/shape with symmetric non-fused commissural angle)	3 (Usually Symmetric)
COMMISSURES	2	2	3 (3 «apparent» commissures where 2 are normal and the third is fused <50%)
RAPHE	Common, visible or not	No	Small, mini-raphe
SPECIFIC PHENOTYPES	**RL cusp fusion** 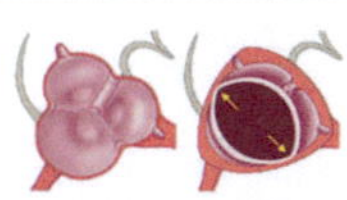	**Latero-lateral** 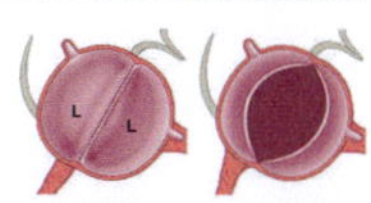	**Partial (<50%) fusion of one commissure** 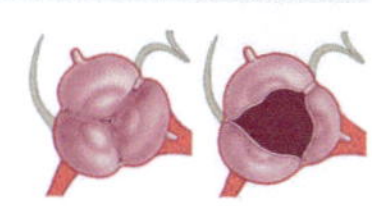
	R-NC cusp fusion 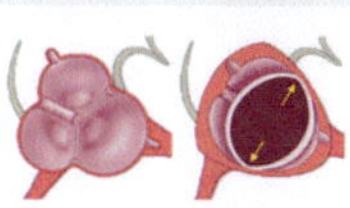	**Antero-posterior** 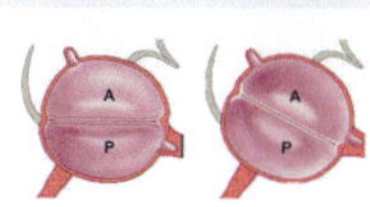	
	L-NC cusp fusion 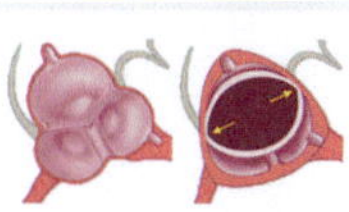		
	Indeterminate cusp fusion 		

Fig. 5.6 Types and specific phenotypes of the congenital BAV according to the International consensus statement on nomenclature and classification of the congenital bicuspid aortic valve and its aortopathy, for clinical, surgical, interventional and research purposes. There are three major types of BAVs and each type has specific phenotypes: fused BAV (RL cusp fusion, R-NC fusion, L-NC fusion and indeterminate phenotypes); 2-sinus BAV (latero-lateral and antero-posterior phenotypes) and partial-fusion BAV or forme fruste BAV (small raphe, single phenotype). Symmetrical or asymmetrical refers to the angle of the commissures of the non-fused cusp as detailed in this figure. (*Modified with permission from Michelena HI et al. International consensus statement on nomenclature and classification of the congenital bicuspid aortic valve and its aortopathy, for clinical, surgical, interventional and research purposes. Eur J Cardiothorac Surg. 2021;60(3):448–76*)

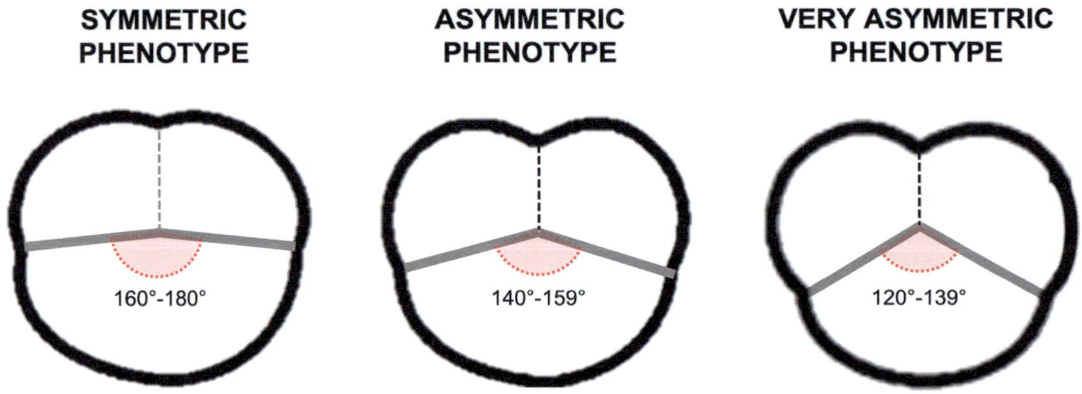

Fig. 5.7 Different commissural angle of the non-fused cusps in the fused type phenotype. The commissural angle of the non-fused cusps in the fused type phenotype can be symmetrical (when the angle is close to 180°) or with different degree of asymmetry (asymmetric when the angle is 140–159° and very asymmetric when is 120–139°)

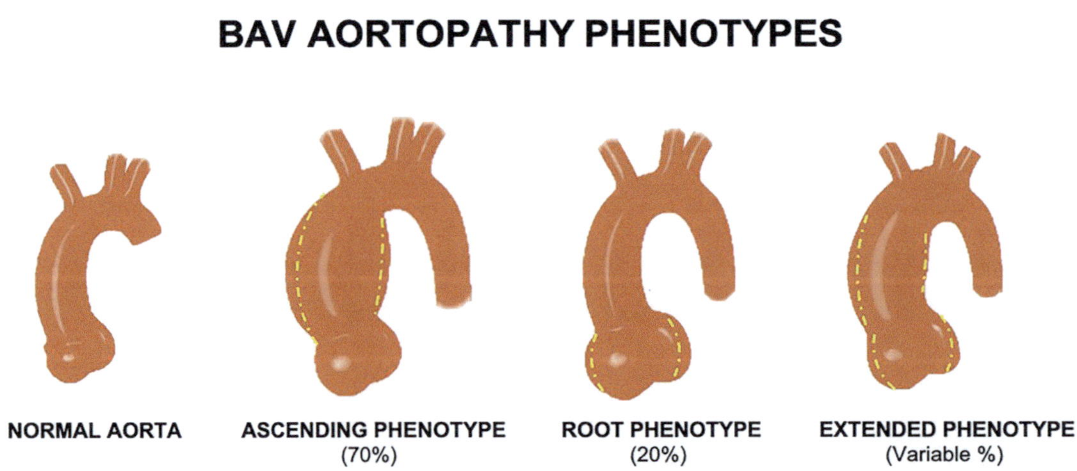

Fig. 5.8 BAV aortopathy phenotypes. On the left is a normal aorta. Moving to the right, the second example refers to the ascending phenotype, where the preferential dilatation is at the level of the tubular ascending aorta. The third example is the root phenotype, which involves preferential dilatation of the root. The last on the right is the extended phenotype, which shows dilatation of different combination of aorta segments including the root, the ascending aorta and the arch

"protect" coronary ostia from the risk of occlusion and facilitate the coronary cannulation [23]. Again, when the angle between 2 coronary ostia is far from the expected 120° and/or when coronary arteries do not originate from the exact center of the AV cusp (bisector), there is an increased risk of partially overlapping the coronary ostium with one of the THV commissures despite commissural alignment techniques [24, 25] (see Chap. 14 for more details).

5.3 The Conduction System in BAV

As far as the conduction system, the AV node is located at the apex of the triangle of Koch, with the His bundle passing under the membranous septum towards the interventricular septum, in correspondence with the interleaflet triangle between the posterior and right anterior semilunar cusps. Aortic valve dysfunction, regardless of

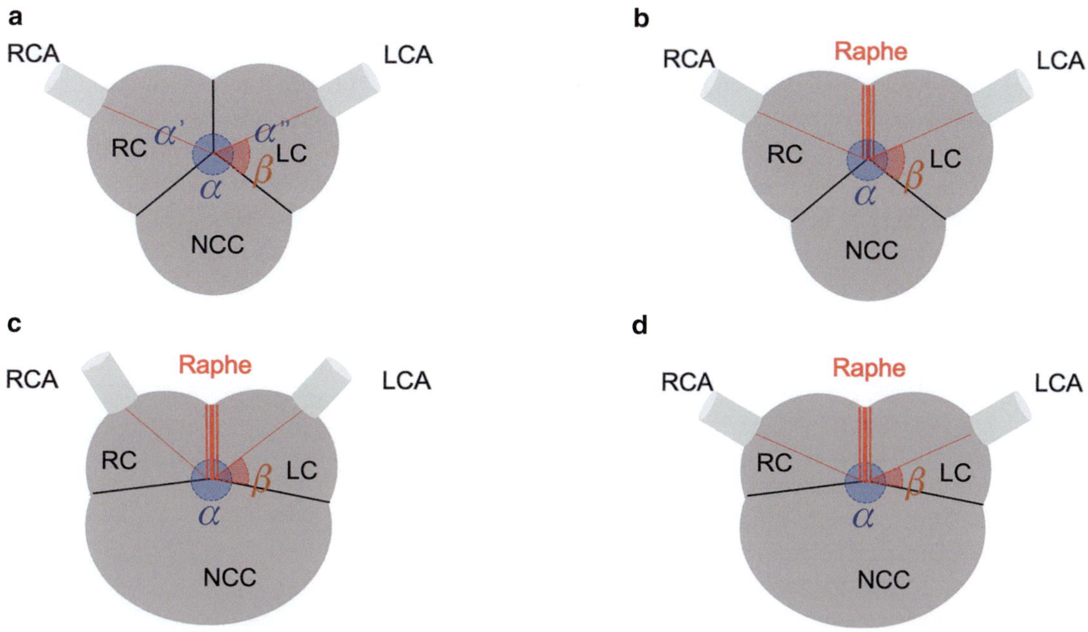

Fig. 5.9 Coronary ostia anatomy in BAV. This figure shows some of the possible coronary ostia configuration in tricuspid aortic valve and in BAV. Panel **a** shows the standard normal anatomy of tricuspid aortic valve, where the three commissures are at 120° (angle $\alpha = \alpha' = \alpha''$) and each coronary ostia are in the bisector, at 60° from the commissures (angle β). Panel **b** shows a tricommissural raphe type BAV, with very asymmetric non fused commissures (angle α 120–139°). In this case, coronary ostia can still be in the bisector, at 60° from the nonfused commissures (angle β). Panel **c** shows another tricommissural raphe type BAV, but with very symmetric non fused commissures (angle α close to 180°). In this example, coronary ostia is again in the bisector of each respective cusp, but the distance from the nonfused commissures (angle β) is inferior than in panel **b**. Finally, panel **c** shows how, in many BAV cases, one or both coronary ostia can be in an eccentric position respect to their respective sinus, close to one commissure (angle β <30°)

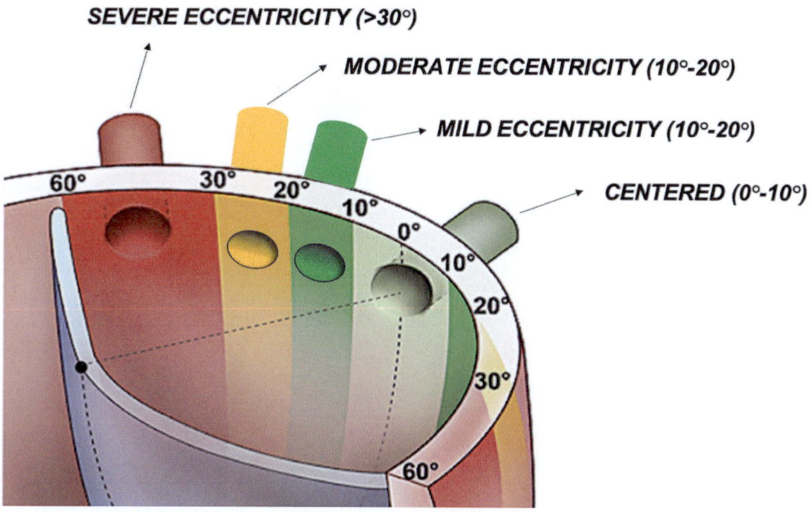

Fig. 5.10 Coronary Ostial Eccentricity. Coronary ostial eccentricity is defined by the angle between the coronary ostium and the center-cusp as follows: centered (0–10°), mild eccentricity (10–20°), moderate eccentricity (20–30°), and severe eccentricity (>30°). (*Modified from Wang, X. et al. J Am Coll Cardiol Intv. 2022;15 (2):123–134*)

the leaflet morphology and burden of calcification, can affect the conduction system, including the AH interval, HV interval, and PR interval [26]. Patients with BAV may be associated with more frequent infra-Hisian conduction disease requiring PPM, as documented by electrophysiologist evaluation [27]. In BAV patients, the spatial relationship between the valve cusps, commissures, raphe, and the AV conduction system is crucial for avoiding conduction disturbances during surgical or transcatheter procedures [28]. Based on the previous studies, having a BAV was—per se—a significant predictor of PPM implantation after aortic valve or aortic root surgery, due to an increased risk of damaging the conduction system during surgical valve manipulation [28–30]. In particular, in cases where there is R-NC cusp fusion, the raphe is located above the membranous septum over the His bundle course, potentially placing the conduction system at even increased risk during surgical excision (Fig. 5.11). This risk is heightened also in cases where the raphe is prominent or where the commissures are close to the conduction tissue, for example in the presence of a short membranous septum length [31]. Even if the association between one particular BAV phenotype and the risk of conduction disorders after surgery is debated, surgeons must be aware of these anatomical relationships to minimize the risk of postoperative complications [8, 30].

In patients undergoing TAVI, the rate of new conduction abnormalities requiring PPM after intervention seems to be similar between patients with TAV or BAV and the depth of valve implantation remains a strong predictors of conduction disturbances after TAVI (Fig. 5.12) [32–34]. However, the BAV phenotype can have an impact (at least theoretically) on the rate of new conduction disorders. Jilaihawi et al. found a trend towards an increased incidence of new PPM after TAVI in patients with fused BAV with LR cusp fusion pattern [11]. The hypothesis was that in patients with LR raphe, this led to a more asymmetrical valve expansion opposite to the fused raphe, enhancing the contact pressure of stent frame against the septum with subsequent injury of the conduction system (Fig. 5.13). In another pooled analysis of association of Sievers BAV

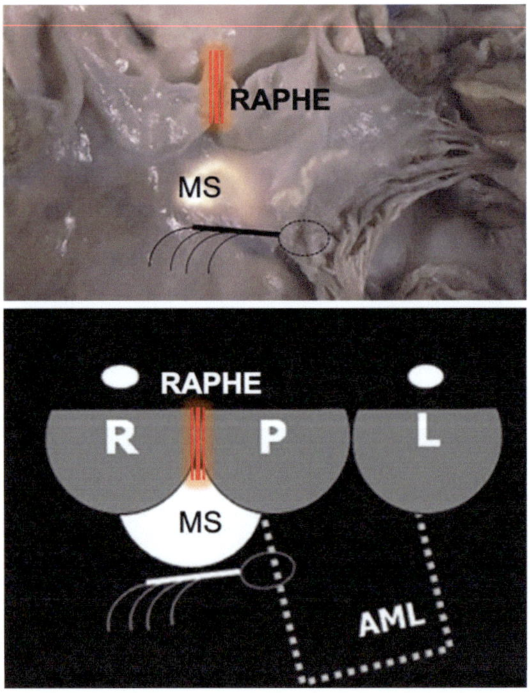

Fig. 5.11 Anatomical relationship between fusion pattern and conduction system. The course of His bundle is underneath the membranous septum. When BAV fusion pattern involves the posterior and right anterior cusps, the raphe is just over the membranous septum and the His bundle. R = right coronary cusp, P = posterior non coronary cusp. L = left coronary cusp, AML = anterior mitral leaflet. (*Modified from Thiene G. et al. Cardiovasc Pathol. 2024 May-Jun:70:107604*)

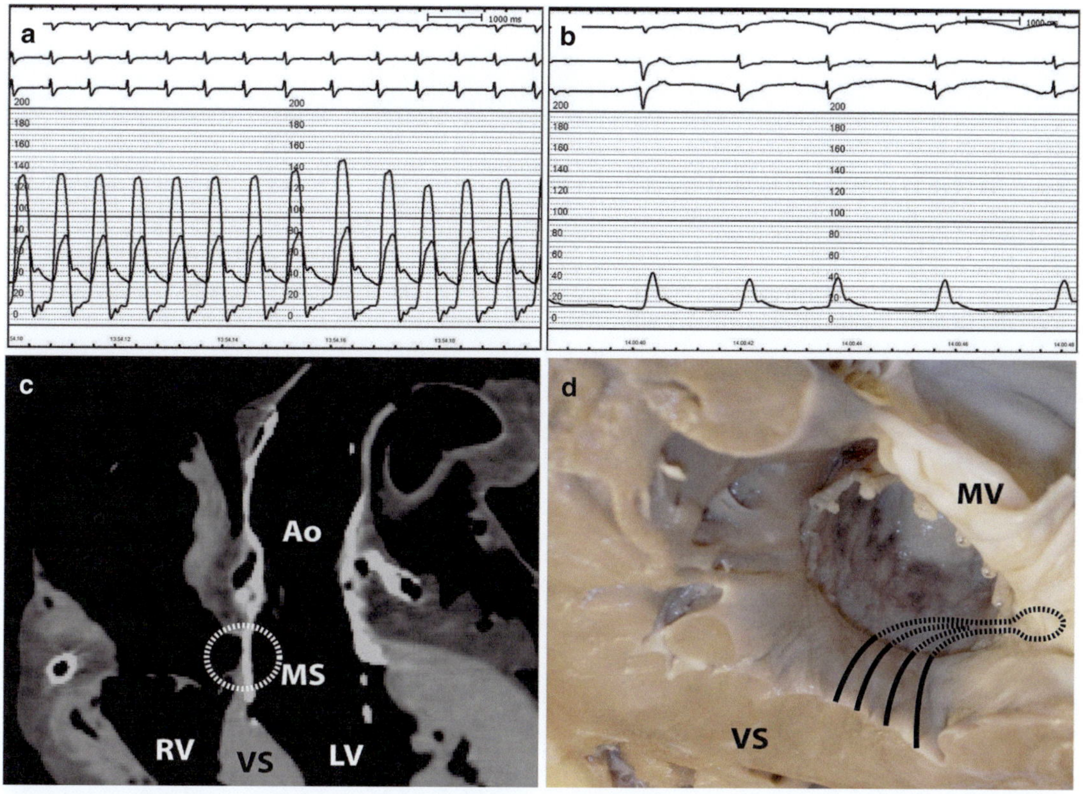

Fig. 5.12 Views of an 85-year-old woman who had developed complete AV block requiring PPM implantation after TAVI who died 13 days after the procedure. (**a**) Electrocardiogram showing right BBB and 60 mmHg transaortic gradient at baseline. (**b**) Electrocardiogram showing complete AV block immediately after TAVI. (**c**) Computed tomography scan of heart explanted at autopsy. Note, deep positioning of CoreValve within left ventricular outflow tract, overlapping membranous septum (*dotted circle*) and crest of interventricular septum. (**d**) Gross anatomic view of left ventricular outflow seen from below. Note, expansion of prosthesis frames in subaortic region compressing ventricular septum and overlapping proximal branching of left bundle branch (*dotted lines*). Ao, aorta; LV, left ventricle; MS, membranous septum; MV, mitral valve; RV, right ventricle; VS, ventricular septum. (*From Chiara Fraccaro, Gianfranco Buja, Giuseppe Tarantini, Valeria Gasparetto, Loira Leoni, Renato Razzolini, Domenico Corrado, Raffaele Bonato, Cristina Basso, Gaetano Thiene, Gino Gerosa, Giambattista Isabella, Sabino Iliceto, Massimo Napodano. Incidence, predictors, and outcome of conduction disorders after transcatheter self-expandable aortic valve implantation. Am J Cardiol. 2011 Mar 1;107(5):747–54. doi: 10.1016/j.amjcard.2010.10.054. Epub 2011 Jan 19*)

morphology with new PPM and conduction abnormalities after TAVI, Zhang et al. found that Sievers type 1 BAV morphology was associated with increased risk of post-TAVI PPM implantation and conduction abnormalities compared to type 0 [35]. L-R fusion as the most prevalent subtype in type 1 BAV morphology could be the driving factor that brought about the association of type 1 BAV morphology with conduction abnormalities and PPI. However, when this LR fusion pattern was compared to the other patterns (L-NC or R-NC pattern), no difference was found and this mechanic theory require more evidences to be confirmed [35].

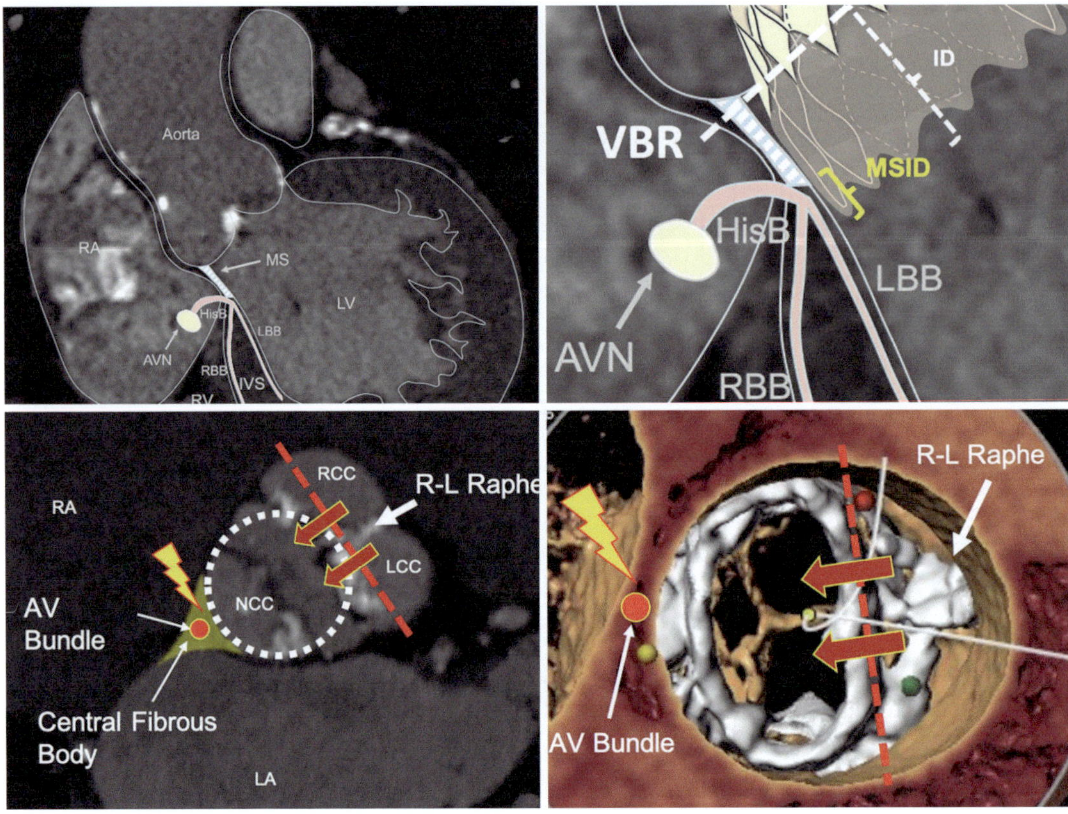

Fig. 5.13 Risk of His conduction system damage in BAV with fusion pattern during TAVI. In patients with fusion of left and right coronary cusp, the raphe is located opposite to the membranous septum. This may cause an elliptical asymmetric expansion of transcatheter valve frame far from the raphe and against the His and the left bundle branch, thus theoretically increasing the risk of conduction system damage and need for PPM

References

1. Thiene G, Rizzo S, Basso C. Bicuspid aortic valve: the most frequent and not so benign congenital heart disease. Cardiovasc Pathol. 2024;70:107604.
2. Fedak PWM, Verma S, David TE, Leask RL, Weisel RD, Butany J. Clinical and pathophysiological implications of a bicuspid aortic valve. Circulation. 2002;106(8):900–4.
3. Roberts WC. The congenitally bicuspid aortic valve. A study of 85 autopsy cases. Am J Cardiol. 1970;26(1):72–83.
4. Brandenburg RO, Tajik AJ, Edwards WD, Reeder GS, Shub C, Seward JB. Accuracy of 2-dimensional echocardiographic diagnosis of congenitally bicuspid aortic valve: echocardiographic-anatomic correlation in 115 patients. Am J Cardiol. 1983;51(9):1469–73.
5. Angelini A, Ho SY, Anderson RH, Devine WA, Zuberbuhler JR, Becker AE, et al. The morphology of the normal aortic valve as compared with the aortic valve having two leaflets. J Thorac Cardiovasc Surg. 1989;98(3):362–7.
6. Sabet HY, Edwards WD, Tazelaar HD, Daly RC. Congenitally bicuspid aortic valves: a surgical pathology study of 542 cases (1991 through 1996) and a literature review of 2,715 additional cases. Mayo Clin Proc. 1999;74(1):14–26.
7. Sievers HH, Schmidtke C. A classification system for the bicuspid aortic valve from 304 surgical specimens. J Thorac Cardiovasc Surg. 2007;133(5):1226–33.
8. Schaefer BM, Lewin MB, Stout KK, Gill E, Prueitt A, Byers PH, et al. The bicuspid aortic valve: an integrated phenotypic classification of leaflet morphology and aortic root shape. Heart. 2008;94(12):1634–8.
9. Kang JW, Song HG, Yang DH, Baek S, Kim DH, Song JM, et al. Association between bicuspid aortic valve phenotype and patterns of valvular dysfunction and bicuspid aortopathy: comprehensive evaluation using MDCT and echocardiography. JACC Cardiovasc Imaging. 2013;6(2):150–61.
10. Michelena HI, Della Corte A, Evangelista A, Maleszewski JJ, Edwards WD, Roman MJ, et al. International consensus statement on nomenclature and classification of the congenital bicuspid aortic valve and its aortopathy, for clinical, surgical, inter-

ventional and research purposes. Eur J Cardiothorac Surg. 2021;60(3):448–76.
11. Jilaihawi H, Chen M, Webb J, Himbert D, Ruiz CE, Rodés-Cabau J, et al. A bicuspid aortic valve imaging classification for the TAVR era. JACC Cardiovasc Imaging. 2016;9(10):1145–58.
12. Sun BJ, Lee S, Jang JY, Kwon O, Bae JS, Lee JH, et al. Performance of a simplified dichotomous phenotypic classification of bicuspid aortic valve to predict type of valvulopathy and combined aortopathy. J Am Soc Echocardiogr. 2017;30(12):1152–61.
13. Murphy IG, Collins J, Powell A, Markl M, McCarthy P, Malaisrie SC, et al. Comprehensive 4-stage categorization of bicuspid aortic valve leaflet morphology by cardiac MRI in 386 patients. Int J Cardiovasc Imaging. 2017;33(8):1213–21.
14. Kusner JJ, Brown JY, Gleason TG, Edelman ER. The natural history of bicuspid aortic valve disease. Struct Heart. 2023;7(2):100119.
15. Michelena HI, Prakash SK, Della Corte A, Bissell MM, Anavekar N, Mathieu P, et al. Bicuspid aortic valve: identifying knowledge gaps and rising to the challenge from the International Bicuspid Aortic Valve Consortium (BAVCon). Circulation. 2014;129(25):2691–704.
16. Johnson AD, Detwiler JH, Higgins CB. Left coronary artery anatomy in patients with bicuspid aortic valves. Br Heart J. 1978;40(5):489–93.
17. Murphy ES, Rösch J, Rahimtoola SH. Frequency and significance of coronary arterial dominance in isolated aortic stenosis. Am J Cardiol. 1977;39(4):505–9.
18. Higgins CB, Wexler L. Reversal of dominance of the coronary arterial system in isolated aortic stenosis and bicuspid aortic valve. Circulation. 1975;52(2):292–6.
19. Hutchins GM, Nazarian IH, Bulkley BH. Association of left dominant coronary arterial system with congenital bicuspid aortic valve. Am J Cardiol. 1978;42(1):57–9.
20. Lerer PK, Edwards WD. Coronary arterial anatomy in bicuspid aortic valve. Necropsy study of 100 hearts. Br Heart J. 1981;45(2):142–7.
21. Rashid A, Saucedo JF, Hennebry TA. Association of single coronary artery and congenital bicuspid aortic valve with review of literature. J Interv Cardiol. 2005;18(5):389–91.
22. Koenraadt WMC, Tokmaji G, DeRuiter MC, Vliegen HW, Scholte AJHA, Siebelink HMJ, et al. Coronary anatomy as related to bicuspid aortic valve morphology. Heart. 2016;102(12):943–9.
23. Tarantini G, Fabris T. Transcatheter aortic valve replacement for bicuspid aortic valve stenosis: a practical operative overview. Circ Cardiovasc Interv. 2021;14(7):e009827.
24. Wang X, De Backer O, Bieliauskas G, Wong I, Bajoras V, Xiong TY, et al. Cusp symmetry and coronary ostial eccentricity and its impact on coronary access following TAVR. JACC Cardiovasc Interv. 2022;15(2):123–34.
25. Tarantini G, Fabris T, Nai Fovino L, Cardaioli F, Pergola V, Montonati C, et al. Definition of trAnscatheter heart valve orIeNtation in biCuspId aortic valve: the DA VINCI pilot study. Front Cardiovasc Med. 2022;9:1056496. https://www.scopus.com/inward/record.uri?eid=2-s2.0-85144964348&doi=10.3389%2ffcvm.2022.1056496&partnerID=40&md5=12d4bf8b0cb05d7260b9153f60bfea23.
26. Friedman HS, Zaman Q, Haft JI, Melendez S. Assessment of atrioventricular conduction in aortic valve disease. Br Heart J. 1978;40(8):911–7.
27. Biswas M, Sandhu A, Kay JD, Khanna AD, Collins KK, Sauer WH, et al. Patients with bicuspid aortic valves may be associated with infra-hisian conduction disease requiring pacemakers. J Interv Card Electrophysiol. 2021;61(1):29–35.
28. Fukuda T, Hawley RL, Edwards JE. Lesions of conduction tissue complicating aortic valvular replacement. Chest. 1976;69(5):605–14.
29. Ribeiro V, Mota Garcia R, Frutuoso C, Melão F, Pereira M, Pinho P, et al. Permanent pacemaker implantation after aortic valve replacement: long-term dependency or rhythm recovery? Rev Port Cardiol. 2015;34(9):529–33.
30. Guglielmetti L, Nazif T, Sorabella R, Akkoc D, Kantor A, Gomez A, et al. Bicuspid aortic valve increases risk of permanent pacemaker implant following aortic root replacement. Eur J Cardiothorac Surg. 2016;50(3):497–503.
31. Nakashima M, Jilaihawi H, He Y, Williams D, Pushkar I, Williams M, et al. Membranous septum length predicts new conduction abnormalities in surgical aortic valve replacement: a novel predictor for permanent pacemaker implantation after surgical aortic valve replacement. J Surg Res. 2024;295:385–92.
32. Yoon SH, Bleiziffer S, De Backer O, Delgado V, Arai T, Ziegelmueller J, et al. Outcomes in transcatheter aortic valve replacement for bicuspid versus tricuspid aortic valve stenosis. J Am Coll Cardiol. 2017;69(21):2579–89.
33. Fraccaro C, Buja G, Tarantini G, Gasparetto V, Leoni L, Razzolini R, et al. Incidence, predictors, and outcome of conduction disorders after transcatheter self-expandable aortic valve implantation. Am J Cardiol. 2011;107(5):747–54.
34. Fraccaro C, Napodano M, Tarantini G. Conduction disorders in the setting of transcatheter aortic valve implantation: a clinical perspective. Catheter Cardiovasc Interv. 2013;81(7):1217–23.
35. Zhang J, Li X, Xu F, Chen Y, Li C. Pooled-analysis of association of sievers bicuspid aortic valve morphology with new permanent pacemaker and conduction abnormalities after transcatheter aortic valve replacement. Front Cardiovasc Med. 2022;9:884911.

Genetic Background and Familiarity

6

Marco Cason, Barbara Bauce, Gaetano Thiene, and Kalliopi Pilichou

The bicuspid aortic valve (BAV) is a congenital heart defect characterized by two cusps instead of the three in the aortic valve. This anatomical anomaly is relatively common, occurring in about 1–2% of the general population [1–5]. The heart develops from mesoderm, while certain components of the cardiac outflow tracts originate from neural crest cells, which are derivatives of ectoderm [6]. The neural crest plays a role in both the septation of the cardiac outflow and the formation of smooth muscle cells in the wall of the ascending aorta. It has been suggested that neural crest abnormalities may be a contributing factor to aortic coarctation and the aortopathy frequently associated with BAV [7]. While often asymptomatic during childhood, BAV is a leading cause of aortic valve disease in adults. Individuals with BAV, frequently affected by coarctation of the aorta [8], are at increased risk of developing complications such as aortic valve stenosis, regurgitation, and ascending aortic aneurysms [8–10]. Furthermore, family members of individuals affected by BAV syndrome may be clinically silent; however they also face significant risks [11, 12].

From a genetic standpoint, BAV is a fascinating condition because of its varied phenotypic presentations and the complex inheritance patterns observed in affected families. This chapter explores the genetic background of BAV, including the key genes involved, mechanisms of inheritance, and how genetics contribute to phenotypes and clinical outcomes.

6.1 Inheritance Patterns and Familial Clustering

Although the majority of BAV occurs sporadically [4], there is significant evidence of familial clustering [13]. The prevalence of familial BAV appears to be in between 4% and 11% of first-degree relatives (FDRs) of affected individuals [14–17]. This suggests an autosomal dominant pattern of inheritance with reduced penetrance and variable expressivity [14, 18] (Fig. 6.1).

A male-to-female predominance has also been demonstrated with a ratio of 3:1 [3, 14, 15, 17]. A 2019 study conducted on 724 FDRs of probands affected by BAV, identified 46 (6.4%) carriers of BAV of which 71.7% were men. According to the fusion pattern of commissures, three morphological types were considered in the study:

M. Cason · G. Thiene · K. Pilichou (✉)
Cardiovascular Pathology Unit, Department of Cardiac, Thoracic, Vascular Sciences and Public Health, School of Medicine, University of Padua, Padua, Italy
e-mail: marco.cason@unipd.it; gaetano.thiene@unipd.it; kalliopi.pilichou@unipd.it

B. Bauce
Cardiology Clinic Unit, Department of Cardiac, Thoracic, Vascular Sciences and Public Health, School of Medicine, University of Padua, Padua, Italy
e-mail: barbara.bauce@unipd.it

G. Thiene et al. (eds.), *Bicuspid Aortic Valve Syndrome*,
https://doi.org/10.1007/978-3-031-85589-4_6

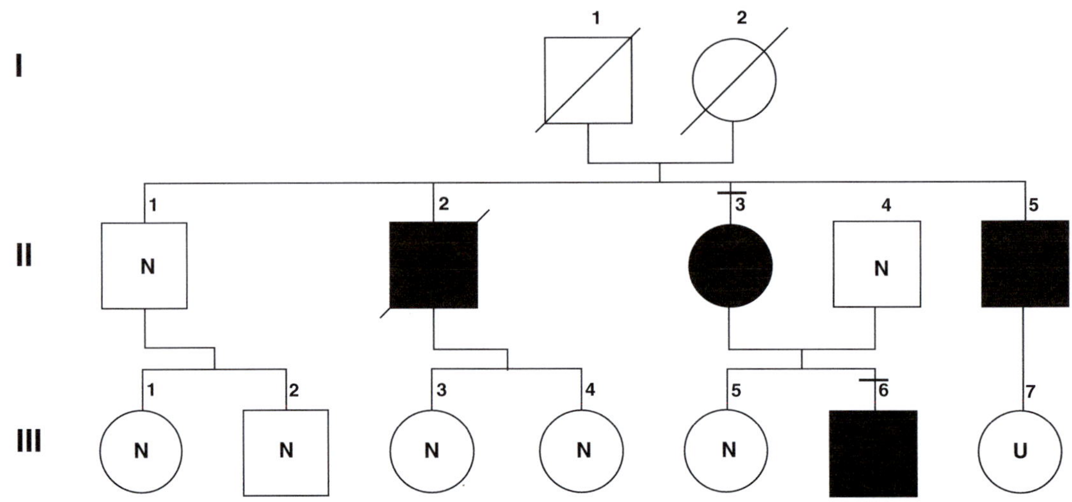

Fig. 6.1 Pedigrees of a family reported by Clementi since 1996 [13] in which 4 members of both sexes in two generations are affected with BAV. N normal at investigation, U: not investigated

Table 6.1 Syndromes with BAV

Syndrome	Gene(s)	Structural features
Loeys-Dietz	*TGFβR1, TGFβR2, TGFβ2, TGFβ3, SMAD3*	Bifid uvula, HTAD
Multisystemic smooth muscle dysfunction	*ACTA2*	Mydriasis, HTAD
Down	47,XY+21; 47,XX+21	Atrioventricular septal defects
Turner	45,X	Short stature, coarctation of the aorta
Marfan	*FBN1*	Aortic root aneurism, dissection
Velocardiofacial	22q11.2 deletion	Truncus arteriosus, tetralogy of Fallot

HTAD hereditary thoracic aortic aneurism and dissection, *TGFβR1 and TGFβR2* Transforming Growth Factor Beta Receptor -1 and -2, *TGFβ2 and TGFβ3* Transforming Growth Factor Beta-2 and -3, *SMAD3* SMAD Family Member 3, *ACTA2* Actin Alpha 2, Smooth Muscle, *FBN1* Fibrillin 1

1. Type R-L with the fusion of the right and left cusps.
2. Type R-N, fusion of the right and non-coronary cusps.
3. Type L-N, fusion of the left and non-coronary cusps.

The distribution of BAV subtypes among family members revealed the following pattern: 34 cases (73.9%) exhibited fusion of the right and left cusps (R-L), 11 cases (23.9%) showed fusion of the right and non-coronary cusps (R-N), while only 1 case (2.2%) displayed fusion of the left and non-coronary cusps (L-N). However, statistical analysis failed to demonstrate a significant correlation between the BAV morphotypes of probands and those of their relatives, with an observed concordance of 56% [17]. This supports the hypothesis of a strong genetic influence, though environmental and epigenetics factors may also play a role during the embryological phase, in modulating the phenotypic expression of the disease.

Inheritance is observed in 47–89% of families [15, 17] suggesting a strong involvement of genetic factors in the etiology of BAV. This vast range of heritability is due to the genetically heterogeneous nature of BAV with a wide spectrum of disease-causing or risk variants. In this context, BAV could be the result of highly penetrant variants in single genes, but also a result of a complex genetic background [19–21]. BAV forms could be syndromic (Table 6.1) or non-syndromic.

Among genetic syndromes, Turner Syndrome (TS) features the highest prevalence of BAV (30% women) usually in association with coarctation of the aorta (CoA) thoracic aortic aneurysm (TAA) [22] and dissection [23]. This observation led to the hypothesis that copy number variation (CNV) of X chromosome genes, including *KDM6A* and *TIMP1*, may predispose to BAV formation [24, 25]. Loeys-Dietz syndrome (LDS) shows lower prevalence of BAV compared to TS, but 10% of patients harbor the congenital defect. LDS is caused by dominantly inherited pathogenic (P) variants in TGF-β pathway genes (*TGFβ2*, *TGFβ3 TGFβR1*, *TGFβR2*, and *SMAD3*).

BAV and aortic dilation are also more prevalent in Velocardiofacial syndrome (VCFS). Large genomic deletions of 22.q11.2 cause VCFS, and haploinsufficiency of *TBX1* is strongly associated with cardiac abnormalities [26]. However, P variants of single VCFS genes do not appear to be prevalent in non -syndromic BAV cases. The prevalence of BAV in Marfan Syndrome (MS) is quite significant (4–13%) [27, 28]. P variants in *fibrillin-1* (*FBN1*) gene, which codes for an extracellular glycoprotein that is secreted by vascular smooth muscle cells and regulates the structural integrity of the aortic media, have been linked to MS. *FBN1* alterations result in progressive disruption of the media by matrix metalloproteinase [29, 30]. Of note, rare FBN1 variants were identified in BAV patients who presented aortic root aneurysm but had no features of MS [31].

Other genetic disorders that include BAV as an occasional feature are Down, Alagille, and Kabuki syndromes [32] (see Table 6.1).

Most patients with BAV do not have syndromic disease but may present other congenital heart and vascular defects with variable disease severity [33]. Among specimens with cardiac malformations from the Congenital CardioVascular Registry at the University of Padua, BAV was found in 51% of ventricular septal defect (VSD) with aortic arch obstruction, in 37% of adult aortic arch coarctation, in 20.5% of isolated VSD, in 7.5% of AV canal, in 2% of Tetralogy of Fallot and only 1% of complete Transposition of great artery. The latter finding would suggest that its occurrence is incidental, similar to the normal population. Overall, BAV was observed in 67 of 474 (14%) CHD cases [33]. Other lesions such as patent ductus arteriosus (8.5%), mitral valve abnormalities (11%), and TAA (50%) have been reported in BAV.

Single P variants in genes were found in monogenic non-syndromic cases of BAV [34–37]. On the other hand, a substantial fraction of BAV heritability is polygenic, with an estimated single nucleotide polymorphism (SNP)-based heritability of 9.55 ± 3.7%. This finding indicates that at least 20% of the heritability of BAV is due to the influence of multiple variants with a relatively high frequency in the population, so-called SNPs, rather than by single P variants [21].

6.2 Key Genes Involved in BAV Development

Cumulative data demonstrate that BAV is caused, in contrast to cardiomyopathies [98], by rare or private variants in many different genes that each contribute to a small proportion of cases (Table 6.2). The first studies, conducted by linkage analysis of single variants in family pedigrees, uncovered multiple genetic loci associated with BAV. In recent years, the advent of Next Generation Sequencing has helped researchers find new genes associated with the condition. Finally, genome-wide association studies (GWAS) have provided valuable insights into the genetic architecture of BAV. In genetic epidemiology, a GWAS is an investigation of the whole genome of different individuals of a particular species to determine gene variations among the individuals under investigation. An attempt is then made to associate the observed differences with particular traits or diseases. These studies have identified several common genetic variants associated with an increased risk of BAV, although most of these variants have small individual effects and most cases of familial BAV remain unsolved to this day (Table 6.2) [19, 21, 38].

Table 6.2 Genes associated with human bicuspid aortic valve and supporting mouse models

Gene	OMIM	Function	Type of study	Associated cardiac defects in addition to BAV in humans	PMID human studies	Mouse models	Valve phenotype in mouse models	Other cardiac defects in mouse models	PMID mouse models study
NOTCH1	190198	Cell specialization and fate determination, cell proliferation and apoptosis	Family base NGS	AVS, CAVD, HLHS, TOF, PVS, VSD, CoA, TAA	16,025100 18593716 16729972 17662764 26820064	Notch1þ/-; Nos3-/-	BAV, thickened AoV and PV	Ascending aortic dilation, VSD, overriding aorta	23583836 27107132
						Notch1+/– mTRG1--3 (mTR-/- generation 1--3)	AVS, Thickened AoV, CAVD, PVS	Ascending aortic dilation, ASD, VSD	28346225
GATA5	611496	Cardiac and aortic development	Family base direct sequencing	ASD, VSD, DORV, TOF	22641149 24796370 24638895	Gata5-/-	BAV	Mild LV hypertrophy	21633169
GATA4	600576	Organogenesis (cardiac development)	GWAS	ASD, VSD, AVSD, PVS, TOF	29,325903 32748548	NA			
GATA6	601656	Organogenesis and cell differentiation	Family base direct sequencing	PTA, TOF	29653232	Gata6 +/-	BAV	None	29567669
NKX2.5	600584	Cardiac development	Family base direct sequencing	ASD, atrioventricular conduction delay, TOF, HLHS	25,438918	Nkx2.5þ/-	BAV	Atrial septal dysmorpho-genesis	11073884
SMAD6	602931	Vascular homeostasis	NGS	AVS, CoA, TAA	22275001 28659821 30796334	NA			
TBX20	606061	Transcription factors involved in the regulation of developmental processes	Targeted genotyping MLPA	TAA	30820038				
ROBO4	607528	Axon guidance and cell adhesion	Family base NGS	AVS, ASD, TAA	30455415	Robo4tm1Lex/tm1Lex	BAV, AVS, thickened AoV	None	30455415
ROBO1	602430	Axon guidance	Family base	NA	35534675	Robo1-/-; Robo2-/-	BAV	VSD	25691540

EXOC4	608185	Vesicle trafficking at neural synapses and vesicle targeting in polarized epithelial cells	GWAS	NA	31387361	NA			
EXOC6	609672	Tubulin polymerization inhibition	GWAS	NA	31387361	NA			
EXOC8	615283	Secretory vesicles targeting (post golgi)	GWAS	NA	31387361	NA			
PALMD	610182	Cytoskeletal regulation	GWAS	NA	35727948	NA			
TEX41			GWAS						
MYH6			GWAS						
MUC4			GWAS						
NOS3	163729	Enzyme involved in nitric oxide production	Family base NGS	NA	26708639 28387797	eNOS-/-	BAV	TAA	10821808 32801116
TAB2	605101	Cardiac development; adaptor involved in activation of JNK and NF-kappa-B signalling	Family base direct sequencing	Aortic stenosis, subaortic stenosis, ASD, TOF, VSD, myxomatous mitral/ tricuspid valves	26139517 20493459	NA			
MAT2A	601468	Enzyme involved in the methionine cycle	Family base direct sequencing	TAA	25557781	NA			
CELSR1	604523	Cell shape and cytoskeletal polarization	Family base NGS	HLHS	35133174	NA			
DSCAM	602523	NA	CNVs analysis	Early onset complications of BAV disease	39240962	NA			

(continued)

Table 6.2 (continued)

Gene	OMIM	Function	Type of study	Associated cardiac defects in addition to BAV in humans	PMID human studies	Mouse models	Valve phenotype in mouse models	Other cardiac defects in mouse models	PMID mouse models study
ADAMTS19	607513	Valve maturation and maintenance	Family base direct sequencing	AVS, aortic valve insufficiency, BAV, subaortic stenosis, PVS, pulmonary valve insufficiency, mitral/tricuspid valve insufficiency	31844321 32323311	Adamts19KO/KO	AVS, AR, thickened AoV, BAV	None	31844321
PCDHA9	604966	Neuronal cell-cell connections		LVOTO malformations	34888534	cdha9m/m	BAV	None	34888534
MIB1	608677	E3 ubiquitin ligase; Notch signalling regulator		LVNC	36325906	Quadruple heterozygous Mib1 Cep192 Tmx3;Bcl7a mice	BAV	LVNC, coronary artery defects	36325906

AoV aortic valve, *AR* aortic regurgitation, *ASD* atrial septal defect, *AVSD* atrioventricular septal defect, *AVS* aortic valve stenosis, *BAV* bicuspid aortic valve, *CAVD* calcific aortic valve disease, *CoA* coarctation of the aorta, *DORV* double outlet right ventricle, *GWAS* genome-wide association study, *HLHS* hypoplastic left heart syndrome, *JNK* c-Jun N-terminal kinase, *LV* left ventricle, *LVNC* left ventricular non-compaction, *LVOTO* left ventricular outflow tract obstruction, *MLPA* multiplex ligation-dependent probe amplification, *NA* not available (or not applicable), *NF-kappa-B* nuclear factor kappa-light-chain-enhancer of activated B cells, *NGS* next generation sequencing, *PMID* PubMed identifier, *PTA* persistent truncus arteriosus, *PV* pulmonary valve, *PVS* pulmonary valve stenosis, *TAA* thoracic aortic aneurysm, *TOF* tetralogy of Fallot, *VSD* ventricular septal defect.

6.2.1 The NOTCH Signaling Pathway

Notch Receptor 1 (*NOTCH1*) is currently the only gene identified as causative for isolated BAV using linkage analysis and positional cloning strategies. However, it is unlikely to account for more than a small proportion of familial cases [34, 39]. In cardiac valve development, Notch signaling provides a major intercellular signaling mechanism that mediates the crosstalk between adjacent cells and is essential in numerous steps: from the initiation of the endocardial-to-mesenchymal transformation (EMT) and endocardial cushion formation to the aortic valve formation and the embryonic pathogenesis of BAV [40].

Together with *RUNX2*, *SOX9* and *XCT790*, *NOTCH1* is also involved in the regulation of osteoblast-cell fate thanks to its repressor activity of *RUNX2*. However, further studies are necessary to define the role of *NOTCH1* genetic variants in Calcific Aortic Valve Disease [41]. In 2005, *NOTCH1* was the first gene associated with inherited aortic valve disease [34]. Authors described the co-segregation of two truncated variants with cardiac lesions (including BAV) in two families [34]. In Fig. 6.2, showing the pedigrees described in the work by Garg and colleagues [34], the extreme phenotypic variability of BAV carriers is highlighted.

Subsequently, *NOTCH1* genetic variants were associated with other cardiac conditions involving left-sided (i.e., coarctation of aorta, hypoplastic left heart syndrome—HLHS) and right-sided (i.e., pulmonary stenosis and atresia) congenital defects as well as tetralogy of Fallot (TOF) and Adams-Oliver syndrome [42–47]. To date, 536 pedigrees with familial congenital cardiac lesions and 9830 sporadic cases of congenital heart disease have been analyzed for genetic variants in the *NOTCH1* gene, showing that the P and likely pathogenic (LP) variants explain only a small proportion of familial (2%) and sporadic (<0.1%) BAV disease. In non-syndromic familial and sporadic forms, *NOTCH1* variants are more commonly associated with more complex congenital phenotypes, particularly TOF and HLHS in particular [48]. Furthermore, in almost half of the analyzed families carrying P *NOTCH1* variants, and almost one-third of carriers of P or LP variants, no clinical phenotype was observed, highlighting the incomplete penetrance of *NOTCH1* variants [48].

6.2.2 GATA Family

The GATA binding protein gene family encodes zinc finger transcription factors that play a crucial role in heart valve differentiation [49]. The name derives from the nucleotide sequence recognized by these transcription factors: (A/T)GATA(A/G) [50]. This family has been highly conserved throughout evolution, and its members are found in plants, animals, and fungi [51]. In vertebrates, a similar amino acid structure is shared by all GATA factors. The proteins consist of an N-terminal transactivation domain involved in the activation or repression of downstream genes, a DNA binding domain consisting of two zinc finger structures, and a nuclear localization sequence at the C-terminal domain [50]. In vertebrates, the family comprises six members (*GATA 1-6*) divided into two subgroups. The first subgroup is mainly expressed in the central nervous and hematopoietic systems (GATA1/2/3), while the second (*GATA4/5/6*) is important for differentiation of embryonic stem cells and in epithelial cells in adults. The first family member for which rare variants were associated with non-syndromic BAV was *GATA5* [52], and subsequent studies have reported genetic variants in GATA5 in syndromic and non-syndromic BAV [53, 54]. Consistent with findings in humans, *Gata5*$^{-/-}$ knockout mice demonstrate partially penetrant (25%) BAV [49].

A GWAS study involving 466 BAV cases and 4660 controls, recently identified *GATA4* as a predisposing gene for BAV (odds ratio ranging from 1.4 to 2.4 depending on the variant) [19]. Mice harboring a human *GATA4* variant display partially penetrant aortic valve stenosis (AVS), supporting this association [55]. *GATA4* risk loci have been confirmed also in a second GWAS

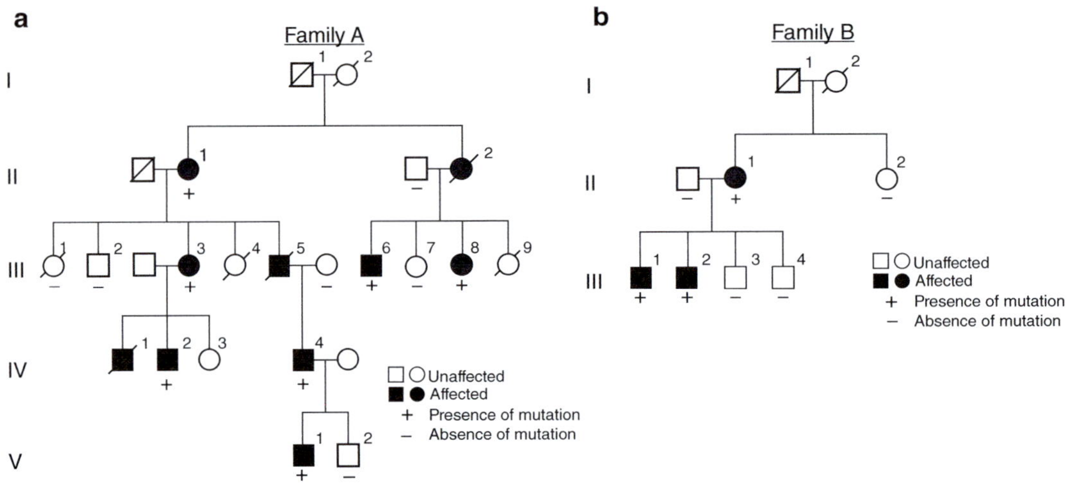

c

Family A	Congenital Heart Disease	Calcification	BAV
II-1	AS	+	–
II-2	AS	+	+
III-3	Dysmorphic AV	+	–
III-4	VSD		
III-5	Severe AS	+	+
III-6	Severe AI	+	+
III-8	Mild AI		+
IV-1	TOF	–	–
IV-2	Mild AS	+	–
IV-4	MS, VSD		+
V-1	Mild AS	+	+

Family B	Congenital Heart Disease	Calcification	BAV
II-1	Severe AS	+	+
III-1	MA, HLV, DORV		+
III-2	Severe AS	+	+

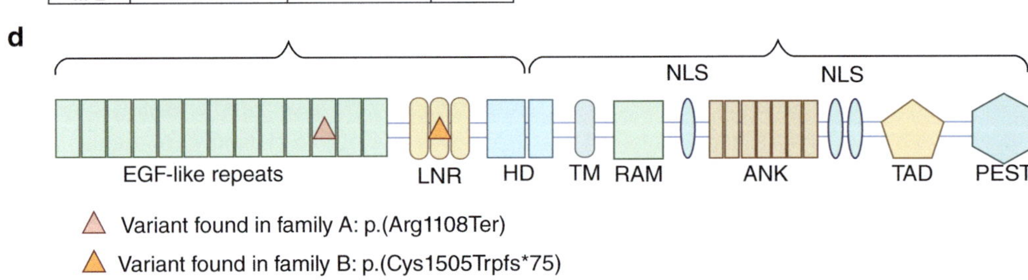

Fig. 6.2 NOTCH1 genetic variations described by Garg et al. in 2005 [34]. The image describes: (**a**) A family tree spanning five generations (labeled with Roman numerals) affected by congenital heart disease and valve calcification. Family members in each generation are numbered. Deceased relatives (marked with a slash) couldn't be tested for variants. Males are represented by squares, females by circles. (**b**) Another family tree showing three members with congenital heart disease. (**c**) A table showing the heart conditions of affected family members. Abbreviations used: AI (aortic insufficiency), AS (aortic stenosis), AV (aortic valve), BAV (bicuspid aortic valve), TOF (tetralogy of Fallot), VSD (ventricular septal defect), DORV (double-outlet right ventricle), HLV (hypoplastic left ventricle), MA (mitral atresia), and MS (mitral stenosis). (**d**) The fully developed NOTCH1 receptor consists of two parts that form a heterodimer: an extracellular portion (NOTCH1-EC) and a section that spans the membrane and extends into the cell (NOTCH1-TMIC). The NOTCH1-EC contains several key components: Epidermal growth factor (EGF)-like repeats, which help with binding to ligands, three LIN-12/NOTCH repeats (LNR), which stop the receptor from activating when no ligands are present, a heterodimerization domain (HD)

study conducted on 1539 BAV patients and 4414 controls [21]. Moreover, CNVs have been reported [56].

Genome-wide association studies have identified several chromosomal regions, such as 18q, which harbors the gene encoding the GATA6 transcription factor [57], and the mechanism by which haploinsufficiency interrupts the aortic valve remodeling and extracellular matrix (ECM) composition was demonstrated in in vitro studies [58], and a loss of function genetic variant was described in a family with non-syndromic BAV [59].

The association of *GATA5* with the pathology is supported by the identification of loss-of-function variants in the *NK2 homeobox 5* (*NKX2.5*) gene in patients with non-syndromic BAV, since these variants interrupt the interaction of *NKX2.5* with *GATA5* [60].

6.2.3 TGF-β Signaling Pathway

The TGF-β family comprises 33 members, which are multifunctional cytokines that interact with TGF-β receptors to trigger signaling cascade responses in cells [61].The TGF-β signaling pathway plays a key role in vascular matrix remodeling and has been linked to connective tissue disorders, tumor angiogenesis and metastasis [62], and cardiac fibrosis [63]. Overexpression of TGF-β factors correlates with the formation and development of fibrosis [64, 65], modulating the transcription of high-affinity DNA-binding factors, such as TCF/LEF and β-catenin [66, 67], and signals of the ERK, JNK and MAPK pathways [68–71], and inducing the expression of interleukin-1 [72]. TGF-β plays an essential role in the expression of ECM proteins and in the differentiation of myofibroblasts via the regulation of connective tissue growth factor (CTGF/CCN2) [73].

Among proteins involved in the TGF-β signaling pathway, *SMAD family member 6* (*SMAD6*) has also been implicated in BAV. This gene encodes a signal transduction protein highly expressed in the embryonic heart and is involved in the inhibition of the TGF-β pathway [74]. Only a very limited number of patients were reported at first [36], but its significant contribution to the etiology of the BAV phenotype was subsequently confirmed in a target resequencing study of individuals with BAV and TAA [75]. In this study, conducted on 441 BAV/TAA patients, authors reported *SMAD6* genetic variants in 2.5% of the cases (two frameshift deletions, two nonsense variants, one in-frame deletion, and six missense variants).

Another TGF- β pathway component recently identified as a possible contributor for BAV is the *T-Box transcription factor 20* (*TBX20*), for which CNVs or single nucleotide variants have been identified in approximately 1% of a cohort of 162 BAV/TAA patients [76]. In vivo studies on Zebrafish animal models showed that the expression of this gene is under the control of SMAD6, and animals in which *TBX20* expression was downregulated developed other congenital cardiac malformations [77]. Of note, other cytokines of the TGF- β pathway (TGF- β2, TGF- β3, TGF-βR1, TGF- βR2 and SMAD3) are involved in syndromic aortopathies where BAV is present in 5 to 30% of the cases [78] (see Table 6.1).

6.2.4 Slit-Robo Signaling Pathway

Experiments conducted on *Drosophila melanogaster* have linked Slit-Robo signaling pathway to heart cell migration, alignment, and lumen formation. In this animal model, a single *slit* (three in mammals) and three *Roundabout homolog - robo* (four in mammals) genes have been identified. Slit proteins are large secreted extracellular

Fig. 6.2 (continued) that creates non-covalent bonds between NOTCH1-EC and NOTCH1-TMIC; the NOTCH1-TMIC is composed of two main parts: A transmembrane domain (TM), an intracellular domain (ICD), also known as NOTCH1-ICD. The NOTCH1-ICD itself has several important regions: an RBPJ-associated molecule (RAM) domain, seven ankyrin (ANK) repeats, nuclear localization signals (NLS), a transactivation domain (TAD), a PEST domain, which plays a role in breaking down active NOTCH1-ICD through the proteasome. Modified from Garg et al. in 2005 [34], with permission molecules that bind to single-pass transmembrane receptors called Robo [79, 80]. *ROBO1*, *ROBO2*, *SLIT2*, and *SLIT3* are expressed in or adjacent to the endocardial cushions and heart valves during mouse development [81]. Targeted homozygous disruption of both Robo1 and Robo2 or Slit3 resulted in highly penetrant septal defects with variably penetrant valve anomalies, including BAV. Robo4 is selectively expressed by coronary vessels, caval veins, aorta, and pulmonary trunk endothelial cells rather than endocardium [82].

Genetic analysis conducted by whole exome sequencing on 736 patients with BAV and AVS or TAAs identified 11 ROBO4 P variants. Genotype-phenotype co-segregation analysis in the family showed that all variant carriers were affected by congenital heart defects, but the phenotypic spectrum was very wide [35]. Recently, a familial study identified a genetic variant also in *ROBO1* that seems to co-segregate with BAV [83], but further in vivo studies are needed to understand the role of the reported variant.

6.2.5 Other Candidate Genes

In recent years, GWAS studies have demonstrated great potential to identify new disease-risk genetic loci. To date, some GWAS studies have been conducted on BAV patients. In addition to the aforementioned work by Yang and colleagues who identified *GATA4* as a BAV-associated gene [19], a second GWAS study has implicated human cilia genes, including *EXOC4*, *EXOC6,* and *EXOC8*. A third GWAS linked BAV to risk loci that include *PALMD*, *TEX41,* and *MYH6* [20] genes.

In 2023, Gehlen and colleagues identified *Mucin 4* (*MUC4)* as a risk locus for BAV [21]. MUC4 is involved in EMT, a cellular process that is of major relevance for cardiac valve development [84–86]. Previously developed *Muc4* knockout mice show no obvious developmental abnormalities; however, mild cardiac valve abnormalities may have been missed [87, 88].

Rare genomic CNVs account for up to 10% of cases and are more likely to be found in patients who present at younger ages or with more severe valve or aortic complications [89]. CNVs found in patients with early onset BAV and TAA that are absent or extremely rare in controls affect genes that interact with each other during heart and vascular development [56, 89].

Several other genes have been linked to BAV, including *NOS3*, *TAB2*, *MAT2A*, *CELSR1* [90–94] and DSCAM [56] (Table 6.2), but more evidence is needed for gene-disease association.

6.3 Clinical Testing

Despite significant advancements in understanding the genetic underpinnings of BAV, the clinical application of genetic testing remains limited, with no official guidelines for clinically actionable genes. The 2022 AHA/ACC Guideline for the Diagnosis and Management of Aortic Disease recommends evaluation by a medical geneticist or genetic aortopathy specialist when the condition is familial [95]. However, implementing these recommendations in routine clinical practice faces several challenges.

The European Society of Cardiology (ESC) Guidelines advocate for echocardiographic screening of first-degree relatives (FDRs) of BAV patients [96], but this practice is not widely adopted. Underutilization of screening may be attributed to various factors, including the underestimation of BAV severity by both patients and healthcare providers, and the relatively low familial occurrence compared to other genetic disorders. Practical obstacles, such as difficulties in recommending screening for unknown family members and potential reluctance of adult relatives to undergo examination, further complicate the issue [97].

Given the variable expressivity of valve and aortic phenotypes in BAV family members, a comprehensive phenotypic analysis of all available family members is crucial before proceeding with genetic testing. It's important to note that parents of affected children, presumed to carry causative genetic variants, may not exhibit BAV themselves but could present other cardiovascular abnormalities.

While genetic testing in BAV families could potentially reduce healthcare costs by eliminating unnecessary surveillance imaging for those who test negative, the long-term cost-benefit analysis remains unclear. There is a lack of robust scientific evidence demonstrating that screening significantly reduces morbidity or mortality rates.

6.4 Conclusion

The inheritance pattern of BAV is characterized by its complexity, typically following an autosomal dominant model with incomplete penetrance and varying degrees of expressivity. While BAV can be associated with certain genetic syndromes and intricate congenital cardiac abnormalities, the majority of cases occur in isolation, without syndromic features. The current understanding of single gene variants falls short in explaining most non-syndromic BAV occurrences. Future research may uncover the role of alternative genetic factors, such as CNVs, non-coding sequences, and epigenetic modifications, which lie beyond the exome. These variants might be identified through advanced techniques like methylation arrays and comprehensive genome sequencing. To further elucidate the genetic basis of BAV, it may be crucial to study family-based groups and populations exhibiting high disease penetrance. In the interim, genetic screening is advisable for BAV patients displaying characteristics of genetic syndromes or hereditary thoracic aortic disease [32].

References

1. Basso C, Boschello M, Perrone C, Mecenero A, Cera A, Bicego D, et al. An echocardiographic survey of primary school children for bicuspid aortic valve. Am J Cardiol. 2004;93(5):661–3.
2. Nistri S, Basso C, Marzari C, Mormino P, Thiene G. Frequency of bicuspid aortic valve in young male conscripts by echocardiogram. Am J Cardiol. 2005;96(5):718–21.
3. Siu SC, Silversides CK. Bicuspid aortic valve disease. J Am Coll Cardiol. 2010;55(25):2789–800.
4. Prakash SK, Bosse Y, Muehlschlegel JD, Michelena HI, Limongelli G, Della Corte A, et al. A roadmap to investigate the genetic basis of bicuspid aortic valve and its complications: insights from the international BAVCon (bicuspid aortic valve consortium). J Am Coll Cardiol. 2014;64(8):832–9.
5. Nistri S, Basso C, Thiene G. Bicuspid aortic valve. G Ital Cardiol. 2012;13(1):25–37.
6. Sieber-Blum M. Cardiac neural crest stem cells. Anat Rec A Discov Mol Cell Evol Biol. 2004;276(1):34–42.
7. Kappetein AP, Gittenberger-de Groot AC, Zwinderman AH, Rohmer J, Poelmann RE, Huysmans HA. The neural crest as a possible pathogenetic factor in coarctation of the aorta and bicuspid aortic valve. J Thorac Cardiovasc Surg. 1991;102(6):830–6.
8. Abbott ME. Coarctation of the aorta of adult type; statistical study and historical retrospect of 200 recorded cases with autopsy; of stenosis or obliteration of discending arch in subjects above age of two years. Am Heart J. 1928;3:574–628.
9. Thiene G, Rizzo S, Basso C. Bicuspid aortic valve: the most frequent and not so benign congenital heart disease. Cardiovasc Pathol. 2024;70:107604.
10. Glotzbach JP, Hanson HA, Tonna JE, Horns JJ, McCarty Allen C, Presson AP, et al. Familial associations of prevalence and cause-specific mortality for thoracic aortic disease and bicuspid aortic valve in a large-population database. Circulation. 2023;148(8):637–47.
11. Prakash SK, Michelena HI, Milewicz DM. Bicuspid aortic valve and thoracic aortic disease: further evidence of clinically silent but deadly risk to family members of affected individuals. Circulation. 2023;148(8):648–50.
12. Chatrath N, Westaby J, Finocchiaro G, Sharma S, Esteban MT, Papadakis M, et al. The role of the bicuspid aortic valve in sudden cardiac death-findings at cardiac autopsy. Cardiovasc Pathol. 2023;65:107527.
13. Clementi M, Notari L, Borghi A, Tenconi R. Familial congenital bicuspid aortic valve: a disorder of uncertain inheritance. Am J Med Genet. 1996;62(4):336–8.
14. Huntington K, Hunter AG, Chan KL. A prospective study to assess the frequency of familial clustering of congenital bicuspid aortic valve. J Am Coll Cardiol. 1997;30(7):1809–12.
15. Cripe L, Andelfinger G, Martin LJ, Shooner K, Benson DW. Bicuspid aortic valve is heritable. J Am Coll Cardiol. 2004;44(1):138–43.
16. Biner S, Rafique AM, Ray I, Cuk O, Siegel RJ, Tolstrup K. Aortopathy is prevalent in relatives of bicuspid aortic valve patients. J Am Coll Cardiol. 2009;53(24):2288–95.
17. Galian-Gay L, Carro Hevia A, Teixido-Tura G, Rodriguez Palomares J, Gutierrez-Moreno L,

Maldonado G, et al. Familial clustering of bicuspid aortic valve and its relationship with aortic dilation in first-degree relatives. Heart. 2019;105(8):603–8.
18. Abdulkareem N, Smelt J, Jahangiri M. Bicuspid aortic valve aortopathy: genetics, pathophysiology and medical therapy. Interact Cardiovasc Thorac Surg. 2013;17(3):554–9.
19. Yang B, Zhou W, Jiao J, Nielsen JB, Mathis MR, Heydarpour M, et al. Protein-altering and regulatory genetic variants near GATA4 implicated in bicuspid aortic valve. Nat Commun. 2017;8:15481.
20. Helgadottir A, Thorleifsson G, Gretarsdottir S, Stefansson OA, Tragante V, Thorolfsdottir RB, et al. Genome-wide analysis yields new loci associating with aortic valve stenosis. Nat Commun. 2018:9.
21. Gehlen J, Stundl A, Debiec R, Fontana F, Krane M, Sharipova D, et al. Elucidation of the genetic causes of bicuspid aortic valve disease. Cardiovasc Res. 2023;119(3):857–66.
22. Niaz T, Poterucha JT, Olson TM, Johnson JN, Craviari C, Nienaber T, et al. Characteristic morphologies of the bicuspid aortic valve in patients with genetic syndromes. J Am Soc Echocardiogr. 2018;31(2):194–200.
23. Carlson M, Silberbach M. Dissection of the aorta in Turner syndrome: two cases and review of 85 cases in the literature. BMJ Case Rep. 2009;2009:bcr0620091998.
24. Corbitt H, Gutierrez J, Silberbach M, Maslen CL. The genetic basis of turner syndrome aortopathy. Am J Med Genet C. 2019;181(1):117–25.
25. Corbitt H, Morris SA, Gravholt CH, Mortensen KH, Tippner-Hedges R, Silberbach M, et al. TIMP3 and TIMP1 are risk genes for bicuspid aortic valve and aortopathy in Turner syndrome. PLoS Genet. 2018;14(10):e1007692.
26. Yagi H, Furutani Y, Hamada H, Sasaki T, Asakawa S, Minoshima S, et al. Role of TBX1 in human del22q11.2 syndrome. Lancet. 2003;362(9393):1366–73.
27. Nistri S, Porciani MC, Attanasio M, Abbate R, Gensini GF, Pepe G. Association of Marfan syndrome and bicuspid aortic valve: frequency and outcome. Int J Cardiol. 2012;155(2):324–5.
28. Monda E, Fusco A, Melis D, Caiazza M, Gragnano F, Mauriello A, et al. Clinical significance of family history and bicuspid aortic valve in children and young adult patients with Marfan syndrome. Cardiol Young. 2020;30(5):663–7.
29. Fedak PW, de Sa MP, Verma S, Nili N, Kazemian P, Butany J, et al. Vascular matrix remodeling in patients with bicuspid aortic valve malformations: implications for aortic dilatation. J Thorac Cardiovasc Surg. 2003;126(3):797–806.
30. Nistri S, Grande-Allen J, Noale M, Basso C, Siviero P, Maggi S, et al. Aortic elasticity and size in bicuspid aortic valve syndrome. Eur Heart J. 2008;29(4):472–9.
31. Pepe G, Nistri S, Giusti B, Sticchi E, Attanasio M, Porciani C, et al. Identification of fibrillin 1 gene mutations in patients with bicuspid aortic valve (BAV) without Marfan syndrome. BMC Med Genet. 2014;15:23.
32. Bravo-Jaimes K, Prakash SK. Genetics in bicuspid aortic valve disease: where are we? Prog Cardiovasc Dis. 2020;63(4):398–406.
33. Duran AC, Frescura C, Sans-Coma V, Angelini A, Basso C, Thiene G. Bicuspid aortic valves in hearts with other congenital heart disease. J Heart Valve Dis. 1995;4(6):581–90.
34. Garg V, Muth AN, Ransom JF, Schluterman MK, Barnes R, King IN, et al. Mutations in NOTCH1 cause aortic valve disease. Nature. 2005;437(7056):270–4.
35. Gould RA, Aziz H, Woods CE, Seman-Senderos MA, Sparks E, Preuss C, et al. ROBO4 variants predispose individuals to bicuspid aortic valve and thoracic aortic aneurysm. Nat Genet. 2019;51(1):42.
36. Tan HL, Glen E, Topf A, Hall D, O'Sullivan JJ, Sneddon L, et al. Nonsynonymous variants in the SMAD6 gene predispose to congenital cardiovascular malformation. Hum Mutat. 2012;33(4):720–7.
37. Wunnemann F, Ta-Shma A, Preuss C, Leclerc S, van Vliet PP, Oneglia A, et al. Loss of ADAMTS19 causes progressive non-syndromic heart valve disease. Nat Genet. 2020;52(1):40–7.
38. Fulmer D, Toomer K, Guo LL, Moore K, Glover J, Moore R, et al. Defects in the exocyst-cilia machinery cause bicuspid aortic valve disease and aortic stenosis. Circulation. 2019;140(16):1331–41.
39. Ellison JW, Yagubyan M, Majumdar R, Sarkar G, Bolander ME, Atkinson EJ, et al. Evidence of genetic locus heterogeneity for familial bicuspid aortic valve. J Surg Res. 2007;142(1):28–31.
40. Wang YD, Fang Y, Lu PF, Wu BR, Zhou B. NOTCH signaling in aortic valve development and calcific aortic valve disease. Front Cardiovasc Med. 2021;8
41. Ackah RL, Yasuhara J, Garg V. Genetics of aortic valve disease. Curr Opin Cardiol. 2023;38(3):169–78.
42. Kerstjens-Frederikse WS, van de Laar IM, Vos YJ, Verhagen JM, Berger RM, Lichtenbelt KD, et al. Cardiovascular malformations caused by NOTCH1 mutations do not keep left: data on 428 probands with left-sided CHD and their families. Genet Med. 2016;18(9):914–23.
43. Page DJ, Miossec MJ, Williams SG, Monaghan RM, Fotiou E, Cordell HJ, et al. Whole exome sequencing reveals the major genetic contributors to nonsyndromic tetralogy of Fallot. Circ Res. 2019;124(4):553–63.
44. Stittrich AB, Lehman A, Bodian DL, Ashworth J, Zong Z, Li H, et al. Mutations in NOTCH1 cause Adams-Oliver syndrome. Am J Hum Genet. 2014;95(3):275–84.
45. Mohamed SA, Aherrahrou Z, Liptau H, Erasmi AW, Hagemann C, Wrobel S, et al. Novel missense mutations (p.T596M and p.P1797H) in NOTCH1 in patients with bicuspid aortic valve. Biochem Biophys Res Commun. 2006;345(4):1460–5.
46. McKellar SH, Tester DJ, Yagubyan M, Majumdar R, Ackerman MJ, Sundt TM 3rd. Novel NOTCH1 mutations in patients with bicuspid aortic valve disease and thoracic aortic aneurysms. J Thorac Cardiovasc Surg. 2007;134(2):290–6.
47. McBride KL, Riley MF, Zender GA, Fitzgerald-Butt SM, Towbin JA, Belmont JW, et al. NOTCH1 muta-

tions in individuals with left ventricular outflow tract malformations reduce ligand-induced signaling. Hum Mol Genet. 2008;17(18):2886–93.
48. Debiec RM, Hamby SE, Jones PD, Safwan K, Sosin M, Hetherington SL, et al. Contribution of NOTCH1 genetic variants to bicuspid aortic valve and other congenital lesions. Heart. 2022;108(14):1114–20.
49. Laforest B, Andelfinger G, Nemer M. Loss of Gata5 in mice leads to bicuspid aortic valve. J Clin Invest. 2011;121(7):2876–87.
50. Lentjes MH, Niessen HE, Akiyama Y, de Bruine AP, Melotte V, van Engeland M. The emerging role of GATA transcription factors in development and disease. Expert Rev Mol Med. 2016;18:e3.
51. Whitcomb J, Gharibeh L, Nemer M. From embryogenesis to adulthood: critical role for GATA factors in heart development and function. IUBMB Life. 2020;72(1):53–67.
52. Padang R, Bagnall RD, Richmond DR, Bannon PG, Semsarian C. Rare non-synonymous variations in the transcriptional activation domains of GATA5 in bicuspid aortic valve disease. J Mol Cell Cardiol. 2012;53(2):277–81.
53. Bonachea EM, Chang SW, Zender G, LaHaye S, Fitzgerald-Butt S, McBride KL, et al. Rare GATA5 sequence variants identified in individuals with bicuspid aortic valve. Pediatr Res. 2014;76(2):211–6.
54. Shi LM, Tao JW, Qiu XB, Wang J, Yuan F, Xu L, et al. GATA5 loss-of-function mutations associated with congenital bicuspid aortic valve. Int J Mol Med. 2014;33(5):1219–26.
55. Li RG, Xu YJ, Wang J, Liu XY, Yuan F, Huang RT, et al. GATA4 loss-of-function mutation and the congenitally bicuspid aortic valve. Am J Cardiol. 2018;121(4):469–74.
56. Carlisle SG, Albasha H, Michelena HI, Sabate-Rotes A, Bianco L, De Backer J, et al. Rare genomic copy number variants implicate new candidate genes for bicuspid aortic valve. PLoS One. 2024;19(9):e0304514.
57. Martin LJ, Ramachandran V, Cripe LH, Hinton RB, Andelfinger G, Tabangin M, et al. Evidence in favor of linkage to human chromosomal regions 18q, 5q and 13q for bicuspid aortic valve and associated cardiovascular malformations. Hum Genet. 2007;121(2):275–84.
58. Gharibeh L, Komati H, Bosse Y, Boodhwani M, Heydarpour M, Fortier M, et al. GATA6 regulates aortic valve remodeling, and its Haploinsufficiency leads to right-left type bicuspid aortic valve. Circulation. 2018;138(10):1025–38.
59. Xu YJ, Di RM, Qiao Q, Li XM, Huang RT, Xue S, et al. GATA6 loss-of-function mutation contributes to congenital bicuspid aortic valve. Gene. 2018;663:115–20.
60. Qu XK, Qiu XB, Yuan F, Wang J, Zhao CM, Liu XY, et al. A novel NKX2.5 loss-of-function mutation associated with congenital bicuspid aortic valve. Am J Cardiol. 2014;114(12):1891–5.
61. Morikawa M, Derynck R, Miyazono K. TGF-beta and the TGF-beta family: context-dependent roles in cell and tissue physiology. Cold Spring Harb Perspect Biol. 2016;8:5.
62. Tie Y, Tang F, Peng D, Zhang Y, Shi H. TGF-beta signal transduction: biology, function and therapy for diseases. Mol Biomed. 2022;3(1):45.
63. Yousefi F, Shabaninejad Z, Vakili S, Derakhshan M, Movahedpour A, Dabiri H, et al. TGF-β and WNT signaling pathways in cardiac fibrosis: non-coding RNAs come into focus. Cell Commun Signal. 2020;18(1):87.
64. Ong CH, Tham CL, Harith HH, Firdaus N, Israf DA. TGF-β-induced fibrosis: a review on the underlying mechanism and potential therapeutic strategies. Eur J Pharmacol. 2021;911
65. Yue YY, Meng K, Pu YJ, Zhang XM. Transforming growth factor beta (TGF-β) mediates cardiac fibrosis and induces diabetic cardiomyopathy. Diabet Res Clin Pract. 2017;133:124–30.
66. Derynck R, Budi EH. Specificity, versatility, and control of TGF-β family signaling. Sci Signal. 2019;12:570.
67. Piersma B, Bank RA, Boersema M. Signaling in fibrosis: TGF-beta, WNT, and YAP/TAZ converge. Front Med (Lausanne). 2015;2:59.
68. Verrecchia F, Tacheau C, Schorpp-Kistner M, Angel P, Mauviel A. Induction of the AP-1 members c-Jun and JunB by TGF-beta/Smad suppresses early Smad-driven gene activation. Oncogene. 2001;20(18):2205–11.
69. Wygrecka M, Zakrzewicz D, Taborski B, Didiasova M, Kwapiszewska G, Preissner KT, et al. TGF-β1 induces tissue factor expression in human lung fibroblasts in a PI3K/JNK/Akt-dependent and AP-1-dependent manner. Am J Resp Cell Mol. 2012;47(5):614–27.
70. Shaulian E, Karin M. AP-1 as a regulator of cell life and death. Nat Cell Biol. 2002;4(5):E131–E6.
71. Cui L, Chen SY, Lerbs T, Lee JW, Domizi P, Gordon S, et al. Activation of JUN in fibroblasts promotes pro-fibrotic programme and modulates protective immunity. Nat Commun. 2020;11(1):2795.
72. Allanki S, Strilic B, Scheinberger L, Onderwater YL, Marks A, Günther S, et al. Interleukin-11 signaling promotes cellular reprogramming and limits fibrotic scarring during tissue regeneration. Sci Adv. 2021;7(37):eabg6497.
73. Yanagihara T, Tsubouchi K, Gholiof M, Chong SG, Lipson KE, Zhou Q, et al. Connective-tissue growth factor contributes to TGF-β1-induced lung fibrosis. Am J Resp Cell Mol. 2022;66(3):260–70.
74. Imamura T, Takase M, Nishihara A, Oeda E, Hanai J, Kawabata M, et al. Smad6 inhibits signalling by the TGF-beta superfamily. Nature. 1997;389(6651):622–6.
75. Gillis E, Kumar AA, Luyckx I, Preuss C, Cannaerts E, van de Beek G, et al. Candidate gene resequencing in a large bicuspid aortic valve-associated thoracic aortic aneurysm cohort: SMAD6 as an important contributor. Front Physiol. 2017;8:400.
76. Luyckx I, Kumar AA, Reyniers E, Dekeyser E, Vanderstraeten K, Vandeweyer G, et al. Copy number variation analysis in bicuspid aortic valve-related

aortopathy identifies TBX20 as a contributing gene. Eur J Human Genet: EJHG. 2019;27(7):1033–43.
77. de Pater E, Ciampricotti M, Priller F, Veerkamp J, Strate I, Smith K, et al. Bmp signaling exerts opposite effects on cardiac differentiation. Circ Res. 2012;110(4):578–U179.
78. Bertoli-Avella AM, Gillis E, Morisaki H, Verhagen JMA, de Graaf BM, van de Beek G, et al. Mutations in a TGF-β ligand cause syndromic aortic aneurysms and dissections. J Am Coll Cardiol. 2015;65(13):1324–36.
79. Kidd T, Brose K, Mitchell KJ, Fetter RD, Tessier-Lavigne M, Goodman CS, et al. Roundabout controls axon crossing of the CNS midline and defines a novel subfamily of evolutionarily conserved guidance receptors. Cell. 1998;92(2):205–15.
80. Seeger M, Tear G, Ferresmarco D, Goodman CS. Mutations affecting growth cone guidance in drosophila—genes necessary for guidance toward or away from the midline. Neuron. 1993;10(3):409–26.
81. Mommersteeg MTM, Yeh ML, Parnavelas JG, Andrews WD. Disrupted slit-Robo signalling results in membranous ventricular septum defects and bicuspid aortic valves. Cardiovasc Res. 2015;106(1):55–66.
82. Mommersteeg MTM, Andrews WD, Ypsilanti AR, Zelina P, Yeh ML, Norden J, et al. Slit-roundabout signaling regulates the development of the cardiac systemic venous return and pericardium. Circ Res. 2013;112(3):465.
83. Jaouadi H, Gérard H, Théron A, Collod-Béroud G, Collart F, Avierinos JF, et al. Identification of non-synonymous variations in ROBO1 and GATA5 genes in a family with bicuspid aortic valve disease. J Hum Genet. 2022;67(9):515–8.
84. Ponnusamy MP, Seshacharyulu P, Lakshmanan I, Vaz AP, Chugh S, Batra SK. Emerging role of mucins in epithelial to mesenchymal transition. Curr Cancer Drug Targets. 2013;13(9):945–56.
85. Xu D, Liu S, Zhang L, Song L. MiR-211 inhibits invasion and epithelial-to-mesenchymal transition (EMT) of cervical cancer cells via targeting MUC4. Biochem Biophys Res Commun. 2017;485(2):556–62.
86. Kostina AS, Uspensky Vcapital Ie C, Irtyuga OB, Ignatieva EV, Freylikhman O, Gavriliuk ND, et al. Notch-dependent EMT is attenuated in patients with aortic aneurysm and bicuspid aortic valve. Biochim Biophys Acta. 2016;1862(4):733–40.
87. Das S, Rachagani S, Sheinin Y, Smith LM, Gurumurthy CB, Roy HK, et al. Mice deficient in Muc4 are resistant to experimental colitis and colitis-associated colorectal cancer. Oncogene. 2016;35(20):2645–54.
88. Rowson-Hodel AR, Wald JH, Hatakeyama J, O'Neal WK, Stonebraker JR, VanderVorst K, et al. Membrane Mucin Muc4 promotes blood cell association with tumor cells and mediates efficient metastasis in a mouse model of breast cancer. Oncogene. 2018;37(2):197–207.
89. Prakash S, Kuang SQ, Regalado E, Guo DC, Milewicz D, Investigators GR. Recurrent rare genomic copy number variants and bicuspid aortic valve are enriched in early onset thoracic aortic aneurysms and dissections. PLoS One. 2016;11(4):e0153543.
90. Dargis N, Lamontagne M, Gaudreault N, Sbarra L, Henry C, Pibarot P, et al. Identification of gender-specific genetic variants in patients with bicuspid aortic valve. Am J Cardiol. 2016;117(3):420–6.
91. Girdauskas E, Geist L, Disha K, Kazakbaev I, Gross T, Schulz S, et al. Genetic abnormalities in bicuspid aortic valve root phenotype: preliminary results. Eur J Cardio-Thorac. 2017;52(1):156–62.
92. Guo DC, Pannu H, Tran-Fadulu V, Papke CL, Yu RK, Avidan N, et al. Mutations in smooth muscle α-actin (acta2) lead to thoracic aortic aneurysms and dissections. Nat Genet. 2007;39(12):1488–93.
93. Weiss K, Applegate C, Wang T, Batista DA. Familial TAB2 microdeletion and congenital heart defects including unusual valve dysplasia and tetralogy of fallot. Am J Med Genet A. 2015;167A(11):2702–6.
94. Guo DC, Gong L, Regalado ES, Santos-Cortez RL, Zhao R, Cai B, et al. MAT2A mutations predispose individuals to thoracic aortic aneurysms. Am J Hum Genet. 2015;96(1):170–7.
95. Isselbacher EM, Preventza O, Black J, Augoustides JG, Beck AW, Bolen MA, et al. 2022 ACC/AHA guideline for the diagnosis and Management of Aortic Disease. J Am Coll Cardiol. 2022;80(24):E223–393.
96. Vahanian A, Beyersdorf F, Praz F, Milojevic M, Baldus S, Bauersachs J, et al. 2021 ESC/EACTS guidelines for the management of valvular heart disease. Eur J Cardiothoracic Surg. 2021;60(4):727–800.
97. Evangelista A, Michelena HI. Family screening for bicuspid aortic valve: indicated, but easy to implement? Eur Heart J. 2023;44(33):3165–7.
98. A Comprehensive Analysis of Non-Desmosomal Rare Genetic Variants in Arrhythmogenic Cardiomyopathy: Integrating in Padua Cohort Literature-Derived Data, Int. J. Mol. Sci., 2024. https://doi.org/10.3390/ijms25116267.

Part II

Natural History and Complications

Aortic Stenosis and Aortic Incompetence

7

Giuseppe Tarantini, Tommaso Fabris, Chiara Fraccaro, and Gaetano Thiene

Aortic valve disease, encompassing both aortic stenosis (AS) and aortic incompetence (AI), represents a significant cause of morbidity and mortality worldwide. These two conditions, while distinct in their pathophysiology, share a common feature of progressive hemodynamic deterioration that leads to left ventricular dysfunction and heart failure if left untreated [1, 2]. Understanding the natural history and complications of AS and AI is essential for optimizing patient outcomes, since timely intervention can prevent irreversible myocardial damage and improve long-term survival.

This chapter provides a comprehensive overview of the natural progression and potential complications of these diseases, with a focus on the pathophysiology, clinical course, and key features of bicuspidity.

G. Tarantini (✉) · T. Fabris · C. Fraccaro
Cardiology Clinic, Department of Cardiac, Thoracic, Vascular Sciences and Public Health, University of Padua, Padua, Italy
e-mail: giuseppe.tarantini.1@unipd.it; tommaso.fabris@aopd.veneto.it; chiara.fraccaro@unipd.it

G. Thiene
School of Medicine, University of Padua, Padua, Italy
e-mail: gaetano.thiene@unipd.it

7.1 Aortic Stenosis

7.1.1 Introduction

AS is a condition characterized by the narrowing of the aortic valve, which obstructs blood flow from the left ventricle to the aorta during systole. Over time, this obstruction increases the workload on the heart, leading to left ventricular hypertrophy and eventually heart failure if left untreated [1, 2]. AS can arise from a variety of causes, including congenital bicuspid aortic valve (BAV), rheumatic heart disease, and degenerative calcific processes (the most prevalent), each with distinct clinical courses and complications (Table 7.1).

The development of calcification in aortic valves is influenced by the valve's etiology. For example, bicuspid valves are more prone to calcification than tricuspid valves [3]. Similarly, the age of onset of symptoms depended on the underlying etiology: 39 ± 18 years for rheumatic AS, 48 ± 6 years for those with bicuspid valves but no history of rheumatic fever, infective endocarditis, or myocarditis, and 66 ± 12 years for degenerative calcific stenosis of tricuspid aortic valves [1].

As the disease progresses, patients typically remain asymptomatic for years until the stenosis becomes critical, at which point symptoms such as angina, syncope, and congestive heart failure indicate a poor prognosis without intervention [4] (Fig. 7.1). The mean survival after the onset

G. Thiene et al. (eds.), *Bicuspid Aortic Valve Syndrome*,
https://doi.org/10.1007/978-3-031-85589-4_7

Table 7.1 Etiology of aortic valve stenosis among surgical pathology studies

Author, year	Country	*N*	TAV, %	BAV, %	RVD, %	Others, %
Subramanian et al. 1984	USA	374	10	46	35	10
Cormier et al., 1988	France	675	45	31	16	8
Passik et al., 1987	USA	646	33	38	24	5
Turri et al., 1990	Italy	140	34	34	38	5

BAV bicuspid aortic valve, *RVD* rheumatic valve disease

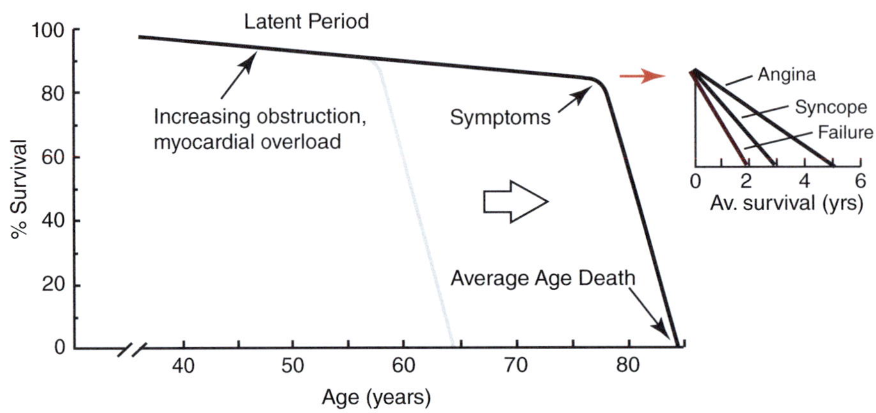

Fig. 7.1 The natural history of AS. After a latent period of gradually increasing valvular stenosis and left ventricular pressure overload, the onset of symptoms heralds an ominous outcome with >50% of patients succumbing over the next 2–3 years. Although Ross and Braunwald [4] in 1968 conceptualized this symptomatic demarcation occurring in middle age (light gray curve), in the current era, the natural history curve often shifts to the right with patients presenting with symptomatic calcific AS in the seventh through ninth decades. Numerous studies confirm the same predictably poor survival after symptom onset in this elderly population unless aortic valve replacement is performed. AS, aortic stenosis. (*Modified with permission from Ross J, Braunwald E. Aortic stenosis. Circulation. 1968;37/38(suppl V): V-61–V-67* [4])

of angina was reported to be 4.5 years, 2.6 years after syncope, and less than 1 year after the onset of heart failure [4]. Mortality seems to increase proportionally to AS severity [5] (Fig. 7.2).

7.1.2 Degenerative (Senile) Calcific Aortic Stenosis

7.1.2.1 Pathophysiology, Progression, and Clinical Course

Degenerative calcific AS is the most common form of stenotic aortic valve, typically affecting individuals over the age of 65 [1, 4, 6]. This form of stenosis is primarily a result of age-related calcification of a previously normal tricuspid aortic valve. The degenerative process involves the deposition of calcium on the valve leaflets, leading to stiffness and restricted motion (Fig. 7.3).

The progression of degenerative AS is slow but relentless, with the degree of calcification increasing over time [1, 4]. The risk factors for developing degenerative calcific AS are similar to those for atherosclerosis, including advanced age, hypertension, hyperlipidemia, smoking, and diabetes mellitus [7]. Some studies suggest that the calcification process in degenerative AS shares common pathways with vascular calcification, including chronic inflammation, lipid deposition, and oxidative stress [8–10]. Most patients with degenerative AS experience a prolonged asymptomatic period, during which the disease progresses silently, followed by rapid clinical deterioration once symptoms emerge [1, 4]. The

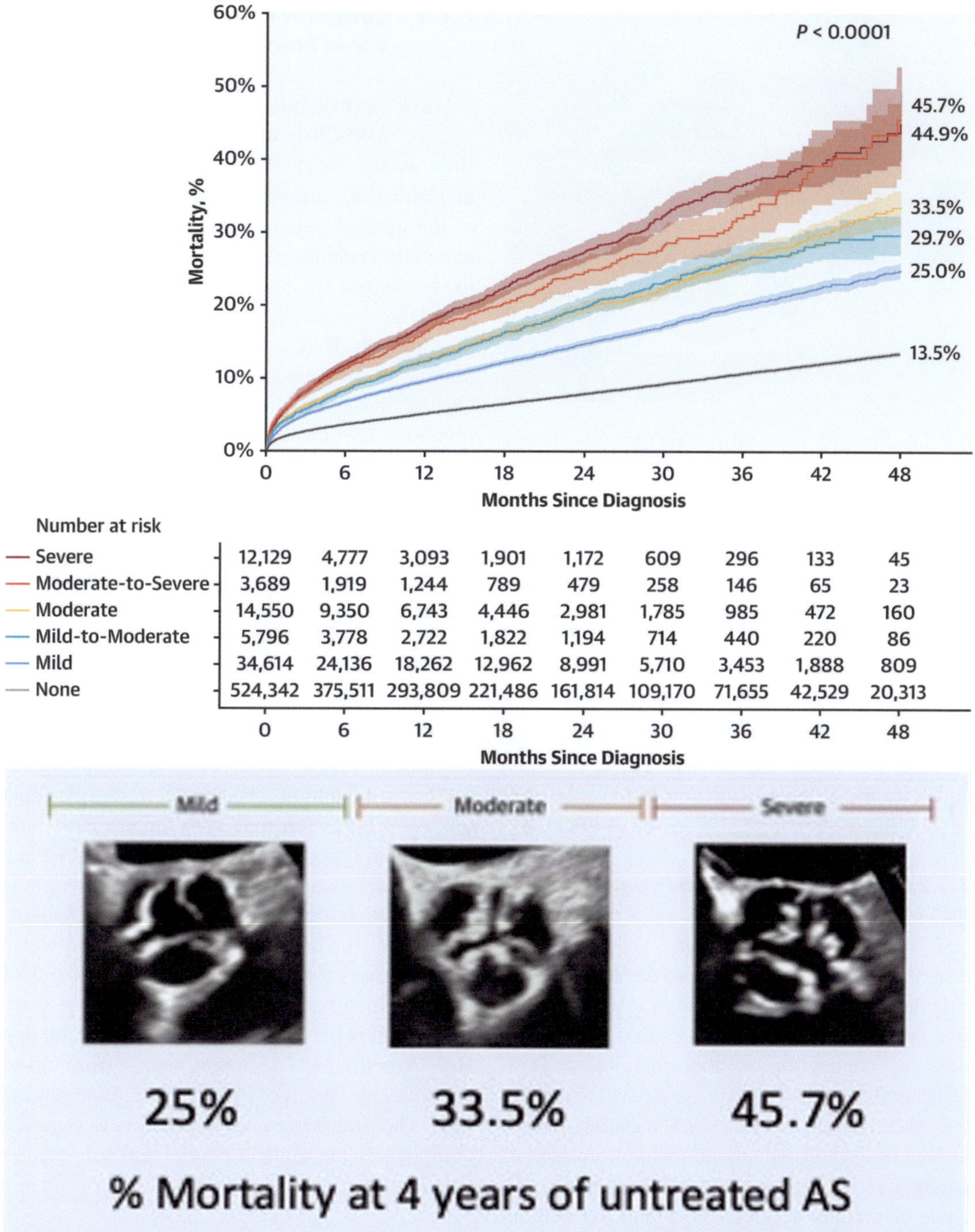

Number at risk	0	6	12	18	24	30	36	42	48
Severe	12,129	4,777	3,093	1,901	1,172	609	296	133	45
Moderate-to-Severe	3,689	1,919	1,244	789	479	258	146	65	23
Moderate	14,550	9,350	6,743	4,446	2,981	1,785	985	472	160
Mild-to-Moderate	5,796	3,778	2,722	1,822	1,194	714	440	220	86
Mild	34,614	24,136	18,262	12,962	8,991	5,710	3,453	1,888	809
None	524,342	375,511	293,809	221,486	161,814	109,170	71,655	42,529	20,313

Fig. 7.2 Mortality with untreated AS over up to 4 years of follow-up after the index documented diagnosis. AS, aortic stenosis. (*Modified from Généreux P,* et al. *The Mortality Burden of Untreated Aortic Stenosis. J Am Coll Cardiol. 2023 Nov 28;82(22):2101–2109* [5])

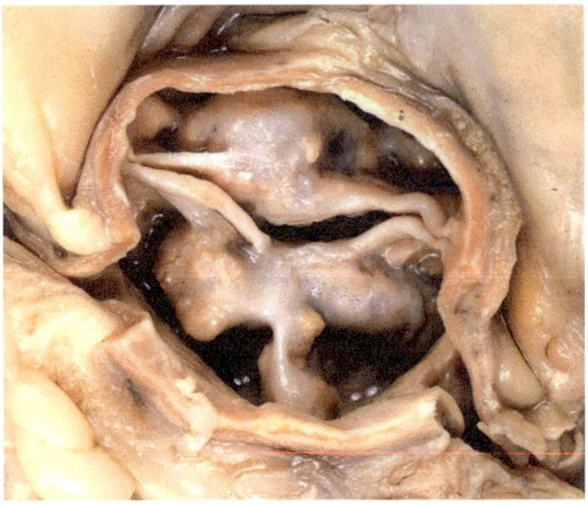

Fig. 7.3 Bicuspid aortic valve (BAV) stenosis. Type 1 autopsy specimen with BAV stenosis by dystrophic calcification in a 65-year-old patient. Note the raphe and the origin of both coronary arteries from the anterior Valsalva sinus. (*from Gaetano Thiene, Stefania Rizzo, Cristina Basso, Bicuspid aortic valve: The most frequent and not so benign congenital heart disease. Cardiovasc Pathol. 2024 May-Jun:70:107604. doi: 10.1016/j.carpath.2024.107604. Epub 2024 Jan 20. PMID: 38253300 DOI: 10.1016/j.carpath.2024.107604* [3])

classic symptoms triad in degenerative AS includes exertional dyspnea, angina, and syncope, which announce the progression to severe stenosis, typically in the seventh or eighth decade of life [1, 4].

7.1.2.2 Complications and Prognosis

Patients with degenerative AS are at risk for several complications, including heart failure, arrhythmias, and sudden cardiac death [1, 4]. The calcific process may also extend to the conduction system, leading to atrioventricular block or other conduction abnormalities [11]. Degenerative AS is also associated with an increased risk of gastrointestinal bleeding, known as Heyde's syndrome, due to the development of angiodysplasia in the gastrointestinal tract [12]. Once symptoms develop, the prognosis without intervention is poor, with an average survival of only 2–3 years [4]. Surgical aortic valve replacement (SAVR) and transcatheter aortic valve implantation (TAVI) are the only definitive treatments for severe AS, and outcomes are excellent when performed in a timely manner [13, 14].

7.1.3 Congenital Bicuspid Aortic Valve Stenosis

7.1.3.1 Pathophysiology, Progression, and Clinical Course

BAV is the most common congenital cardiac malformation, present in approximately 0.5–2% of the general population [6, 15, 16]. However, our understanding of actual BAV prevalence is likely skewed by the fact that many individuals with BAV remain asymptomatic.

Although BAV is present at birth, it may remain asymptomatic in infancy and childhood. During these early years, the valve typically functions adequately, although mild stenosis or regurgitation may already be present [17]. The progression of BAV stenosis is variable but often faster than tricuspid aortic valves [18]. In fact, BAV associated structural abnormalities and disrupted hemodynamics lead to early calcification, often present by the age of 40 [19–21]. Consequently, the progression of AS in BAV frequently leads to valve replacement earlier in life. A study by Michelena et al. [17] found that among 212 asymptomatic BAV patients, 12.3% required AVR for severe AS by a median age of 49. Sievers subtype classification suggests that disease progression may vary among BAV subtypes, with type 1 right-non coronary showing a higher incidence of AS, while type 1 left-right is more often associated with AI [22, 23]. Patients with BAV who had isolated severe AS had worse preoperative LV function and an increased risk of postoperative heart failure hospitalization compared with patients who had TAV, suggesting that BAV patients with AS might benefit from closer surveillance and possibly earlier intervention [24]. The presence of AI, in addition to stenosis, complicates the clinical course and can exacerbate left ventricular dysfunction [25].

7.1.3.2 Complications, Aortopathy, and Prognosis

One of the distinguishing features of BAV is its association with sinusal aortic root and tubular ascending aortic pathology, known as BAV aortopathy, which show a prevalence ranging from

20% to 84% [26–29]. Patients with BAV are at increased risk of ascending aortic aneurysm and aortic dissection due to inherent weaknesses in the aortic wall. This phenomenon is thought to be related to underlying connective tissue abnormalities, with changes in the aortic media, such as cystic medial necrosis [30] (Fig. 7.4). Accelerated degeneration of the tunica media, with progressive aortic dilation, can lead to intimal tears, dissection, and rupture, possibly precipitated by hypertensive episodes (Fig. 7.5). Autopsy studies have found the prevalence of BAV in individuals who suffered aortic dissection ranges from 9% to 13%, which is approximately 10 times higher than the rate of BAV in the general population [3, 31] (Table 7.2). The main risk factors associated with aortic dissection in BAV are uncontrolled hypertension, a personal history of coarctation, a family history of aortic dissection or early unexplained sudden cardiac death, and a confirmed aortic diameter increase of >3 mm/year [13, 14, 31]. In these high-risk cases, prophylactic ascending aorta replacement is recommended at a lower threshold (aortic diameter of ≥50 mm) compared to the generally recommended value (≥55 mm) [13, 14].

The prognosis for BAV stenosis is generally favorable if diagnosed early and managed appropriately. However, once symptoms appear, untreated BAV stenosis can rapidly lead to heart failure or sudden cardiac death [20, 21]. Surgery or transcatheter interventions are typically required in most symptomatic patients. While several registries have reported excellent outcomes of TAVI in patients with a bicuspid valve, SAVR remains the first choice, especially in bicuspid patients with associated aortic root and ascending aortic pathology requiring a surgical approach [13, 14].

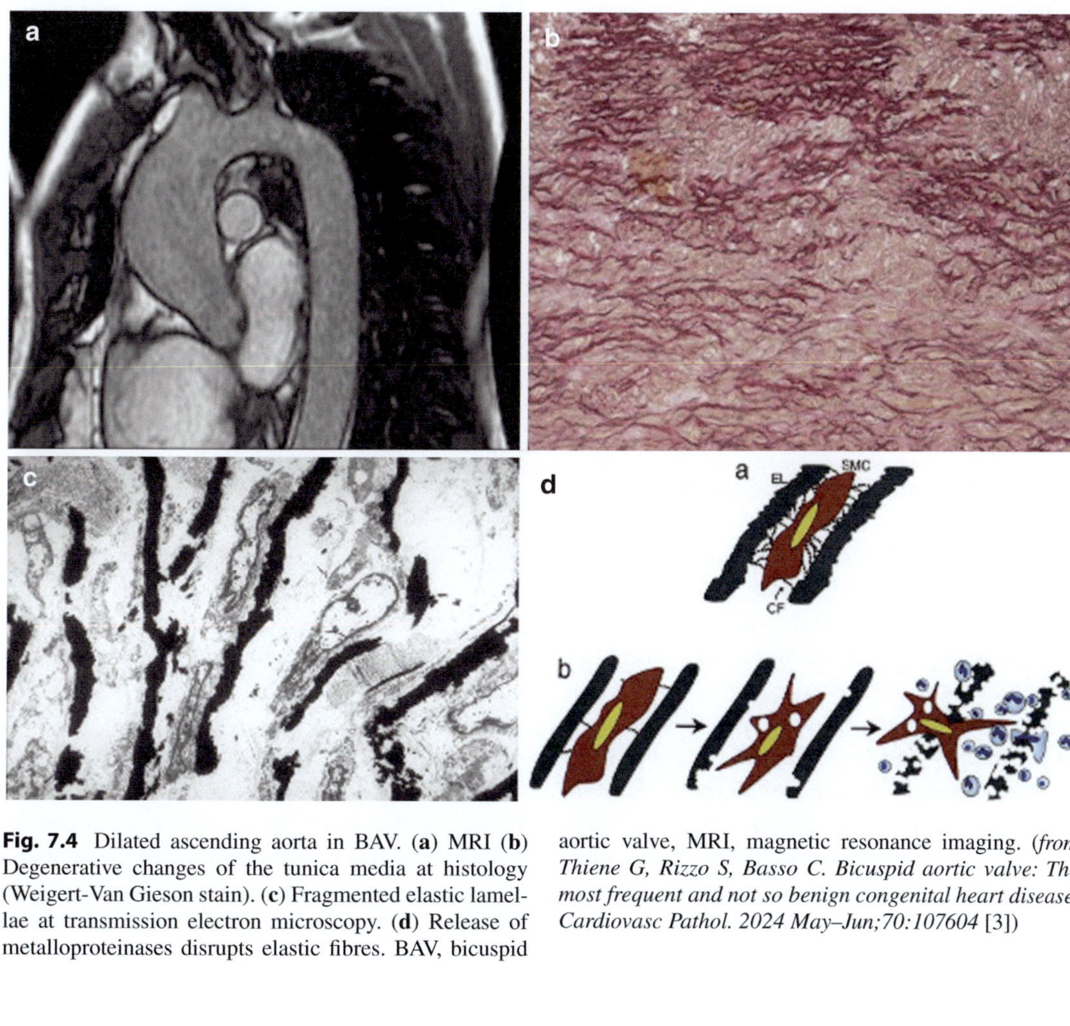

Fig. 7.4 Dilated ascending aorta in BAV. (**a**) MRI (**b**) Degenerative changes of the tunica media at histology (Weigert-Van Gieson stain). (**c**) Fragmented elastic lamellae at transmission electron microscopy. (**d**) Release of metalloproteinases disrupts elastic fibres. BAV, bicuspid aortic valve, MRI, magnetic resonance imaging. (*from Thiene G, Rizzo S, Basso C. Bicuspid aortic valve: The most frequent and not so benign congenital heart disease. Cardiovasc Pathol. 2024 May–Jun;70:107604* [3])

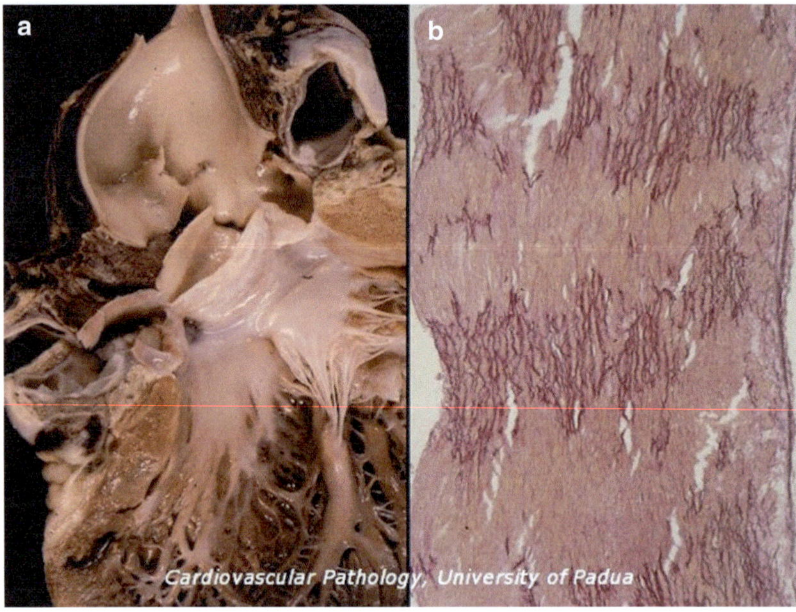

Fig. 7.5 Aortic dissection in a 35-year-old male affected by BAV: (**a**) Gross view of the "explosion"of the aorta with intimal tear and wall dissection. The diameter of ascending tubular aorta was 3.8 cm. (**b**) Extensive loss of elastic fibres in the tunica media. Weigert-Van Gieson stain. BAV, bicuspid aortic valve. (*From Thiene G, Rizzo S, Basso C. Bicuspid aortic valve: The most frequent and not so benign congenital heart disease. Cardiovasc Pathol. 2024 May–Jun;70:107604* [3])

Table 7.2 Frequency of BAV in aortic dissection. BAV from autopsy series of aortic dissection. The rate of BAV varies from 9% to 13%

Author, year	*N*	BAV (%)
Gore, 1952	85	13
Edwards, 1978	119	9
Larson, 1984	161	11
Giusti, 1990	87	12

7.1.4 Rheumatic Aortic Stenosis

7.1.4.1 Pathophysiology, Progression, and Clinical Course

Rheumatic AS results from rheumatic fever, a post-streptococcal autoimmune response that leads to chronic inflammation and scarring of the heart valves. Although rheumatic heart disease most commonly affects the mitral valve, the aortic valve is frequently involved, often leading to combined stenosis and regurgitation. Rheumatic AS is associated with fibrosis and fusion of the valve commissures, which limits valve opening and progressively narrows the valve orifice. Unlike degenerative calcific AS, which is primarily a disease of older adults, rheumatic AS can present in younger patients, with symptoms emerging in middle age, especially in regions where rheumatic fever remains endemic. In developed countries, where the incidence of rheumatic fever has significantly decreased, rheumatic AS is less common but still seen in migrate people. The progression of rheumatic AS is marked by the gradual thickening and calcification of the valve leaflets. Patients may experience a prolonged asymptomatic phase, but once stenosis becomes severe, symptoms such as dyspnea, angina, and syncope emerge. Rheumatic stenosis often affects both the aortic and mitral valves, leading to combined valvular disease, which complicates the clinical course [6].

7.1.4.2 Complications and Prognosis

The presence of combined mitral and aortic valve disease in rheumatic heart disease significantly increases the risk of heart failure. Mitral stenosis, in particular, exacerbates left atrial pressure and pulmonary congestion, compounding the hemodynamic burden. Atrial fibrillation, a common complication of mitral stenosis, further complicates the management of these patients. Additionally, rheumatic AS is associated with an increased risk of infective endocarditis, particularly in individuals with damaged or scarred valve tissue. Regular monitoring and early intervention are essential to prevent complications, particularly in younger patients who may experience rapid deterioration [6].

The prognosis of rheumatic AS is closely tied to the extent of valve involvement and the presence of other rheumatic valve lesions. In the absence of timely intervention, severe stenosis can lead to heart failure and premature death. SAVR or TAVI are the definitive treatments for advanced disease, and outcomes are generally favorable. However, patients with combined mitral and aortic disease often require more complex interventions, and their long-term prognosis is less favorable than those with isolated AS.

7.1.5 Other Less Common Etiologies: Radiation-Induced Aortic Stenosis

In rare cases, AS can develop as a late complication of radiation therapy to the chest, typically decades after treatment for conditions such as Hodgkin lymphoma or breast cancer.

7.2 Aortic Incompetence

7.2.1 Introduction

Aortic incompetence occurs when the aortic valve fails to close properly, allowing blood to flow back into the left ventricle during diastole.

The clinical course of AI varies significantly depending on the chronicity of the condition and underlying etiologies, such as rheumatic heart disease, infective endocarditis, BAV, or diseases of the aortic root like Marfan syndrome or aortic dissection (Table 7.3) [32]. The progression of chronic AI is typically slow and insidious, with patients remaining asymptomatic for prolonged periods before symptoms emerge [33].

7.2.2 Chronic Aortic Incompetence: Pathophysiology, Clinical Course, and Complications

In chronic AI, the volume overload initially leads to compensatory left ventricular dilatation and hypertrophy, enabling the heart to maintain adequate forward flow for many years [32]. However, as the condition progresses, these compensatory mechanisms become inadequate, leading to elevated left ventricular end-diastolic pressure, reduced cardiac output, and eventual heart failure [32].

The clinical course of chronic AI is typically protracted, with patients remaining asymptomatic for many years despite significant regurgitation [33]. This long latent phase can last several decades, during which time patients experience little to no limitation in daily activities [33]. Symptoms usually become evident in the fourth or fifth decade of life, with dyspnea on exertion and fatigue being the most common initial signs [32, 33]. In the later stages, left ventricular dysfunction is evident, with often the onset of secondary mitral regurgitation due to left ventricle dilatation [33]. Without surgical intervention, this stage often culminates in irreversible myocardial damage, which can persist even after valve replacement [34]. In patients with associated aortic root disease, such as Marfan syndrome

Table 7.3 Etiology of aortic valve regurgitation among surgical pathology studies

Author, year	Country	*N*	TAV, %	BAV, %	RVD, %	IE, %	Others, %
Olsen et al. 1984	USA	225	21	20	46	9	4
Guiney et al., 1987	UK	72	34	9	26	21	10
Turri et al., 1990	Italy	254	59	7	20	11	3
Michel et al., 1991	France	313	35	5	38	11	11

BAV bicuspid aortic valve, *IE* infective endocarditis, *RVD* rheumatic valve disease, *TAV* tricuspid aortic valve or BAV, AI may be complicated by aortic dissection or aneurysm formation [32]. These conditions significantly increase the risk of sudden death and require close surveillance and timely surgical intervention.

7.2.3 Bicuspid Aortic Valve and Aortic Incompetence

In BAV patients, the prevalence of AI ranges from 26.7% to 47.2% [17, 31, 35–37]. Significant AI, marked by New York Heart Association (NYHA) class III or higher symptoms, often presents at a younger age than AS, with a mean onset around 45 years, compared to 54 years for severe AS [35]. Altered cusp and aortic wall composition in BAV frequently lead to cusp prolapse and immobility, causing AI [38] (Fig. 7.6). Acute AI in younger BAV patients can also arise from the rupture of a cord connecting the raphe to the aortic wall at the sinotubular junction (Fig. 7.7), endocarditis, or complications from balloon val-

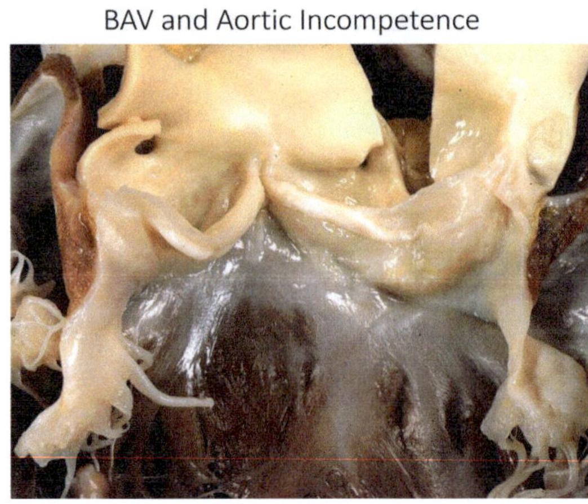

Fig. 7.6 Bicuspid aortic valve (BAV) incompentence. Autopsy specimen with BAV incompetence due to malformed cusps. Note thickened free margin of the cusps, probably due also to longstanding wear and tear. Male, 52 years old. (*from Gaetano Thiene, Stefania Rizzo, Cristina Basso, Bicuspid aortic valve: The most frequent and not so benign congenital heart disease. Cardiovasc Pathol. 2024 May-Jun:70:107604. doi: 10.1016/j.carpath.2024.107604. Epub 2024 Jan 20. PMID: 38253300 DOI: 10.1016/j.carpath.2024.107604* [3])

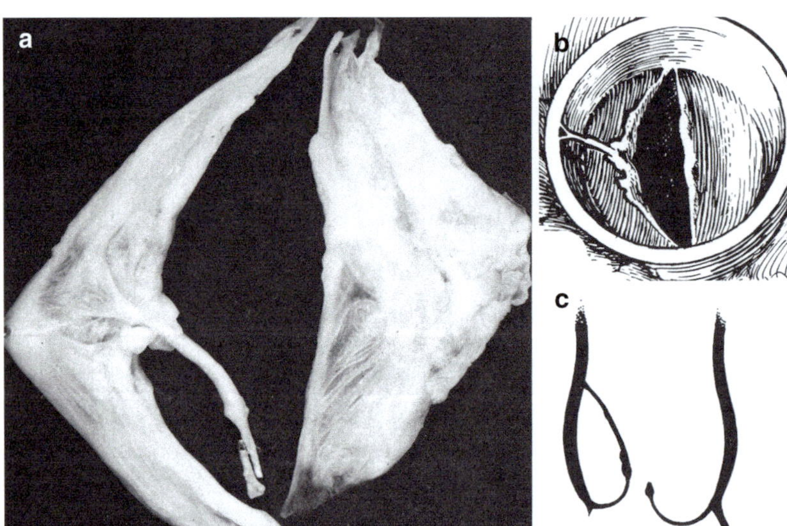

Fig. 7.7 Panel a, b and c show different views and representations of a congenitally bicuspid aortic valve devoid of calcific deposits containing an anomalous cord attached to the conjoined cusp. (*From Travis J Vowels, Gonzalo V Gonzalez-Stawinski, Jong M Ko, Gregory D Trachiotis, Brad J Roberts, Charles S Roberts, William C Roberts. Anomalous cord from the raphe of a congenitally bicuspid aortic valve to the aortic wall producing either acute or chronic aortic regurgitation. J Am Coll Cardiol. 2014 Jan 21;63(2):153–7. doi: 10.1016/j.jacc.2013.09.030. Epub 2013 Oct 16* [39])

vuloplasty [3, 39]. In older patients, AI may develop from aortic dilation at the sinus segments or sinotubular junction, affecting cusp coaptation [40]. Sievers type 1 BAV is more frequently associated with AI than type 0, possibly due to differences in the underlying aortic wall pathology [22, 37, 41].

References

1. Horstkotte D, Loogen F. The natural history of aortic valve stenosis. Eur Heart J. 1988;9(Suppl E):57–64. https://doi.org/10.1093/eurheartj/9.suppl_e.57.
2. Turina J, Hess O, Sepulcri F, Krayenbuehl HP. Spontaneous course of aortic valve disease. Eur Heart J. 1987;8(5):471–83. https://doi.org/10.1093/oxfordjournals.eurheartj.a062307.
3. Thiene G, Rizzo S, Basso C. Bicuspid aortic valve: The most frequent and not so benign congenital heart disease. Cardiovasc Pathol. 2024;70:107604. https://doi.org/10.1016/j.carpath.2024.107604. Epub 2024 Jan 20
4. Ross J, Braunwald E. Aortic stenosis. Circulation. 1968;37/38(suppl V):V-61–7.
5. Généreux P, Sharma RP, Cubeddu RJ, Aaron L, Abdelfattah OM, Koulogiannis KP, Marcoff L, Naguib M, Kapadia SR, Makkar RR, Thourani VH, van Boxtel BS, Cohen DJ, Dobbles M, Barnhart GR, Kwon M, Pibarot P, Leon MB, Gillam LD. The Mortality Burden of Untreated Aortic Stenosis. J Am Coll Cardiol. 2023;82(22):2101–9. https://doi.org/10.1016/j.jacc.2023.09.796. Epub 2023 Oct 24
6. Passik CS, Ackermann DM, Pluth JR, Edwards WD. Temporal changes in the causes of aortic stenosis: a surgical pathologic study of 646 cases. Mayo Clin Proc. 1987;62(2):119–23. https://doi.org/10.1016/s0025-6196(12)61880-1.
7. Peter M, Hoffmann A, Parker C, Lüscher T, Burckhardt D. Progression of aortic stenosis: Role of age and concomitant coronary artery disease. Chest. 1993;103(6):1715–9.
8. Mohler ER 3rd, Gannon F, Reynolds C, Zimmerman R, Keane MG, Kaplan FS. Bone formation and inflammation in cardiac valves. Circulation. 2001;103(11):1522–8. https://doi.org/10.1161/01.cir.103.11.1522.
9. Rajamannan NM, Gersh B, Bonow RO. Calcific aortic stenosis: from bench to the bedside-emerging clinical and cellular concepts. Heart. 2003;89:801–5.
10. Freeman RV, Otto CM. Spectrum of calcific aortic valve disease: pathogenesis, disease progression, and treatment strategies. Circulation. 2005;111:3316–26.
11. Prihadi EA, Leung M, Vollema EM, Ng AC, Ajmone Marsan N, Bax JJ, Delgado V. Prevalence and prognostic relevance of ventricular conduction disturbances in patients with aortic stenosis. Am J Cardiol. 2017;120:2226–32. https://doi.org/10.1016/j.amjcard.2017.08.046.
12. Hudzik B, Wilczek K, Gasior M. Heyde syndrome: gastrointestinal bleeding and aortic stenosis. CMAJ. 2016;188(2):135–8. https://doi.org/10.1503/cmaj.150194. Epub 2015 Jun 29. PMID: 26124230; PMCID: PMC4732965
13. Vahanian A, Beyersdorf F, Praz F, Milojevic M, Baldus S, Bauersachs J, Capodanno D, Conradi L, De Bonis M, De Paulis R, Delgado V, Freemantle N, Gilard M, Haugaa KH, Jeppsson A, Jüni P, Pierard L, Prendergast BD, Sádaba JR, Tribouilloy C, Wojakowski W, ESC/EACTS Scientific Document Group, ESC Scientific Document Group. 2021 ESC/EACTS Guidelines for the management of valvular heart disease. Eur Heart J. 2021;28:ehab395. https://doi.org/10.1093/eurheartj/ehab395.
14. Otto CM, Nishimura RA, Bonow RO, Carabello BA, Erwin JP 3rd, Gentile F, Jneid H, Krieger EV, Mack M, McLeod C, O'Gara PT, Rigolin VH, Sundt TM 3rd, Thompson A, Toly C. 2020 ACC/AHA guideline for the management of patients with valvular heart disease: executive summary: a report of the American College of Cardiology/American Heart Association Joint Committee on Clinical Practice Guidelines. J Am Coll Cardiol. 2021;77(4):450–500. https://doi.org/10.1016/j.jacc.2020.11.035. Epub 2020 Dec 17. Erratum in: J Am Coll Cardiol. 2021 Mar 9;77(9):1276
15. Cripe L, Andelfinger G, Martin LJ, Shooner K, Benson DW. Bicuspid aortic valve is heritable. J Am Coll Cardiol. 2004;44(1):138–43. https://doi.org/10.1016/j.jacc.2004.03.050.
16. Sillesen AS, Vøgg O, Pihl C, Raja AA, Sundberg K, Vedel C, Zingenberg H, Jørgensen FS, Vejlstrup N, Iversen K, Bundgaard H. Prevalence of bicuspid aortic valve and associated aortopathy in newborns in Copenhagen. Denmark JAMA. 2021;325(6):561–7. https://doi.org/10.1001/jama.2020.27205. PMID: 33560321; PMCID: PMC7873775
17. Michelena HI, Desjardins VA, Avierinos JF, Russo A, Nkomo VT, Sundt TM, Pellikka PA, Tajik AJ, Enriquez-Sarano M. Natural history of asymptomatic patients with normally functioning or minimally dysfunctional bicuspid aortic valve in the community. Circulation. 2008;117(21):2776–84. https://doi.org/10.1161/CIRCULATIONAHA.107.740878. PMID: 18506017; PMCID: PMC2878133
18. Shen M, Tastet L, Capoulade R, Bédard É, Arsenault M, Clavel MA, Pibarot P. Determinants of Aortic Stenosis Progression in Bicuspid and Tricuspid Aortic Valves. CJC Pediatr Congenit Heart Dis. 2022;1(4):184–92. https://doi.org/10.1016/j.cjcpc.2022.06.004. PMID: 37969929; PMCID: PMC10642113
19. Yetkin E, Waltenberger J. Molecular and cellular mechanisms of aortic stenosis. Int J Cardiol. 2009;135(1):4–13. https://doi.org/10.1016/j.ijcard.2009.03.108.
20. Siu SC, Silversides CK. Bicuspid aortic valve disease. J Am Coll Cardiol. 2010;55(25):2789–800. https://doi.org/10.1016/j.jacc.2009.12.068.

21. Kusner JJ, Brown JY, Gleason TG, Edelman ER. The natural history of bicuspid aortic valve disease. Struct Heart. 2022;7(2):100119. https://doi.org/10.1016/j.shj.2022.100119. PMID: 37275583; PMCID: PMC10236785
22. Sievers HH, Schmidtke C. A classification system for the bicuspid aortic valve from 304 surgical specimens. J Thorac Cardiovasc Surg. 2007;133(5):1226–33. https://doi.org/10.1016/j.jtcvs.2007.01.039.
23. Mai Z, Guan L, Mu Y. Association between bicuspid aortic valve phenotype and patterns of valvular dysfunction: a meta-analysis. Clin Cardiol. 2021;44(12):1683–91. https://doi.org/10.1002/clc.23736.
24. Wedin JO, Vedin O, Rodin S, Simonson OE, Malmborg JH, Pallin J, James SK, Flachskampf FA, Ståhle E, Grinnemo K-H. Patients with bicuspid aortic stenosis demonstrate adverse left ventricular remodeling and impaired cardiac function before surgery with increased risk of postoperative heart failure. Circulation. 2022;146(17):1310–22. https://doi.org/10.1161/CIRCULATIONAHA.122.060125.
25. Ngiam JN, Chew NWS, Pramotedham T, Tan BY, Sia CH, Loh PH, Ruan W, Tay E, Kong WKF, Yeo TC, Poh KK. Implications of coexisting aortic regurgitation in patients with aortic stenosis. JACC Asia. 2021;1(1):105–11. https://doi.org/10.1016/j.jacasi.2021.05.004. PMID: 36338366; PMCID: PMC9627873
26. Nistri S, Sorbo MD, Marin M, Palisi M, Scognamiglio R, Thiene G. Aortic root dilatation in young men with normally functioning bicuspid aortic valves. Heart. 1999;82:19–22.
27. Novaro GM, Griffin BP. Congenital bicuspid aortic valve and rate of ascending aortic dilatation. Am J Cardiol. 2004;93(4):524–5. https://doi.org/10.1016/j.amjcard.2003.07.051.
28. Verma S, Siu S. Aortic dilatation in patients with bicuspid aortic valve. N Engl J Med. 2014;370(20):1920–9.
29. Li Y, Wei X, Zhao Z, Liao Y, He J, Xiong T, Xu Y, Lv W, Ou Y, Tang H, Feng Y, Chen M. Prevalence and complications of bicuspid aortic valve in Chinese according to echocardiographic database. Am J Cardiol. 2017;120(2):287–91. https://doi.org/10.1016/j.amjcard.2017.04.025. Epub 2017 Apr 27
30. Halushka MK, Angelini A, Bartoloni G, Basso C, Batoroeva L, Bruneval P, Buja LM, Butany J, d'Amati G, Fallon JT, Gallagher PJ, Gittenberger-de Groot AC, Gouveia RH, Kholova I, Kelly KL, Leone O, Litovsky SH, Maleszewski JJ, Miller DV, Mitchell RN, Preston SD, Pucci A, Radio SJ, Rodriguez ER, Sheppard MN, Stone JR, Suvarna SK, Tan CD, Thiene G, Veinot JP, van der Wal AC. Consensus statement on surgical pathology of the aorta from the Society for Cardiovascular Pathology and the Association For European Cardiovascular Pathology: II. Noninflammatory degenerative diseases—nomenclature and diagnostic criteria. Cardiovasc Pathol. 2016;25(3):247–57. https://doi.org/10.1016/j.carpath.2016.03.002. Epub 2016 Mar 12
31. Michelena HI, Khanna AD, Mahoney D, Margaryan E, Topilsky Y, Suri RM, Eidem B, Edwards WD, Sundt TM 3rd, Enriquez-Sarano M. Incidence of aortic complications in patients with bicuspid aortic valves. JAMA. 2011;306(10):1104–12. https://doi.org/10.1001/jama.2011.1286.
32. Goldschlager N, Pfeifer J, Cohn K, Popper R, Selzer A. The natural history of aortic regurgitation. A clinical and hemodynamic study. Am J Med. 1973;54(5):577–88. https://doi.org/10.1016/0002-9343(73)90115-0.
33. Bonow RO, Lakatos E, Maron BJ, Epstein SE. Serial long-term assessment of the natural history of asymptomatic patients with chronic aortic regurgitation and normal left ventricular systolic function. Circulation. 1991;84(4):1625–35. https://doi.org/10.1161/01.cir.84.4.1625.
34. Gaasch WH, Andrias CW, Levine HJ. Chronic aortic regurgitation: the effect of aortic valve replacement on left ventricular volume, mass and function. Circulation. 1978;58(5):825–36. https://doi.org/10.1161/01.cir.58.5.825.
35. Masri A, Svensson LG, Griffin BP, Desai M. Contemporary natural history of bicuspid aortic valve disease: a systematic review. Heart. 2017;103:1323–30.
36. Basso C, Boschello M, Perrone C, Mecenero A, Cera A, Bicego D, Thiene G, De Dominicis E. An echocardiographic survey of primary school children for bicuspid aortic valve. Am J Cardiol. 2004;93(5):661–3. https://doi.org/10.1016/j.amjcard.2003.11.031.
37. Kong WK, Delgado V, Poh KK, Regeer MV, Ng AC, McCormack L, Yeo TC, Shanks M, Parent S, Enache R, Popescu BA, Liang M, Yip JW, Ma LC, Kamperidis V, van Rosendael PJ, van der Velde ET, Ajmone Marsan N, Bax JJ. Prognostic implications of Raphe in bicuspid aortic valve anatomy. JAMA Cardiol. 2017;2(3):285–92. https://doi.org/10.1001/jamacardio.2016.5228.
38. Roberts WC. Congenitally bicuspid aortic valve causing severe, pure aortic regurgitation without superimposed infective endocarditis. Am J Cardiol. 1961;47:206–9.
39. Vowels TJ, Gonzalez-Stawinski GV, Ko JM, Trachiotis GD, Roberts BJ, Roberts CS, Roberts WC. Anomalous cord from the raphe of a congenitally bicuspid aortic valve to the aortic wall producing either acute or chronic aortic regurgitation. J Am Coll Cardiol. 2014;63(2):153–7. https://doi.org/10.1016/j.jacc.2013.09.030. Epub 2013 Oct 16
40. Roman MJ, Devereux RB, Niles NW, Hochreiter C, Kligfield P, Sato N, Spitzer MC, Borer JS. Aortic root dilatation as a cause of isolated, severe aortic regurgitation. Prevalence, clinical and echocardiographic patterns, and relation to left ventricular hypertrophy and function. Ann Intern Med. 1987;106(6):800–7. https://doi.org/10.7326/0003-4819-106-6-800.
41. Zhang Y, Choi BH, Chee HK, Kim JS, Ko SM. Aortic Valve Dysfunction and Aortopathy Based on the Presence of Raphe in Patients with Bicuspid Aortic Valve Disease. J Cardiovasc Dev Dis. 2023;10(9):372. https://doi.org/10.3390/jcdd10090372.

8 Aortopathy, Coarctaction, and Aortic Dissection

Cristina Basso, Stefania Rizzo, Monica De Gaspari, and Gaetano Thiene

Association of bicuspid aortic valve (BAV) with aortopathy [1] was not recognized until 1928, when Maude Abbott reported 200 autopsy cases with aortic arch coarctation [2]. She found that BAV was frequently associated with aortic coarctation and dissection of the ascending aorta (Fig. 8.1), which led her to argue "for a congenital weakness of the (aortic) wall, put above the cusps, as a probable cause of rupture." Indeed, a significant number of people affected by BAV undergo with time to progressive ectasia of ascending aorta, both sinusal and tubular, due to degenerative changes in the tunica media: medionecrosis with disappearance of elastic lamellae units and loss of elastic properties with accumulation of mucoid substance in the extracellular matrix (Fig. 8.2), known in the past as Erdheim's medionecrosis idiopathic of the aorta [3] (Fig. 8.3).

The severity of aortic wall abnormalities is related to the aortic diameter and age.

However, patients with aneurysm of the ascending aorta have less aortic wall alterations in BAV than in non-rheumatic tricuspid aortic valve (TAV) incompetence.

Even a normally functioning BAV shows a progressive dilatation of the ascending aorta, beginning in childhood [4]. Half of 20 years old conscripts with BAV showed dilated ascending aorta at echo screening [5].

Progression of the aorta enlargement is reported to be 1–2 mm per year, faster growth occurring occasionally. A rapid progression of >5 mm per year is associated with increased risk of aortic dissection and rupture as to require preventive surgery [6].

Metanalysis of family screening for BAV revealed an overall occurrence rate of BAV by 7.3% among relatives, 23.6% if limited to family and prevalence of aortic dilatation of 0.4% among relatives [7].

Ascending aorta in BAV patients, without yet aortopathy dilatation, is characterized by a differention defect of vascular smooth muscle cells of the tunica media and a thinner intimal layer [8].

An association between BAV and Erdheim's cystic medial necrosis was first observed by McKusick VA in 1956 [9] and confirmed in 1972 [10]. These findings raised the suspect that BAV might be a development defect, explaining both arch coarctation (Fig. 8.1) and cystic medial necrosis of Erdheim's in the aorta [3] (Fig. 8.2).

C. Basso (✉) · S. Rizzo · M. De Gaspari · G. Thiene
Cardiovascular Pathology Unit, Department of Cardiac, Thoracic, Vascular Sciences and Public Health, School of Medicine, University of Padua, Padua, Italy
e-mail: cristina.basso@unipd.it; s.rizzo@unipd.it; monica.degaspari@unipd.it; gaetano.thiene@unipd.it

G. Thiene et al. (eds.), *Bicuspid Aortic Valve Syndrome*,
https://doi.org/10.1007/978-3-031-85589-4_8

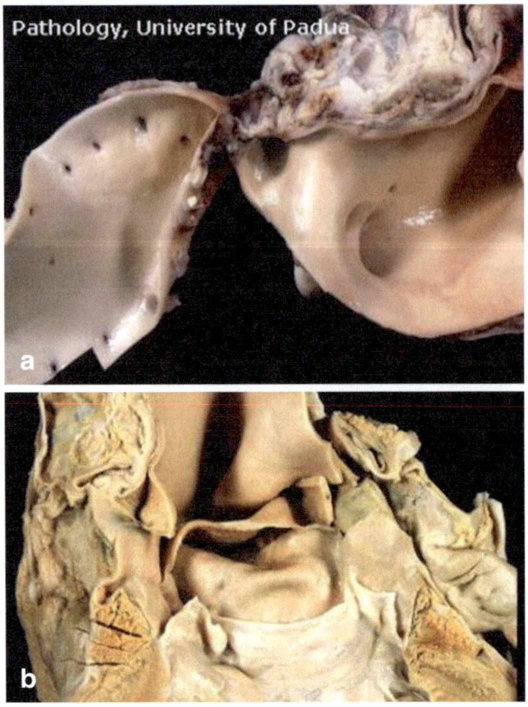

Fig. 8.1 BAV with coarctation of the aortic arch (**a**) and dissection (**b**) with intimal tear, just above the sino-tubular junction of the ascending aorta

Proposal of prophylactic replacement of a dilated ascending aorta at the time of surgery by BAV dysfunction [11] has been advanced by the late Bill Roberts (Fig. 8.3).

Young men with BAV and dilated proximal aorta manifest systemic endothelium dysfunction and high plasma levels of metalloproteinase 2 [12].

Hemodynamic theory of abnormalities in aortopathy of BAV patients has been postulated by Robicsek F et al. in 2004 [13], through an experiment with cryo preserved human artic roots including BAV vs TAV. BAV function was characterized by excessive folding and creasing with extended area of cusp contact, eventually complicated by calcification and stenosis. The abnormal flow patterns would produce dilatation and dissection of the ascending aorta, due to higher and uneven stress.

An epidemiology study performed by Mayo Clinic in the Olmsted Country on BAV carriers, devoid of valve stenosis or incompetence, showed that 23% developed an aneurysm of the ascending aorta within 22 years follow-up, with aortic dissection occurring much more frequent than in general population without BAV [14].

BAV is more frequent in Marfan patients (4.7% vs 0.5% of general population) and BAV with Marfan develops more severe aortopathy [15].

Deficient fibrillin content, not genetically determined as in Marfan syndrome, was advanced as a trigger of metalloproteinases release with disruption of lamellae units [16].

Neural crest stem cells play an important role in the development of heart outflow tracts and their deficiency is probably the biological mechanism of the BAV syndrome [17, 18].

The Society for Cardiovascular Pathology and the Association for European Cardiovascular Pathology put forward a new nomenclature and histologic diagnostic criteria for degenerative non-inflammatory alterations of aortic tunica media. The term "Medionecrosis aorta idiopathica" introduced by Erdheim in 1929 [3] has been replaced by "mucoid extracellular matrix accumulation" (MEMA) in 2016 [19].

Drug therapy to treat BAV aortopathy is currently tested in Clinical Trials on BAV patients, similarly to Marfan patients. β-blockers and angiotensin receptors blockers (Losartan), through direct and indirect inhibition of transforming growth factor (TGF β), have been shown to reverse aorta root dilatation, in analogy with Marfan syndrome. Although BAV and Marfan are distinct morbid entities (aortic valve in Marfan is mostly tricuspid), β-blockers and angiotensin receptor blockers may work in a similar fashion in the two entities [20].

Knowledge of aortic dimension in normal aortic valves is mandatory to evaluate aorta dilatation over time in the follow-up of the disease and to plan appropriate therapeutic strategy [21]. Don't forget that BAV is associated by progressive dilatation of the ascending aorta, beginning in childhood [4, 5].

Pregnancy has been recognized as predisposing factor of aortic dissection. It is associated with histopathologic alterations in the tunica media, most probably related to immune influ-

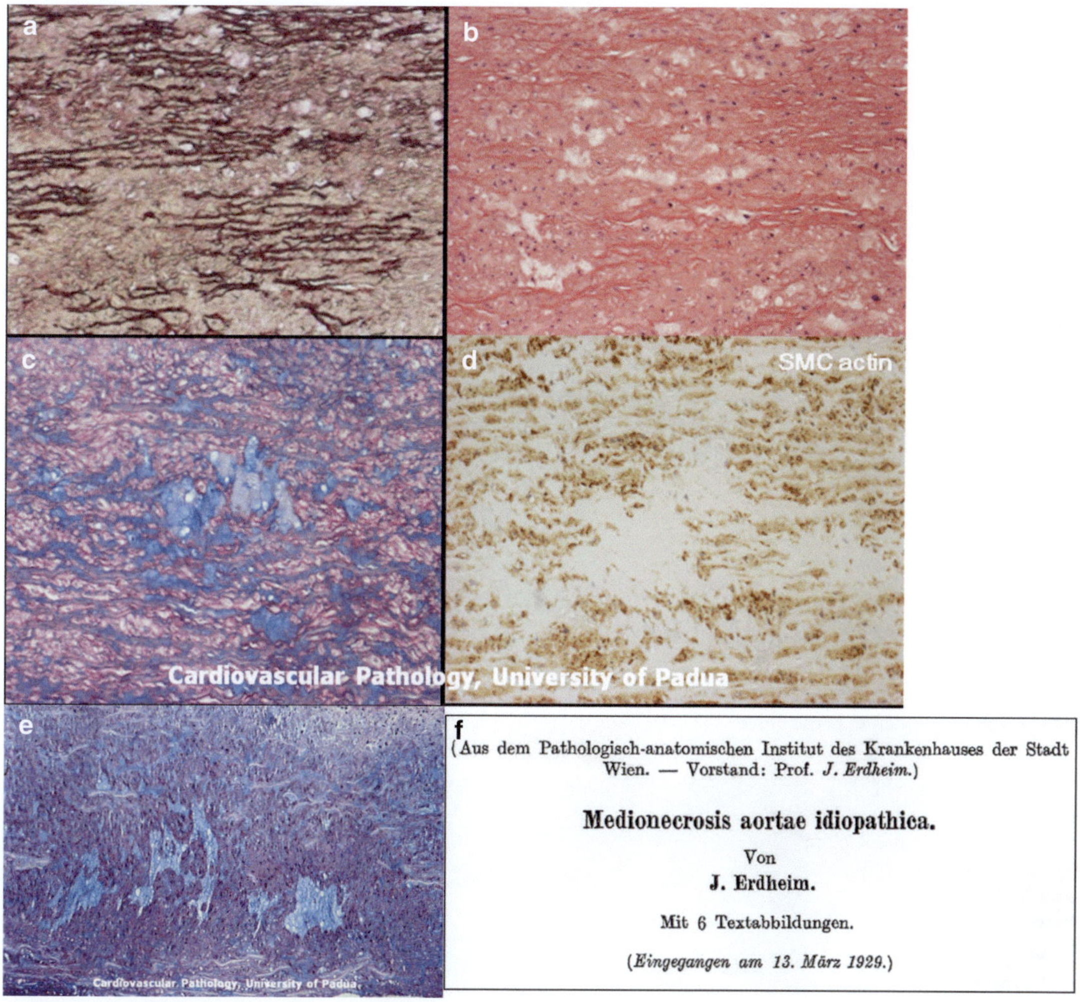

(Aus dem Pathologisch-anatomischen Institut des Krankenhauses der Stadt Wien. — Vorstand: Prof. *J. Erdheim.*)

Medionecrosis aortae idiopathica.

Von

J. Erdheim.

Mit 6 Textabbildungen.

(*Eingegangen am 13. März 1929.*)

Fig. 8.2 Histological alterations within the tunica media of the dilated ascending aorta in a patient with BAV aortic wall, operated for both replacement of stenotic aortic valve and the ascending aorta. (**a**) Disruption of elastic lamellae (Weigert-Van Gieson stain). (**b**) Cysts of mucoid substance (hematoxylin and eosin stain). (**c**) Spots of mucoid substance deposits (Alcian-Pas stain). (**d**) Medionecrosis by loss of smooth muscle cells at immunohistochemistry. (**e**) Cysts of mucoid substance (Alcian-Pass stain). (**f**) Original paper by J. Erdheim, entitled "*Medionecrosis aortae idiopathica*"

ence on the connective tissue. Pregnant women affected by BAV are at a major risk to develop aortic aneurysm and rupture [22–27].

Ascending aorta size threshold of 5.5 cm appears among criteria for surgical replacement of aortic aneurysm also for BAV patients [6]. A threshold of 5 cm has been postulated [28], especially if the dilatation growth is fast. However the lowering of the threshold as prophylactic aortic surgery has been questioned [29, 30].

Aortic dissection of the ascending aorta with external rupture, hemopericardium, and cardiac tamponade is the main cause of sudden death in BAV syndrome (Figs. 8.4 and 8.5).

Incidence of BAV in aortic dissection is much more frequent than in normal population (Table 8.1), so that the occurrence of aortic dissection in BAV cannot be considered incidental.

The dilation may accelerate in the case of progressive medionecrosis of the tunica media. A

Fig. 8.3 Photo of the late Bill Roberts, a first class pathologist, researcher on aortic valve

hypertension attach in BAV patients may trigger the intimal tear and the aortic dissection (Fig. 8.1).

At autopsy we experienced cases of BAV and aortic dissection with diameter of the ascending aorta quite less than the 5 cm threshold (Figs. 8.5 and 8.6). This suggests to perform echocardiography regularly, to check the growth speed of the aorta diameter. Moreover, since aortic histology cannot be performed, except at the time of surgery, the poor systole-diastole excursion of the aortic wall at echo may help to detect increased aortic stiffness, due to degenerative changes of the tunica media at risk of impeding dissection [35, 36].

Aortic dissection in BAV is a significant cause of sudden death in the young (Fig. 8.7).

Fig. 8.4 Dissection of the ascending aorta in a 25-year-old male patient with concealed bicuspid aortic valve. (**a**) Gross view with huge dilatation (diameter 7 cm). (**b**) The aortic wall of the tubular ascending aorta show dissection in the tunica media (Weigert-Van Gieson stain). (**c**) Spots of cystic mucoid deposits (MEMA) (Alcian-Pas stain)

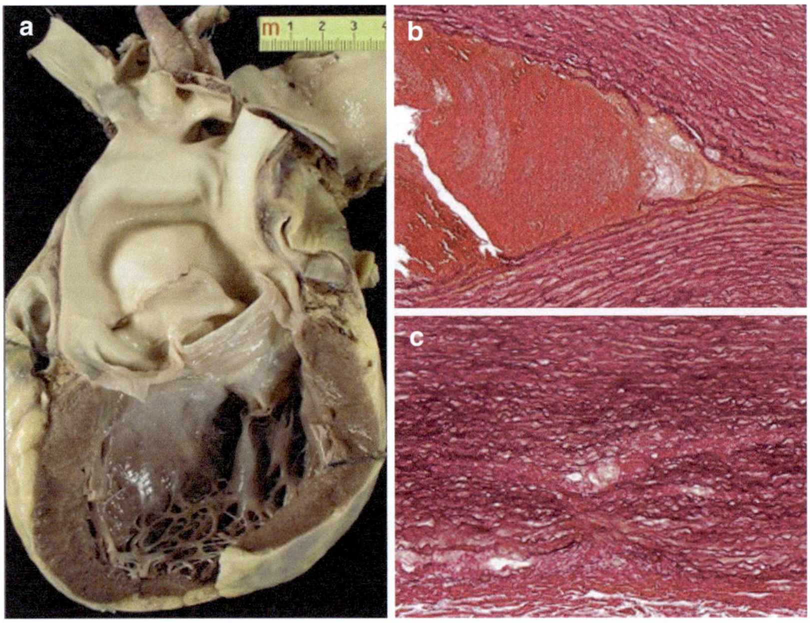

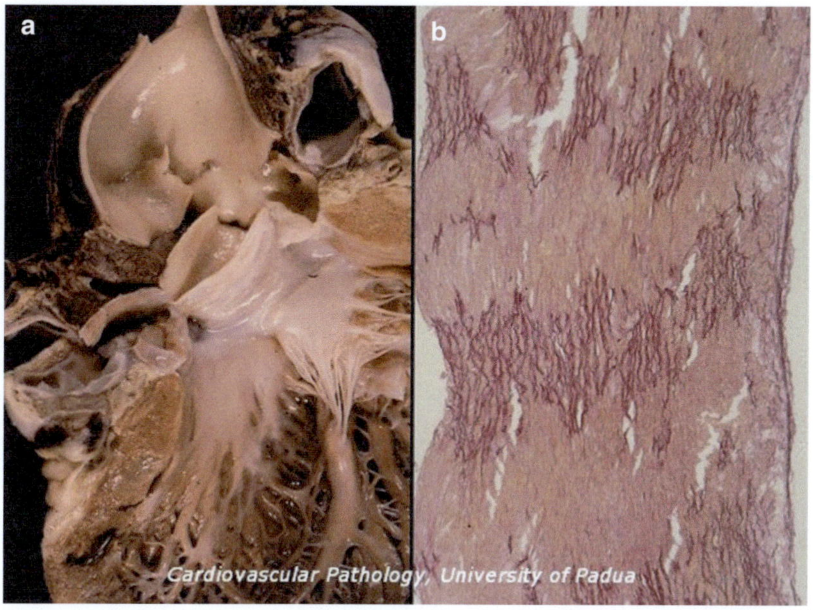

Fig. 8.5 Dissection of the ascending aorta occurred in a 38-year-old patient with BAV. (**a**) Gross view with burst of the ascending aorta. The 38 mm diameter was not considered at risk of dissection and therefore not operated on. (**b**) Massive disappearance of elastic fibers in the tunica media (Weigert-Van Gieson stain)

Table 8.1 Incidence of BAV in aortic dissection

Author	Year	N. dissection	% BAV
Gore [31]	1952	85	13%
Edwards et al. [32]	1978	119	9%
Larson et al. [33]	1984	161	11%
Giusti et al. [34]	1990	87	12%

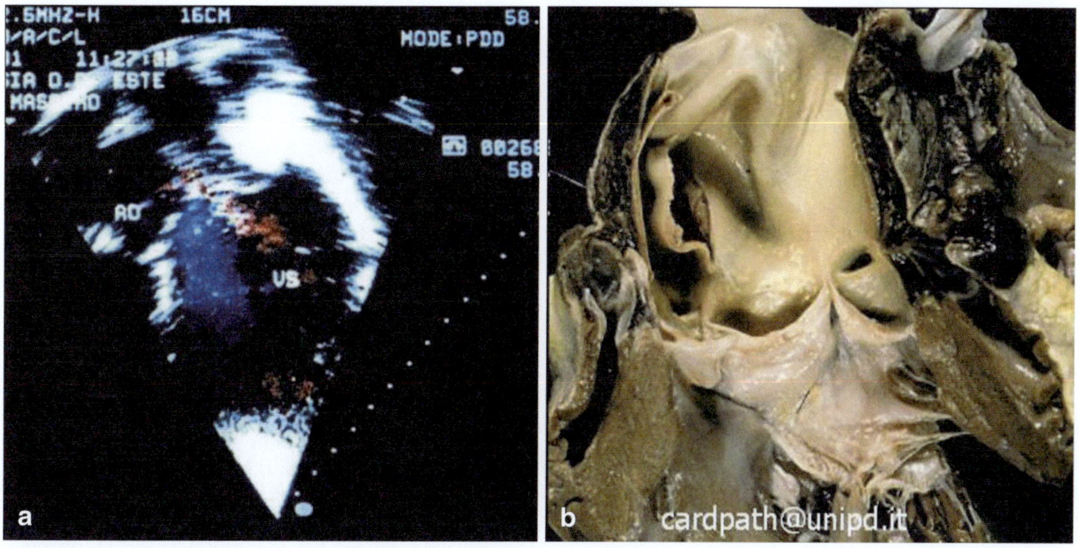

Fig. 8.6 Male, 31 years 2D echo diagnosis of bicuspid aortic valve. Last echo follow-up: mild aortic incompetence, ascending aorta diameter 38 mm (**a**). Ten months later (**b**)

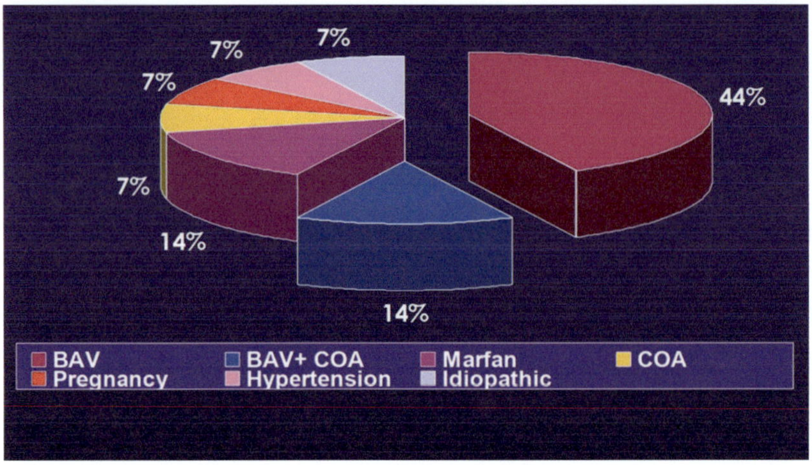

Fig. 8.7 Sudden cardiac death in the young (<35 years) due to aortic dissection. *No 15 out 450 cases (3.5%)*

References

1. Michelena HI. The bicuspid aortic valve aortopathy mystery continues: are we that mediocre? Trends Cardiovasc Med. 2015;25(5):452–5.
2. Abbott ME. Coartation of the aorta of adult type; statistical study and historical retrospect of 200 recorded cases with autopsy of stenosis or obliteration of descending arch in subjects above age of two years. Am Heart J. 1928;3:574.
3. Erdheim J. Medionecrosis aortae idiopathica. Virchows Arch Path Anat. 1929;273:454–79.
4. Beroukhim RS, Kruzick TL, Taylor AL, Gao D, Yetman AT. Progression of aortic dilation in children with a functionally normal bicuspid aortic valve. Am J Cardiol. 2006;98(6):828–30.
5. Nistri S, Sorbo MD, Marin M, Palisi M, Scognamiglio R, Thiene G. Aortic root dilatation in young men with normally functioning bicuspid aortic valves. Heart. 1999;82(1):19–22.
6. Erbel R, Aboyans V, Boileau C, Bossone E, Bartolomeo RD, Eggebrecht H, Evangelista A, Falk V, Frank H, Gaemperli O, Grabenwöger M, Haverich A, Iung B, Manolis AJ, Meijboom F, Nienaber CA, Roffi M, Rousseau H, Sechtem U, Sirnes PA, Allmen RS, Vrints CJ, ESC Committee for Practice Guidelines. ESC Guidelines on the diagnosis and treatment of aortic diseases: document covering acute and chronic aortic diseases of the thoracic and abdominal aorta of the adult. The task force for the diagnosis and treatment of aortic diseases of the European Society of Cardiology (ESC). Eur Heart J. 2014;35(41):2873–926. Erratum in: Eur Heart J. 2015 Nov 1;36(41):2779
7. Bray JJH, Freer R, Pitcher A, Kharbanda R. Family screening for bicuspid aortic valve and aortic dilatation: a meta-analysis. Eur Heart J. 2023;44(33):3152–64.
8. Grewal N, Girdauskas E, Idhrees M, Velayudhan B, Klautz R, Driessen A, Poelmann RE. Structural abnormalities in the non-dilated ascending aortic wall of bicuspid aortic valve patients. Cardiovasc Pathol. 2023;62:1074–8.
9. McKusick VA, Logue RB, Bahnson HT. Association of aortic valvular disease and cystic medial necrosis of the ascending aorta; report of four instances. Circulation. 1957;16(2):188–94. https://doi.org/10.1161/01.cir.16.2.188.
10. McKusick VA. Association of congenital bicuspid aortic valve and Erdheim's cystic medial necrosis. Lancet. 1972;1(7758):1026–7.
11. Roberts WC. Prophylactic replacement of a dilated ascending aorta at the time of aortic valve replacement of a dysfunctioning congenitally unicuspid or bicuspid aortic valve. Am J Cardiol. 2011;108(9):1371–2.
12. Tzemos N, Lyseggen E, Silversides C, et al. Endothelial function, carotid-femoral stiffness, and plasma matrix metalloproteinase-2 in men with bicuspid aortic valve and dilated aorta. J Am Coll Cardiol. 2010;55(7):660–8.
13. Robicsek F, Thubrikar MJ, Cook JW, Fowler B. The congenitally bicuspid aortic valve: how does it function? Why does it fail? Ann Thorac Surg. 2004;77(1):177–85.
14. Yang LT, Ye Z, Wajih Ullah M, Maleszewski JJ, Scott CG, Padang R, Pislaru SV, Nkomo VT, Mankad SV, Pellikka PA, Oh JK, Roger VL, Enriquez-Sarano M, Michelena HI. Bicuspid aortic valve: long-term morbidity and mortality. Eur Heart J. 2023;44(43):4549–62.
15. Nistri S, Porciani MC, Attanasio M, Abbate R, Gensini GF, Pepe G. Association of Marfan syndrome and bicuspid aortic valve: frequency and outcome. Int J Cardiol. 2012;155(2):324–5. https://doi.org/10.1016/j.ijcard.2011.12.009. Epub 2012 Jan 5
16. Fedak PW, de Sa MP, Verma S, Nili N, Kazemian P, Butany J, Strauss BH, Weisel RD, David TE. Vascular matrix remodeling in patients with bicuspid aortic valve malformations: implications for aortic dilatation. J Thorac Cardiovasc Surg. 2003;126(3):797–806. https://doi.org/10.1016/s0022-5223(03)00398-2.

17. Sieber-Blum M. Cardiac neural crest stem cells. Anat Rec A Discov Mol Cell Evol Biol. 2004;276(1):34–42.
18. Brown CB, Baldwin HS. Neural crest contribution to the cardiovascular system. Adv Exp Med Biol. 2006;589:134–54.
19. Halushka MK, Angelini A, Bartoloni G, Basso C, Batoroeva L, Bruneval P, Buja LM, Butany J, d'Amati G, Fallon JT, Gallagher PJ, Gittenberger-de Groot AC, Gouveia RH, Kholova I, Kelly KL, Leone O, Litovsky SH, Maleszewski JJ, Miller DV, Mitchell RN, Preston SD, Pucci A, Radio SJ, Rodriguez ER, Sheppard MN, Stone JR, Suvarna SK, Tan CD, Thiene G, Veinot JP, van der Wal AC. Consensus statement on surgical pathology of the aorta from the Society for Cardiovascular Pathology and the Association For European Cardiovascular Pathology: II. Noninflammatory degenerative diseases—nomenclature and diagnostic criteria. Cardiovasc Pathol. 2016;25(3):247–57.
20. Broberg CS, Therrien J. Understanding and treating aortopathy in bicuspid aortic valve. Trends Cardiovasc Med. 2015;25(5):445–51.
21. Vriz O, Aboyans V, D'Andrea A, Ferrara F, Acri E, Limongelli G, Della Corte A, Driussi C, Bettio M, Pluchinotta FR, Citro R, Russo MG, Isselbacher E, Bossone E. Normal values of aortic root dimensions in healthy adults. Am J Cardiol. 2014;114(6):921–7.
22. De Martino A, Morganti R, Falcetta G, Scioti G, Milano AD, Pucci A, Bortolotti U. Acute aortic dissection and pregnancy: review and meta-analysis of incidence, presentation, and pathologic substrates. J Card Surg. 2019;34(12):1591–7.
23. Manalo-Estrella P, Barker AE. Histopathologic findings in human aortic media associated with pregnancy. Arch Pathol. 1967;83(4):336–41.
24. Zhu JM, Ma WG, Peterss S, Wang LF, Qiao ZY, Ziganshin BA, Zheng J, Liu YM, Elefteriades JA, Sun LZ. Aortic dissection in pregnancy: management strategy and outcomes. Ann Thorac Surg. 2017;103(4):1199–206.
25. Al-Hussaini A. Pregnancy and aortic dissections. Eur Heart J. 2020;41(44):4243–4. https://doi.org/10.1093/eurheartj/ehaa754.
26. Beyer SE, Dicks AB, Shainker SA, Feinberg L, Schermerhorn ML, Secemsky EA, Carroll BJ. Pregnancy-associated arterial dissections: a nationwide cohort study. Eur Heart J. 2020;41(44):4234–42.
27. Immer FF, Bansi AG, Immer-Bansi AS, McDougall J, Zehr KJ, Schaff HV, Carrel TP. Aortic dissection in pregnancy: analysis of risk factors and outcome. Ann Thorac Surg. 2003;76(1):309–14.
28. Wu J, Zafar MA, Liu Y, Chen JF, Li Y, Ziganshin BA, Ellauzi H, Mukherjee SK, Rizzo JA, Elefteriades JA. Fate of the unoperated ascending thoracic aortic aneurysm: three-decade experience from the Aortic Institute at Yale University. Eur Heart J. 2023;44(43):4579–88.
29. Milleron O, Eliahou L, Jondeau G. There is insufficient evidence to lower the threshold for prophylactic aortic surgery. Eur Heart J. 2024;45(8):632.
30. Hultgren R, Sakalihasan N. Do we need new thresholds for surgical repair in patients with ascending thoracic aortic aneurysm disease? Eur Heart J. 2023;44(43):4589–91.
31. Gore I. Pathogenesis of dissecting aneurysm of the aorta. AMA Arch Pathol. 1952;53(2):142–53.
32. Edwards WD, Leaf DS, Edwards JE. Dissecting aortic aneurysm associated with congenital bicuspid aortic valve. Circulation. 1978;57(5):1022–5.
33. Larson EW, Edwards WD. Risk factors for aortic dissection: a necropsy study of 161 cases. Am J Cardiol. 1984;53(6):849–55.
34. Giusti S, Cocco P, Thiene G. Valvola aortića biscuspide: una cardiopatia congenita "minore" a rischio di gravi complicanze [the bicuspid aortic valve: a "minor" congenital cardiopathy at risk of serious complications]. G Ital Cardiol. 1991;21(2):189–201. Italian
35. Nistri S, Sorbo MD, Basso C, Thiene G. Bicuspid aortic valve: abnormal aortic elastic properties. J Heart Valve Dis. 2002;11(3):369–73.
36. Nistri S, Grande-Allen J, Noale M, Basso C, Siviero P, Maggi S, Crepaldi G, Thiene G. Aortic elasticity and size in bicuspid aortic valve syndrome. Eur Heart J. 2008;29(4):472–9.

Endocarditis

9

Gaetano Thiene and Cristina Basso

9.1 Infective Endocarditis

Infective endocarditis (IE) on bicuspid aortic valve (BAV) (Fig. 9.1) occurs in the setting of bacteremia which promotes deposits of microorganisms (bacteria and fungi upon the native pathologic substrate)[1] (Table 9.1). In the past rheumatic valve disease and congenital heart diseases were the main morbid entities, prone to endocarditis [1]. They are still occurring in third world countries (Table 9.2) [2]. Nowadays, according to surgical pathology data (tissue removed at the time of valve replacement), BAV represents the most frequent predisposing factor (25%) at risk of IE, followed by mitral valve prolapse, rheumatic valve disease, congenital heart disease, and hypertrophic cardiomyopathy (Tables 9.3 and 9.4) [3].

Sterile, tiny fibrin or platelets aggregates serve as milieu for bacteria adhesion on the valve surface [4]. Endothelial injury by surface erosion accounts for thrombi deposits as nidus for infection. Mechanical stress by leaflet coaptation on line of closure with continuous trauma explains the prevalent location of endocarditis on left sided valves (aortic and mitral). Take into account that the number of systoles—diastoles, with valve opening and closure during life, are billions.

The peculiar sources of microorganisms entering into circulation and causing bacteriemia are infections of gums, urinary and colic tracts. Gram positive bacteria (streptococci and staphylococci) show more propensity to adhere to valve surface than gram negative bacteria (Table 9.5) [4]. Decreased host defence may play a role.

Microorganisms proliferation gives origin to huge septic thrombotic vegetation (vegetative endocarditis) (Fig. 9.2), and tearing with perforations of the cusps (ulcero-vegetative endocarditis) (Fig. 9.3). Septic vegetations may detach giving origin to systemic embolism with organ infarcts (Figs. 9.4 and 9.5); at spleen levels, the embolic infarct may lead to rupture.

Patients with BAV have a very high risk of native valve infective endocarditis [5]. BAV is a frequent substrate in patients with isolated aortic valve IE and is associated with several perivalvular complications and neurological events. The better survival with tricuspid aortic valve

[1] Viruses never reported

G. Thiene (✉) · C. Basso
Cardiovascular Pathology Unit, Department of Cardiac, Thoracic, Vascular Sciences and Public Health, School of Medicine, University of Padua, Padua, Italy
e-mail: gaetano.thiene@unipd.it; cristina.basso@unipd.it

G. Thiene et al. (eds.), *Bicuspid Aortic Valve Syndrome*,
https://doi.org/10.1007/978-3-031-85589-4_9

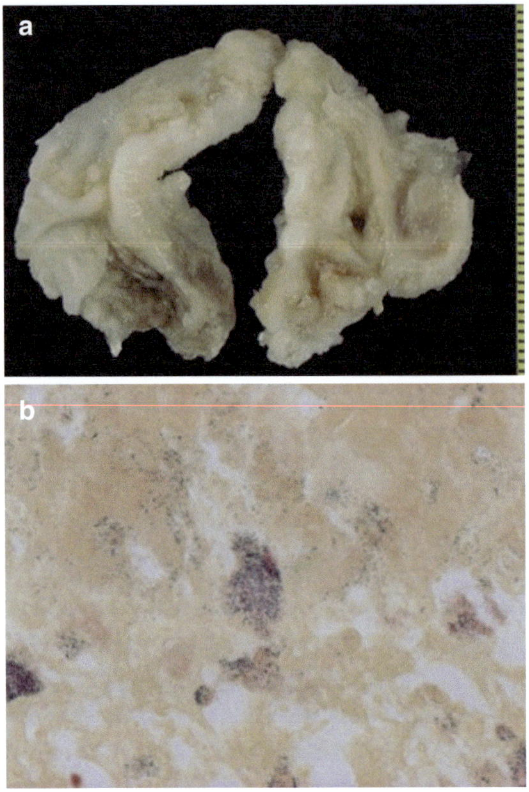

Fig. 9.1 Surgical pathology specimen with infective endocarditis on a bicuspid valve. (**a**) Gross view showing tiny vegetations, upon calcified cusps, with the perforation of a cusp (on the right cusp). (**b**) At histology colonies of gram-positive cocci, surrounded by a necrotic tissue (Gram stain)

Table 9.1 Cardiac substrates at risk of infective endocarditis [1]

Congenital heart disease	Bicuspid aortic valve
Acquired valve disease	Rheumatic, mitral valve prolapse
Cardiomyopathies	Hypertrophic cardiomyopathy
Prosthetic valves	Biological, mechanical prostheses
Cardiac catheters	Pacemaker, defibrillator, synchronization

(TAV-IE) are probably related to the age and different comorbidities between these two populations [6].

The attention was recently paid on the occurrence of bacterial endocarditis in patients with apparently innocent BAV [7] as first underlined by Osler in 1886 [8].

Valve disease, endocarditis, and aortopathy are unavoidable by age 90. The lifetime incidence of IF is higher than that of aortic dissection [9].

IE can lead to valve perforation or destruction (Fig. 9.6). If this occurs acutely, it is poorly tolerated and fatal [10]. Risk of endocarditis, originally estimated to range between 10% and 30%, was estimated much lower recently, between 2% or 0.3%/year [11]. Because the risk of endocarditis in BAV is felt to be low, the ACC/AHA practice guidelines no longer suggest bacterial prophylaxis in patients with BAV, except in patients with a prior history of endocarditis [12], a questionable recommendation in clinical practice.

Huge vegetations, larger than 10–11 mm, should be removed by surgery to prevent devastating embolism [13].

Extension of infection to sinus aortic wall can create aortic aneurysms, tunnels, fistulae to the atrial cavity (Table 9.6) (Fig. 9.7), and even involvement of the av. conduction system with av. block (Fig. 9.7) [14]. Extension to mitral valve with the mechanism "kissing lesion" may give onset of mitral aneurysm (Fig. 9.8).

BAV, despite being a left side valve, may be the target of infection in drug addicts by intravenous injection (Fig. 9.9).

As far as primary prevention, the topic is the source of controversy. According to Maude Abbott classification [15], BAV belongs to benign non cyanotic congenital heart diseases, which should not deserve antibiotic prophilaxis. However, 2004 guidelines of the European Society of Cardiology recommended patients with BAV to undergo to antibiotic prophylaxis, when submitted to procedures at risk of bacteremia, like dental maneuvers or colonoscopy [16].

Embolism of septic vegetation may occur also along coronary circulation bed, with even occlusion of subepicardical arteries, precipitating myocardial infarcts (Fig. 9.10a) and sudden death.

Septic dissemination may also involve microcirculation of the heart, giving origin to septic myocarditis (Fig. 9.10b).

Table 9.2 Cardiac heart substrates, currently predisposing to infective endocarditis, in Indian patients [2]

	$n°$	%
Rheumatic valve disease	79	42%
Congenital heart disease	62[a]	33%
Normal valve	17	9%
Uncertain etiology	24	13%
Mitral valve prolapse	2	1%
Prosthetic valves	2	1%
	186	

[a]25 bicuspid aortic valve (13%)

Table 9.3 Current structural cardiac disease at risk of infective endocarditis according to surgical pathology findings [Modified from Castonguay MC. et al. [3]]

Risk factor/comorbidity	$n°$	%
Bicuspid aortic valve	71	25%
Mitral valve prolapse	52	18%
Rheumatic valve disease	22	8%
Other congenital heart disease	16	5.5%
Hypertrophic cardiomyopathy	7	2.5%
	168	–

Table 9.4 Predisposing valve disease for native infective endocarditis

Author, year	*N*	BAV (%)	RVD (%)	MPV (%)	Others (%)	Normal (%)
McKinsey, 1987	63	10	6	29	28	27
Muller-Haache, 1987	39	20	8	18	31	23
Dressler, 1981	80	9	10	–	–	81
Maizza, 1990	131	27	16	13	9	35

BAV bicuspid aortic valve, *RVD* rheumatic valve disease, *MPV* mitral valve prolapse

Table 9.5 Microorganisms of infective endocarditis (from 3 published experiences). The leading infective microorganisms are Streptococcus viridans and Staphilococcus aureus. However blood culture may be frequently negative

	Müller (1964–82)	Johnson (1950–79)	Awadallah (1970–90)
Streptococcus viridans	35%	28%	29%
Streptococci, others	10%	–	–
Staphilococcus aureus	20%	38%	27%
Staphilococcus epidermidis	3%	6%	–
Fungi	5%	2%	6%
Others	14%	14%	27%
Negative culture	23%	12%	11%

In vivo diagnosis of endocarditis can be achieved by the combination of positive blood culture and detection by eco-doppler valve vegetations and incompetence, as established by "Duke criteria" (Table 9.7) [9].

Infective endocarditis was reported in several autopsy studies on BAV [17–22].

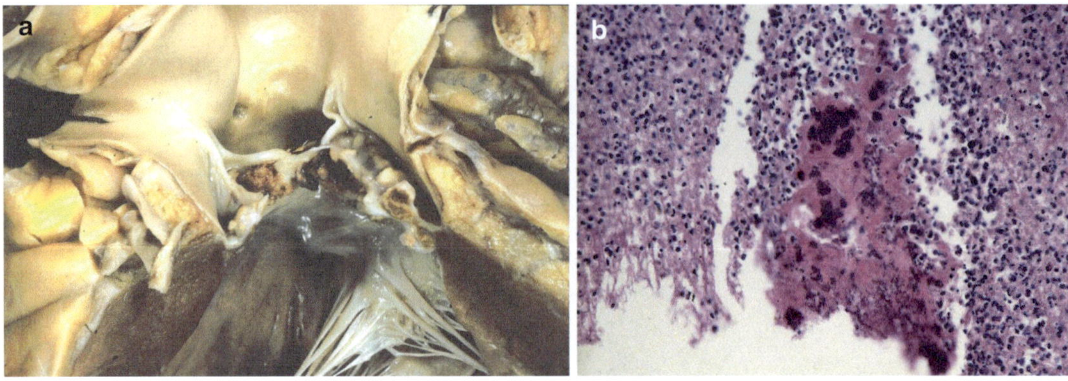

Fig. 9.2 (**a**) Gross view of infective endocarditis of aortic valve with coarse, huge vegetations at impeding detachment with embolism. (**b**) At histology, vegetations include colonies of microorganisms surrounded by neutrophil inflammatory infiltrates and necrosis (Haematoxylin-Eosin stain)

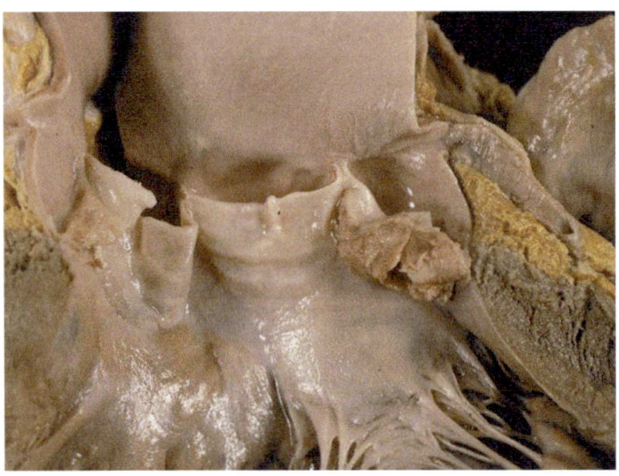

Fig. 9.3 Gross view of ulcerative endocarditis of the aortic valve

Fig. 9.4 Large ischemic infarct of the spleen, following embolism due to vegetative endocarditis

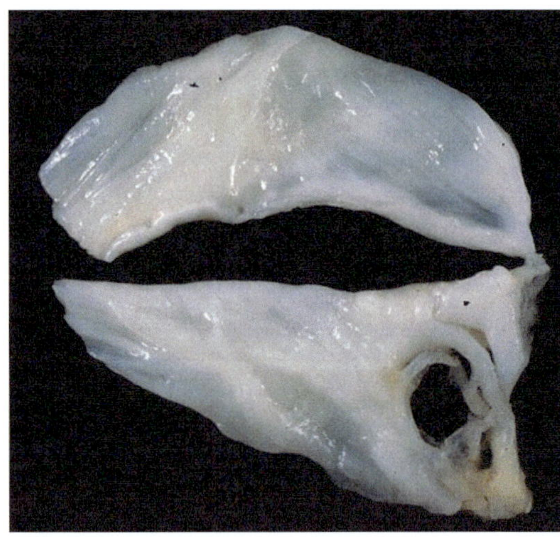

Fig. 9.5 Perforation of bicuspid valve by infective endocarditis with chronic valve incompetence

Fig. 9.6 (**a**) Abscess of the aortic valve annulus complicating infective endocarditis of aortic valve. (**b**) The abscess is penetrating the right atrium

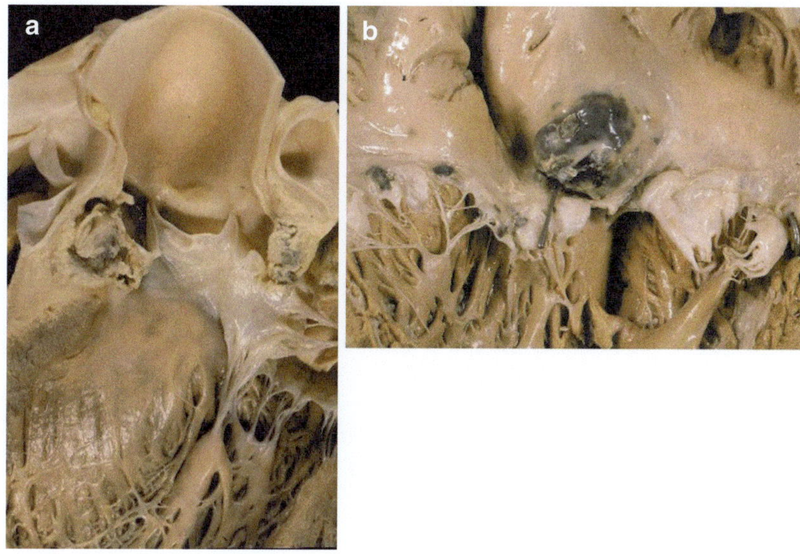

Table 9.6 Infective endocarditis of aortic valve: complications

Local	Distal
Vegetations	Infarction of systemic organ
Perforations	Stroke
Abscess with aneurysm, tunnel, fistulae	Myocarditis, pericarditis
AV block	Conduction tissue
	Glomerulonephritis
	Splenic abscess
	Arterial mycotic aneurysms

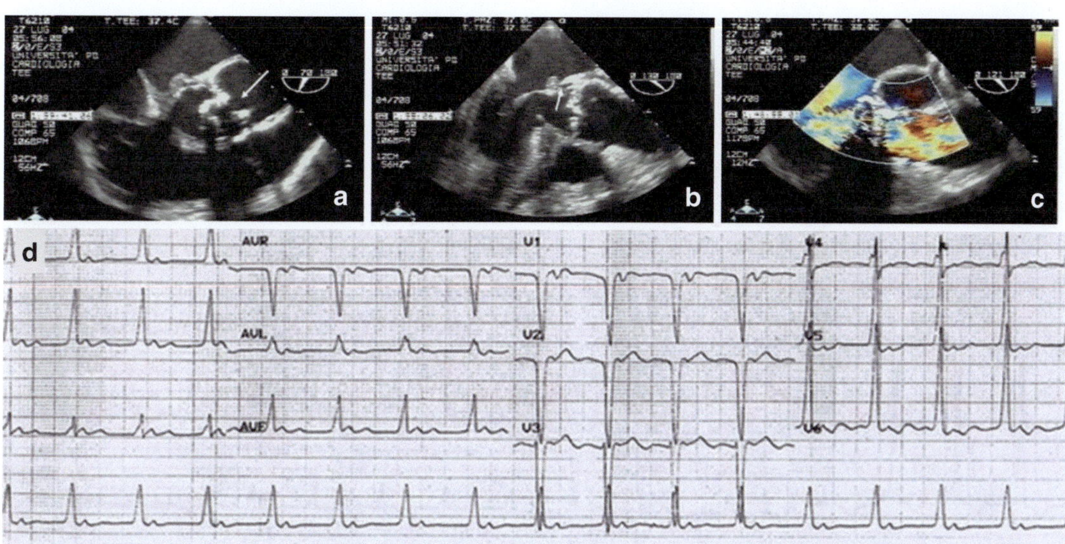

Fig. 9.7 Transesophageal 2D echo in a patient with involvement of the conduction system by infective endocarditis of a bicuspid aortic valve (same patient of Fig. 9.1). (**a**) Short axis view: note the tunnel from the aortic orifice. (**b**) Long axis view: note the abscess nearly the mitroaortic conduit. (**c**) Severe aortic regurgitation valve at the color-Doppler. (**d**) Ecg with first degree av. block and complete left bundle branch block

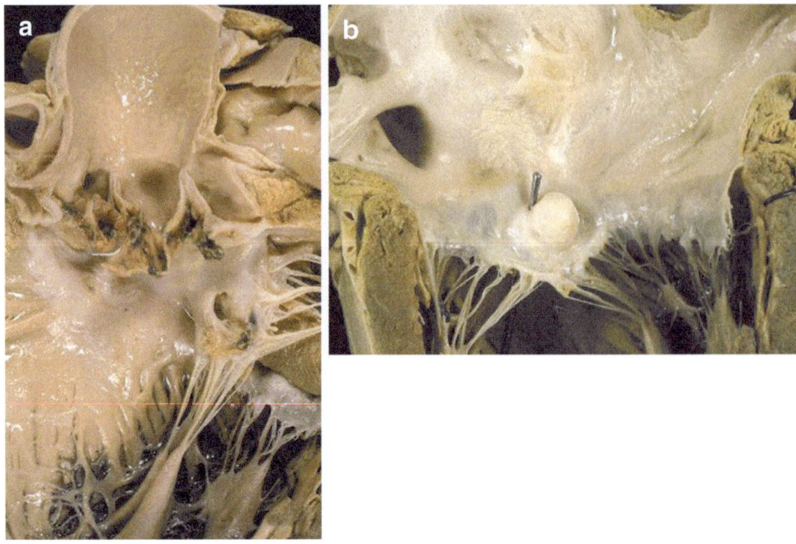

Fig. 9.8 (**a**) Gross view of infective endocarditis of the aortic valve. (**b**) Aneurysm of the mitral anterior leaflet by "kissing lesion" of infective aortic valve vegetation

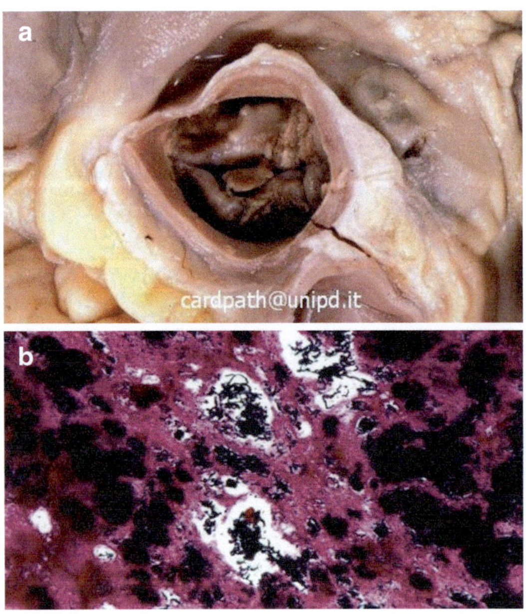

Fig. 9.9 Infective vegetative endocarditis of bicuspid aortic valve in a drug addict. (**a**) Gross view of coarse vegetations. (**b**) Cocci microorganism at histology (Gram stain)

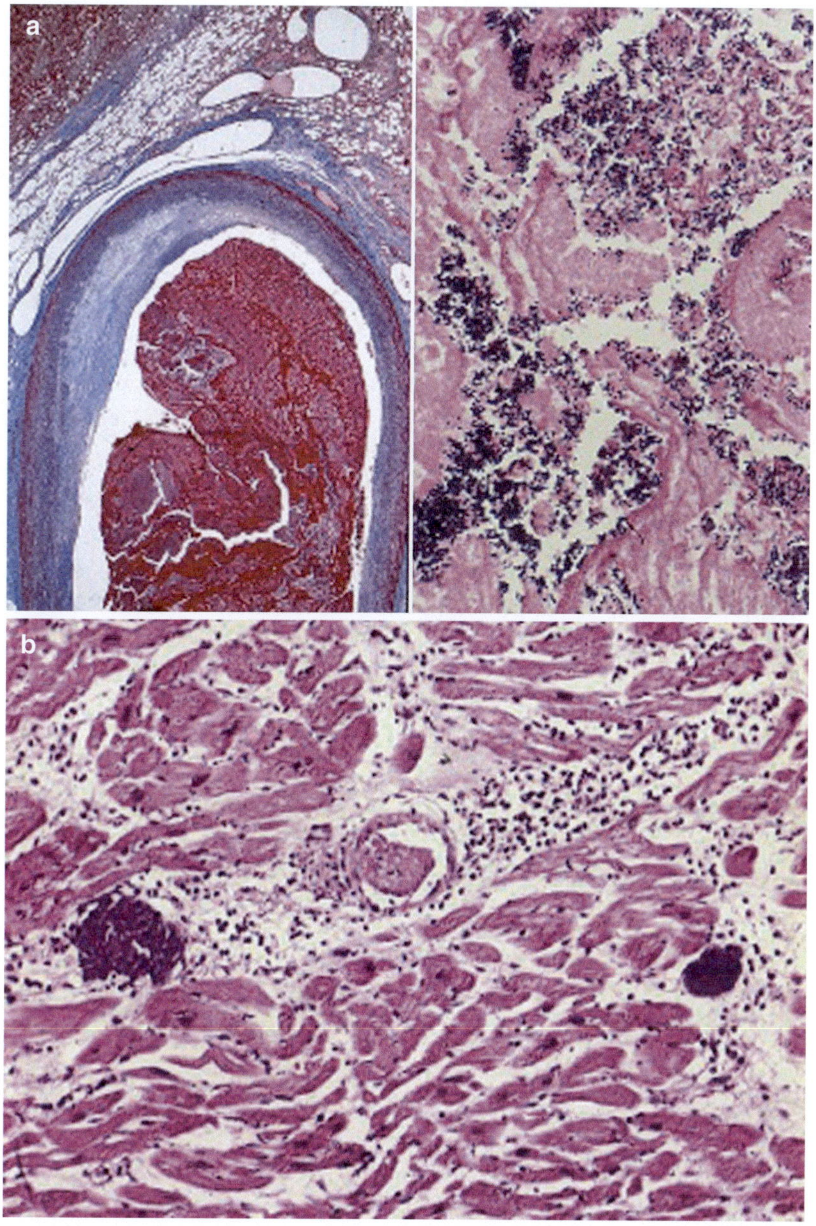

Fig. 9.10 (**a**) Coronary embolism in infective vegetative endocarditis of the aortic valve with microorganisms inside the occlusive coronary embolism. (**b**) The distal dissemination of septic vegetations accounts for onset of infective myocarditis (Hematoxylin-Eosin)

Table 9.7 According to Duke Criteria for diagnosis of infective myocarditis, two major or one major + three minor or five minor criteria are required

Major	Minor
– Positive blood culture – Evidence of valve vegetation or perforation at echo – Vegetations – Regurgitation	– Predisposing heart conditions – Intravenous drug use – Fever >38° – Vascular phenomena – Microbiological evidence – Echocardiogram valve alteration

References

1. Michel PL, Acar J. Native cardiac disease predisposing to infective endocarditis. Eur Heart J. 1995;16(Suppl B):2–6.
2. Choudhury R, Grover A, Varma J, Khattri HN, Anand IS, Bidwai PS, Wahi PL, Sapru RP. Active infective endocarditis observed in an Indian hospital 1981–1991. Am J Cardiol. 1992;70(18):1453–8.
3. Castonguay MC, Burner KD, Edwards WD, Baddour LM, Maleszewski JJ. Surgical pathology of native valve endocarditis in 310 specimens from 287 patients (1985–2004). Cardiovasc Pathol. 2013;22:19–27.
4. Thiene G, Basso C. Pathology and pathogenesis of infective endocarditis in native heart valves. Cardiovasc Pathol. 2006;15(5):256–63.
5. Pereira SC, Abrantes AL, António PS, Morais P, Sousa C, David C, Pinto FJ, Almeida AG, Caldeira D. Infective endocarditis risk in patients with bicuspid aortic valve: systematic review and meta-analysis. Int J Cardiol Heart Vasc. 2023;28(47):101249.
6. Bohbot Y, Denev S, Benvenga RM, Philip M, Michelena HI, Citro R, Habib G, Tribouilloy C. Characteristics and prognosis of isolated aortic valve infective endocarditis in patients with bicuspid aortic valves: a propensity matched study. Front Cardiovasc Med. 2023;10:1304957.
7. Mills P, Leech G, Davies M, Leathan A. The natural history of a non-stenotic bicuspid aortic valve. Br Heart J. 1978;40(9):951–7.
8. Osler W. The bicuspid condition of the aortic valve. Trans Assoc Am Physicians. 1986;2:185–92.
9. Yang LT, Ye Z, Wajih Ullah M, Maleszewski JJ, Scott CG, Padang R, Pislaru SV, Nkomo VT, Mankad SV, Pellikka PA, Oh JK, Roger VL, Enriquez-Sarano M, Michelena HI. Bicuspid aortic valve: long-term morbidity and mortality. Eur Heart J. 2023;44(43):4549–62.
10. Siu SC, Silversides CK. Bicuspid aortic valve disease. J Am Coll Cardiol. 2010;55(25):2789–800.
11. Ward C. Clinical significance of the bicuspid aortic valve. Heart. 2000;83(1):81–5.
12. Nishimura RA, Carabello BA, Faxon DP, Freed MD, Lytle BW, O'Gara PT, O'Rourke RA, Shah PM, Bonow RO, Carabello BA, Chatterjee K, de Leon AC Jr, Faxon DP, Freed MD, Gaasch WH, Lytle BW, Nishimura RA, O'Gara PT, O'Rourke RA, Otto CM, Shah PM, Shanewise JS, Smith SC Jr, Jacobs AK, Buller CE, Creager MA, Ettinger SM, Krumholz HM, Kushner FG, Lytle BW, Nishimura RA, Page RL, Tarkington LG, Yancy CW Jr, American College of Cardiology/American Heart Association Task Force. ACC/AHA 2008 guideline update on valvular heart disease: focused update on infective endocarditis: a report of the American College of Cardiology/American Heart Association Task Force on Practice Guidelines: endorsed by the Society of Cardiovascular Anesthesiologists, Society for Cardiovascular Angiography and Interventions, and Society of Thoracic Surgeons. Circulation. 2008;118(8):887–96.
13. Steckelberg JM, Murphy JG, Ballard D, Bailey K, Tajik AJ, Taliercio CP, Giuliani ER, Wilson WR. Emboli in infective endocarditis: the prognostic value of echocardiography. Ann Intern Med. 1991;114(8):635–40.
14. Bacchion F, Cukon S, Rizzoli G, Gerosa G, Daliento L, Thiene G, Basso C. Infective endocarditis in bicuspid aortic valve: atrioventricular block as sign of perivalvular abscess. Cardiovasc Pathol. 2007;16(4):252–5.
15. Abbott ME. Atlas of congenital cardiac disease. New York: American Heart Association Publisher; 1936. Chart I
16. Horstkotte D, Follath F, Gutschik E, Lengyel M, Oto A, Pavie A, Soler-Soler J, Thiene G, von Graevenitz A, Priori SG, Garcia MA, Blanc JJ, Budaj A, Cowie M, Dean V, Deckers J, Fernández Burgos E, Lekakis J, Lindahl B, Mazzotta G, Morais J, Oto A, Smiseth OA, Lekakis J, Vahanian A, Delahaye F, Parkhomenko A, Filipatos G, Aldershvile J, Vardas P, Task Force Members on Infective Endocarditis of the European Society of Cardiology, ESC Committee for Practice Guidelines (CPG), Document Reviewers. Guidelines on prevention, diagnosis and treatment of infective endocarditis executive summary; the task force on infective endocarditis of the European society of cardiology. Eur Heart J. 2004;25(3):267–76.
17. Lewis T, Grant RT. Observations relating to subacute infective endocarditis. 1. Notes on the normal structure of the aortic valve. 2. Bicuspid aortic valves of congenital origin. 3. Bicuspid aortic valves in subacute infective endocarditis. Heart. 1923;10:21–99.
18. Grant RT, Wood GE Jr, Jones TD. Heart valve irregularities in relation to subacute bacterial endocarditis. Heart. 1928;14:247–55.
19. Gross L. So-called congenital bicuspid aortic valve. Arch Pathol. 1937;23:350–62.

20. Roberts WC. The congenitally bicuspid aortic valve. A study of 85 autopsy cases. Am J Cardiol. 1970;26(1):72–83.
21. Datta BN, Bhusnurmath B, Khattri HN, Sapru RP, Bidwai PS, Wahi PL. Anatomically isolated aortic valve disease. Morphologic study of 100 cases at autopsy. Jpn Heart J. 1988;29(5):661–70.
22. Pauperio HM, Azevedo AC, Ferreira CS. The aortic valve with two leaflets—a study in 2,000 autopsies. Cardiol Young. 1999 Sep;9(5):488–98.

Part III

Imaging and Diagnosis

Echocardiography

10

Stefano Nistri, Riccardo Beccari, and Donato Mele

10.1 Introduction

Bicuspid aortic valve (BAV) is a syndrome with a wide range of phenotypic and clinical manifestations based on a heterogeneous, yet largely undefined, genetic background with multiple prognostic implications. To reconcile this diverseness, a three-partite nosological approach has been proposed including (a) complex valvulo-aortopathy, where concomitant or associated disorders overcome in severity the BAV condition; (b) typical valvulo-aortopathy, the most common group, with progressive BAV dysfunction and/or aortic dilatation without major associated disorders; and (c) undiagnosed or uncomplicated BAV, a life-long silent condition with mild or non-progressing valvulo-aortopathy which may come to light incidentally by imaging planned for other situations or even at autopsy [1–8].

Transthoracic echocardiography (TTE) plays a pivotal role in diagnosing BAV, to detect those congenital heart disease (CHD) whose prevalence is significantly higher in BAV patients (i.e., aortic coarctation, patent ductus arteriosus, atrial and ventricular septal defects, coronary anomalies and mitral valve prolapse) and assess presence, severity, and progression of its more commonly associated complications, i.e., valvular impairment and aortopathy. Moreover, TTE is the perfect tool for performing family screenings of relatives of BAVs or of patients with other congenital left-ventricular outflow tract obstructions, who are at an increased risk of being carrier of BAV or of aortopathy [1–11].

When performed with methodological care, TTE can answer the vast majority of clinical questions regarding BAV syndrome, needing of multimodality imaging only when acoustic window is suboptimal, or for specific problems such as defining aortic valve (AV) morphology when heavily calcified, or visualize region of the thoracic aorta (TA) which cannot be fully visualized by TTE [12–15]. Since TTE is widely available, feasible, and repeatable, its accurate performance, together with a learned interpretation of its results, makes TTE not only the first-line approach but often the only imaging tool needed for a substantial number of patients and, more importantly, for a substantial part of their natural history. TTE must thus be performed comprehensively, and reported precisely to allow appropriate spreading of the most important imaging data. This is particularly important, since BAVs

S. Nistri (✉)
Department of Cardiac Thoracic Vascular Sciences and Public Health, University of Padua, Padua, Italy

Cardiology Service, CMSR Veneto Medica, Altavilla Vicentina (VI), Italy

R. Beccari · D. Mele
Department of Cardiac Thoracic Vascular Sciences and Public Health, University of Padua, Padua, Italy
e-mail: Riccardo.beccari@studenti.unipd.it; donato.mele@unipd.it

G. Thiene et al. (eds.), *Bicuspid Aortic Valve Syndrome*,
https://doi.org/10.1007/978-3-031-85589-4_10

may be long-term cared in primary care settings where communication gaps between general practitioner and specialists may persist, potentially resulting in care fragmentation [16]. It should be also be emphasized that any imaging report (including that from routine TTE) should be used as an integration of a thorough clinical evaluation and not as a stand-alone procedure, in order to plan appropriate follow-up strategies, and provide accurate clinical framing and decision-making.

10.2 Phenotyping and Diagnosing

10.2.1 Classification and Nomenclature

Multiple classifications and nomenclatures have been proposed for the BAV condition [4]. Numbers and letters for the BAV and aorta phenotypes have been used by some, while others either proposed five numerical types of BAV phenotypes associated with four aortic phenotypes, or oversimplified using a dichotomous approach. A further combination of some of these schemes has also been provided [revised in [4]]. These numerous classifications were based on different clinical needs mostly generated in tertiary referral centers. Moreover, they have also created confusion in clinical practice and have negatively affected the possibility to compare the results of multiple studies in which different classifications were adopted [4].

These problems have been overcome by an international consensus statement encompassing both the nomenclature and the classification of the BAV and its aortopathy. Since its publication, we adopt it systematically, classifying BAV morphology into three types: (a) the fused BAV, (b) the 2-sinus BAV, and (c) the partial-fusion BAV, each with specific sub phenotypes [4] (Fig. 10.1). Possible limitations in applying this classification may be present on a single-patient basis. For instance, it may be difficult to visualize the number of Valsalva's sinuses, in which case terms such as "possible *or* probable 2-sinus BAV" may be used, without appealing to other imaging modalities (e.g., trans-esophageal echocardiography (TEE), computed tomography (CT) or magnetic resonance (MR)) if the congenital bicuspid nature of the valve is recognizable and if not otherwise indicated. Moreover, it is occasionally impossible to clearly discern the fusion phenotype in which case using "BAV with indeterminate cusp fusion" is appropriate. On the other side, the description of the fused BAV type may be enriched by the definition of the symmetry of the cusps (defined as the angle between the commissures of the non-fused cusp) since this parameter has recently become a critical aspect in the planning and performance of BAV repair for pure aortic regurgitation [9, 17].

10.2.2 Echocardiographic Technique

Introduction of 2-dimensional echocardiography, further potentiated by harmonic imaging, has superseded the original m-mode evaluation of the AV morphology [4, 8–10]. While BAV may be suspected by TTE parasternal long-axis view, based on the systolic doming of the conjoined cusps, the diagnosis of BAV (whatever the imaging modality) is performed by short-axis views crossing the base of the heart, unequivocally demonstrating the presence of only two commissures demarcating two valve cusps in systole. It should also be considered that, due to the systolic apically-oriented movement of the base of the heart (which is in turn affected by age, sex, heart rate, blood pressure and left ventricular size), the ultrasonic beam does not cross the same point(s) of the heart in general, and of the AV in particular, in systole and in diastole [18]. This, might constitute a significant problem when the remodeling of the inter-leaflet triangle underlying the fused cusps is limited in extension, as in partial fusion (forme fruste) BAV (Fig. 10.2). Thus, a slight upward sweeping of the ultrasonic beam or even an upward displacement movement of the probe, with the patient recumbent on the left side, sometimes with further rotation toward the bed, at end-expiration, may be necessary to disclose the presence of BAV.

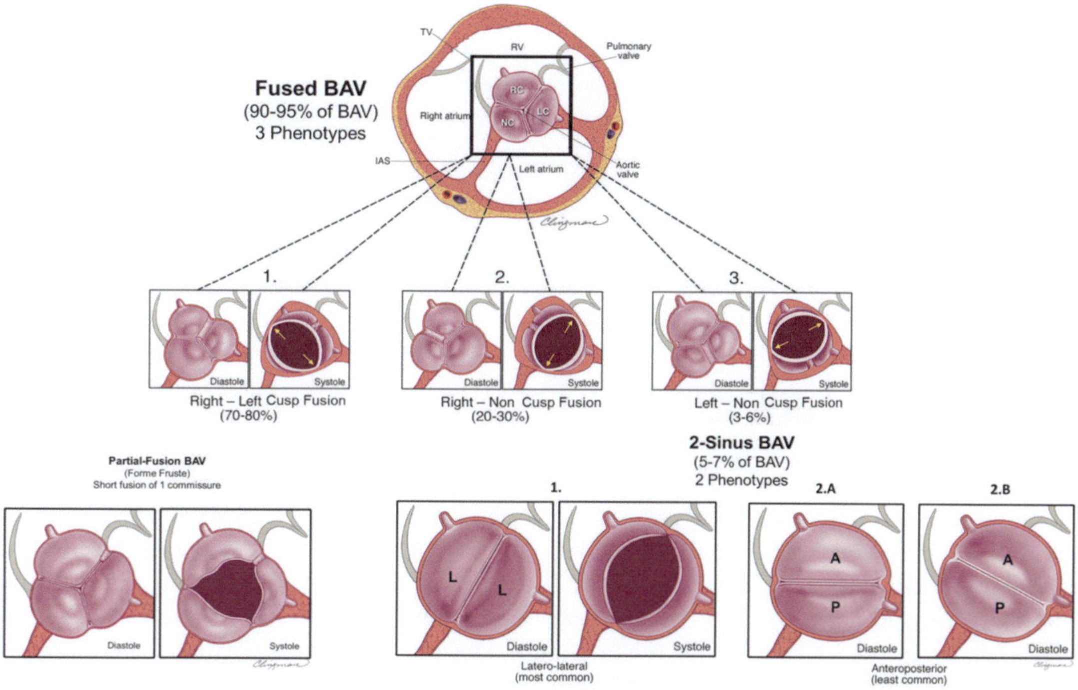

Fig. 10.1 Schematic BAV phenotypes as seen by parasternal short-axis transthoracic echocardiography as classified in the Consensus. *Upper panel*: Fused BAV phenotypes; *Lower right panel:* 2-sinus BAV phenotypes; *Lower left panel:* Partial-fusion BAV (Forme fruste). With permission from reference [4]

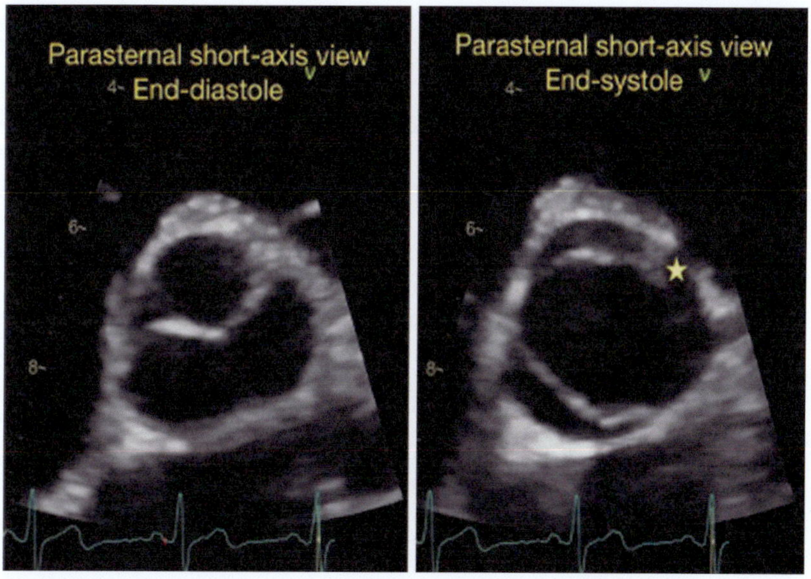

Fig. 10.2 Partial fusion phenotype. A 16-year-old male was referred to echocardiography for the detection of a systolic ejection click. The appearance of the valve at end-diastole (left) is that of a tricuspid valve, while the end-systolic frame (right) shows a cusp fusion of less than 50% at the base of the commissures between the two coronary cusps (star), consistent with a "mini-raphe"

Originally, a high diagnostic accuracy of TTE in BAV detection was shown, mostly in ideal conditions, by expert, dedicated researchers, ruling out sub-optimal TTE exams, and with high-tech echocardiographic machines. The issue of the diagnostic accuracy of routine TTE for BAV has thus been assessed by a retrospective, cross-sectional study of 216 adults (108 BAV and 108 age-matched individuals with tricuspid aortic valve (TAV)) with all diagnoses confirmed at surgery by Hillebrand et al. [10]. TTE accuracy was assessed (I) in all 216 individuals, according to the original diagnoses by the primary investigators during original TTE examination, and (II) according to expert re-evaluation of 158 TTE with availability of original recordings. Sensitivity, specificity, and accuracy of (I) primary investigators were 46.3%, 97.2%, and 71.8% as compared to (II) expert re-evaluation with 59.7%, 93%, and 77.8%, respectively. A systematic review and meta-analysis [10] of 9 studies identified a pooled sensitivity of 87.7% (95% CI, 85.0–90.5%); and a pooled specificity of 88.3% (95% CI, 85.6–90.9%). Furthermore, in a separate meta-analysis in three studies including non-diagnostic TTE results, the pooled sensitivity and specificity for BAV were 59.2% (95% CI, 53.2–65.1%) and 71.3% (95% CI, 66.4–76.2%) respectively. Performing TTE at a non-tertiary care center ($P = 0.012$), presence of aortic aneurysm ($P = 0.001$) and presence of severe AV calcification ($P = 0.003$) predicted an inaccurate diagnosis of BAV [10]. It thus seems clear that TTE visualization of BAV is simple but neither easy or straightforward, particularly in busy clinical facilities when time restraint and a multitude of either doctor-sonographer-related or machine and/or patient-related issues can affect the quality of TTE. TEE is not generally required to diagnose BAV. However, when it is performed and significant valvular calcification is absent, it has a high sensitivity for detecting the BAV (approaching 100%). Because of the better spatial resolution and the unobstructed visualization, TEE also provides a more detailed evaluation of the AV compared to TTE and facilitate direct planimetry of the valve orifice. To characterize the AV using TEE, the valve should be imaged in both short- and long-axis views, with and without color flow Doppler. Two-dimensional (2D) TEE images fail in recognizing BAV phenotype in about 10% of BAV patients, while three-dimensional (3D) TEE seems more effective, especially in experienced hands and in patients with short raphe or calcification of the aortic leaflets. 3D TEE is helpful to describe size and features (thickening, calcification) of the cusps, fusion of the commissures, and the orientation of the coronary ostia [19].

10.2.3 Echocardiographic Phenotyping

Observational studies reported an empirical association between BAV cusp fusion pattern and the associated aortopathy as well as with valvular dysfunction [20, 21]. RN phenotype is often associated with dilation of the ascending aorta and is more prevalent in patients with aortic stenosis (AS), while raphe is more prevalent (or visible?) in patients with significant valvular impairment.

However, (a) these results were not invariably reproducible in any study [4]; (b) aortic size was assessed based on different measurement methods and using different thresholds defining aortic dilatation; (c) the potential role for progression of valve impairment, as well as time-dependent changes in aortic dilatation, could not be assessed; (d) while experimental studies support different mechanisms underlying different BAV phenotypes, the observation that different BAV phenotype can coexist in the same family makes a dichotomous interpretation of different BAV morphologies somehow disappointing. Thus, as underscored by Krieger and Hung [22], despite the statistically significant associations between demographics, valve morphology, valve dysfunction, and aortic dilation, these differences are relatively small, with too much overlap, thus not providing enough discrimination to offer different surveillance strategies to patients based on valve morphology. Notwithstanding this, a precise description of BAV phenotype, together with accurate hemodynamic characterization and detailed assessment of the TA in every patient should be considered a quality marker of any TTE examination for allowing proper staging and favor appropriate follow-up.

10.3 Valvular Heart Disease and Associated Congenital Heart Disease

Overall, the most common complication of BAV is progressive valvular dysfunction eventually leading to aortic valve replacement (AVR) or repair even in patients aged ≥80 years. In the community, long-term risk of AVR is about 50% 25 years after BAV diagnosis, more frequently due to AS, which is 2 times more frequent than significant regurgitation (AR). Significant AR, more commonly encountered in men, occurs about two decades earlier than in patients with TAV (TAV) independently from AS [2, 3, 23–27]. AR is determined by multiple potential mechanisms including cusp prolapse of the larger of unequal cusps, myxoid degeneration of the valve, cusp restriction/retraction, or may be the result of a former infective endocarditis. Moreover, the degree of AR may be further aggravated by the BAV-associated aortopathy, in particular dilatation of annulus and/or sinotubular junction (root complex) [27, 28]. Interestingly, surgical repair is an option on these patients, more successful than in those with TAV, and echocardiography plays a determinant role in assessing reparability. Commissural orientation and symmetry, annular size, geometric and effective height [4, 17] are all determinants of repair durability, and should be systematically assessed when considering these surgical techniques.

Since stenosis of the BAV is uncommon in young patients without congenital AS [8], the relevant incidence of acquired AS in BAVs underscores the need to early detect AV calcification and hemodynamic progression. The prognostic value of calcific AV sclerosis in the general population has been extensively demonstrated [29, 30]. When using a visual score system, after proper gain setting, integrating echogenicity, thickening, localization, and mobility, the assessment of AV calcification is reproducible and correlates with calcium score assessed by CT [31]. Michelena et al. [26], in BAV patients with no or only minimal valvular dysfunction, utilized a comparable degeneration score. After 20 years of follow-up, valve degeneration at diagnosis with a score ≥ 3 univariately predicted all cardiovascular events and remained so after adjustment for age and sex. In the multivariate analysis, baseline predictors of all cardiovascular events were age ≥ 50 years and valve degeneration. Initiation and progression of AV calcification implicates a myriad of potential pathways with multifaceted entanglements [32]. This is even more concerning for BAVs, that are significantly more prone to severe AV calcification that individuals with TAV for reasons that remain largely unclear, and suffer from accelerated progression of disease when compared with those with the standard configuration of a TAV [33–35].

When grading valvular heart disease (VHD) in BAV, the same comprehensive approach suggested by current guidelines should be utilized both for estimating the severity of valve impairment, and for the assessment of the adaptive consequences on the whole heart in patients with TAV [36–38]. Of particular relevance is the need of utilizing multiple echocardiographic windows, also in patients with "apparent" mild AS. Importantly, the view with the highest aortic velocity/gradient should be described in the report and re-utilized both at follow-up and, after AVR, preferred to assess the hemodynamic of the AV prosthesis [39, 40] (Fig. 10.3).

Grading the severity of AR should not rely on any single echocardiographic parameter. AR severity cannot be reliably assessed simply based on the length and area of the regurgitant jet into the left ventricle, thus, both semi-quantitative and quantitative methods should be systematically attempted (i.e., pulsed wave Doppler with the sample volume in the descending aorta, volumetric and proximal isovelocity surface area (PISA) measurements allowing calculation of the effective regurgitant orifice area (EROA) and regurgitant volume (RVol) and regurgitant fraction). Moreover, there is no clear hierarchy in weighting the different parameters used in the comprehensive assessment of AR severity, although ERO and RVol provide the strongest imaging predictors of outcome in AR (Fig. 10.4). Inconsistencies of different measures (e.g., aortic valve area and gradients in AS patients with normal ejection fraction) or impossibilities to obtain

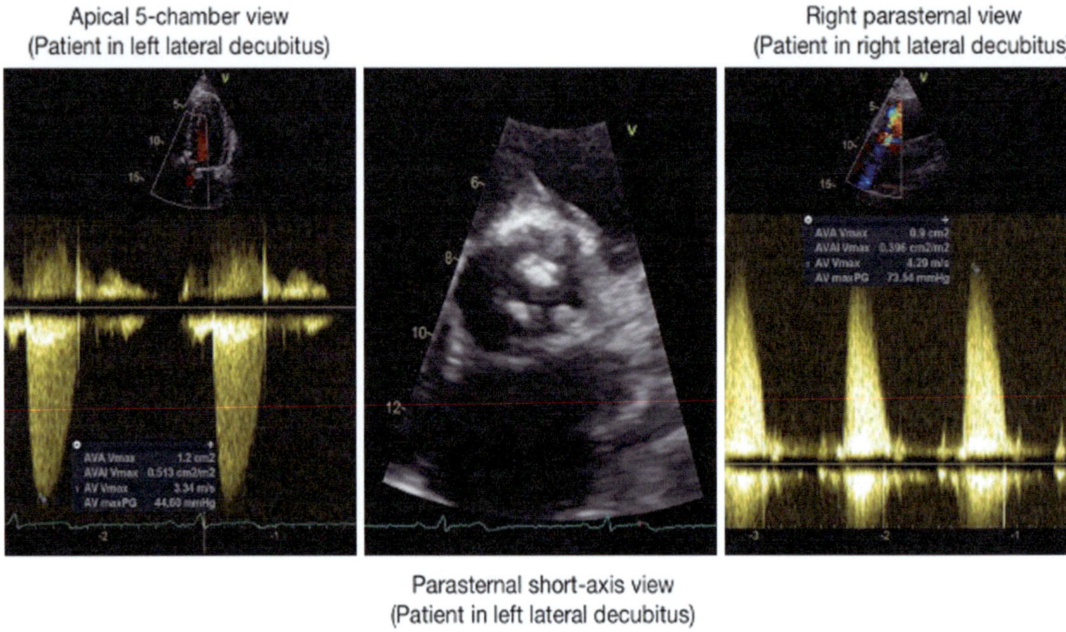

Fig. 10.3 Assessment of the severity of aortic stenosis (AS). A 43-year-old asymptomatic male with a severely calcified bicuspid aortic valve (confirmed as aortic valve calcium score 3214 AU by computed tomography) *(middle panel)*. Continuous wave Doppler from apical view *(left panel)* measured 3.34 m/s of peak aortic velocity, consistent with moderate AS. Interrogating the aortic valve from the right parasternal view *(right panel)* resulted in a peak aortic velocity of 4.3 m/s, thus reclassifying this AS as severe

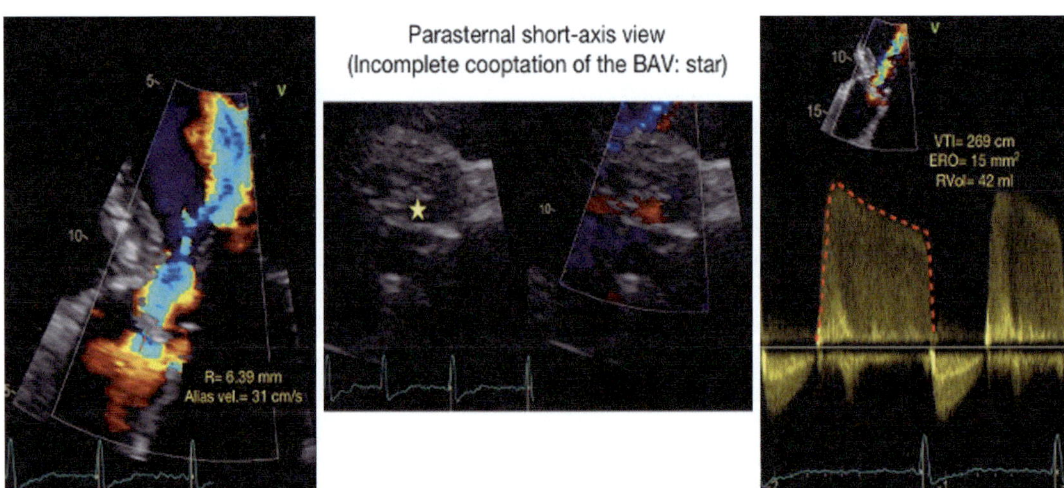

Fig. 10.4 Assessment of the severity of aortic regurgitation (AR). A 81-year-old male with chronic atrial fibrillation, in whom echocardiography showed a moderately calcified bicuspid aortic valve with central cusp retraction causing incomplete coaptation (star) *(middle panel)*. Color coded echocardiography, 5-chamber apical view, with aliasing velocity adjusted at 31 cm/s allowed measuring the radium (R) of the proximal isovelocity surface are *(left panel)*. After tracing the velocity-time-integral obtained in the same view by continuous wave Doppler *(right panel)*, the effective regurgitant orifice (ERO) and regurgitant volume (RVol) were calculated. AR was thus graded as moderate

accurate measures (i.e., effective regurgitant orifice area and/or regurgitant volume and fraction in AR patients) benefit from a multimodality approach using either CT or MR [27, 28, 37, 38, 41–43].

LV remodeling and subclinical functional impairment may occur early in life, even in BAVs without VHD but with clinical aortopathy, as captured by deformation abnormalities by speckle-tracking TTE, and are linked to changes in loading conditions due to alterations in aortic elasticity. Conflicting data are available regarding myocardial consequences of BAV in AS patients. Some studies reported that BAV was protective toward cardiac damage while others demonstrated adverse LV remodeling and impaired cardiac function in BAV-AS patients in comparison with TAV-AS patients. It should be reminded that, independently from aortic valve morphology, the risk of adverse clinical outcomes increases significantly even when LV ejection fraction falls below 60% in AS patients. Since assessment of subendocardial dysfunction is one of the theoretical advantages of studying longitudinal systolic function, global longitudinal strain (GLS) is more sensitive in detecting early myocardial functional abnormalities even when LVEF is normal, and it is useful in the risk stratification of patients with AR/AS. To match quantitative assessment of the severity of VHD and of their consequences on LV size and function, with a comprehensive clinical evaluation, integrated with biomarkers and stress testing as needed, is critical to schedule the frequency of follow-up evaluations and to optimize the decision-making process [37–43]. The BAV is at risk for infective endocarditis, possibly related to abnormal leaflet structure or turbulent flow. About 1/4 of aortic valve endocarditis is due to BAV, and in children BAV is one of the most common underlying lesions associated with endocarditis. A previously unrecognized BAV may be first detected in the course of an episode of endocarditis. Signs of infective endocarditis in TTE include vegetation, valvular perforation, and perivalvular abscess. TEE has higher accuracy for detecting aortic valve vegetation and abscesses. Given the higher frequency of perivalvular abscesses in BAV than in TAV patients with infective endocarditis, systematic use of TEE could be advantageous for prognosis. It has also been reported that BAV endocarditis requires surgery more often than trileaflet aortic valve endocarditis [19, 44].

10.4 Aortopathy

Dilatation of the TA is one of the component of the BAV-related aortopathy together with impaired aortic elasticity and dissection [1, 4, 45]. Aortic dilatation is prevalent at every age, increasingly incident with aging, and is independent form the presence of significant VHD. Interestingly, aortic dilatation is also exceedingly prevalent in first degree relatives of BAV patients having normal TAV, more than would be expected in the general population [45–47].

The diagnosis of aortic dilatation implies both methodological and technical issues, regarding the conceptual framework of "aortic normalcy" and the way TA is measured by TTE. Though many studies explored the normal limits of TA, multiple issues (regarding interfaces, timing and TTE modes) make findings from these previous studies not comparable [48](Fig. 10.5).

To date, only two studies provide calculators to assess aortic normalcy at each aortic level (see below) in a wide age range. One is exclusively based on linear regression modeling via ordinary least squares, providing a Z-score, the other one provides both a Z-score and another score (Q-score) based on a machine learning technique known as one-class support vector machine [48, 49]. The aortic root, the sino-tubular junction, and ascending aorta should be measured at end-diastole, using the leading-to-leading method and the aortic annulus in mid-systole by the inner-to-inner convention, in parasternal long-axis views. When proper methodology is utilized, intraclass coefficient for all segments of the ascending aorta, by TTE is higher than by CT and MRI, with smaller mean of differences TTE is a feasible and accurate for the assessment and follow-up of thoracic aortic diameters in patients with

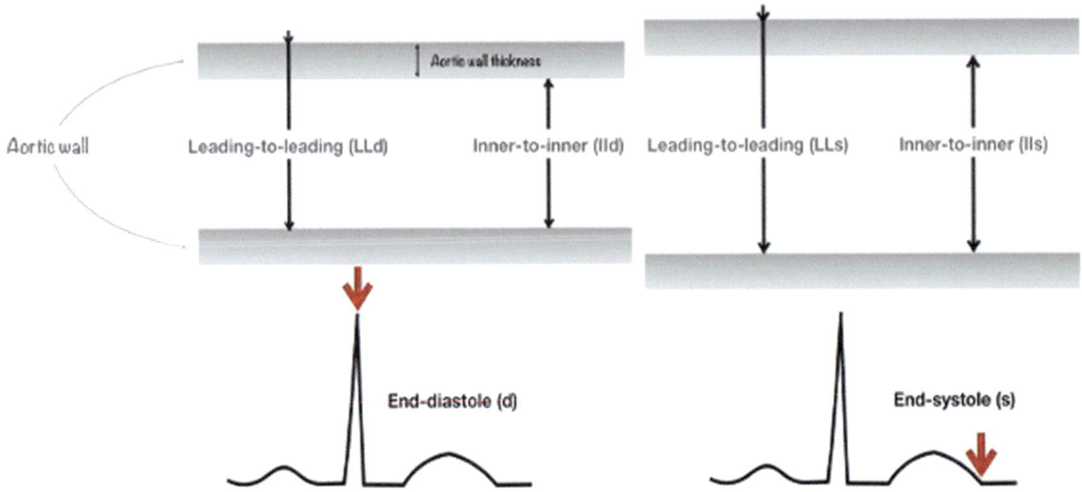

Fig. 10.5 Diagrammatic representation of different methods to measure aortic diameter by echocardiography. *Left panel*: end-diastolic measurement using either the leading-to-leading (LLd) convention or the inner-to-inner (IId) conventions. Differences are related to inclusion/exclusion of the aortic wall. *Right panel*: same measurements but at end-systole, thus further adding changes due to aortic elasticity. On average, LLs is the largest diameter, IId the smallest, while LLd and IIs are quite comparable

ascending aortic aneurysms, and an unbiased correlations between CT and TTE measurements can be obtained when appropriate methods are used for the sinuses and the ascending aorta [50–53]. Regrettably, in clinical practice, these discrepancies can be hardly reconciled, since measurements (by TTE, CT, MR) are often performed and repeated in different institutions, by different practitioners and methods, without the possibility to check for previous images for consistency. Importantly, the aortic root morphology in patients with BAV can be predicted by BAV phenotype: the largest root diameter is roughly perpendicular to the orientation of the valve opening [53]. Therefore, TTE measurements of the aortic root in patients with BAV-NR should be performed in parasternal short-axis view, to avoid underestimation of the maximal diameter.

Aortic dilatation BAV phenotypes are to date [4] classified as (a) the ascending phenotype (70% of cases: dilatation preferentially located at the tubular ascending tract beyond the STJ); (b) the root phenotype [20% of case: dilatation preferentially located at the root (sinuses of Valsalva), possibly involving also the ventriculoaortic junction/annulus]; and (c) an extended phenotype when the dilation of the aorta does not significantly prevail at 1 segment (Fig. 10.6). Association of the root phenotype with acute aortic events after isolated AVR as well as with potentially aortopathy-related genetic variants has been reported. Consistently, the root phenotype has been recently included among the adjunctive risk factors to consider when indicating earlier elective surgery for BAV aortopathy [54–57]. However, in one preliminary study aimed at suggesting a clinic-echocardiographic score to predict the presence of a major connective tissue in BAV, Pepe et al. [58] demonstrated that BAV patients with characteristics suggesting significant connective tissue disorder were not distinguishable from the other BAVs based on either aortic valve hemodynamic, aortic size or root phenotype. Discrepancies between those findings and those supporting the peculiarities of the root-phenotype should be seen cautiously, considering differences of clinical settings possibly resulting into relevant differences among study populations. In patients with Marfan and Loeys-Dietz syndromes, recently reported very detailed definition of root phenotypes demonstrated anatomic patterns at different aortic levels irrespective of aortic dilation and disease severity. These features may represent the expression

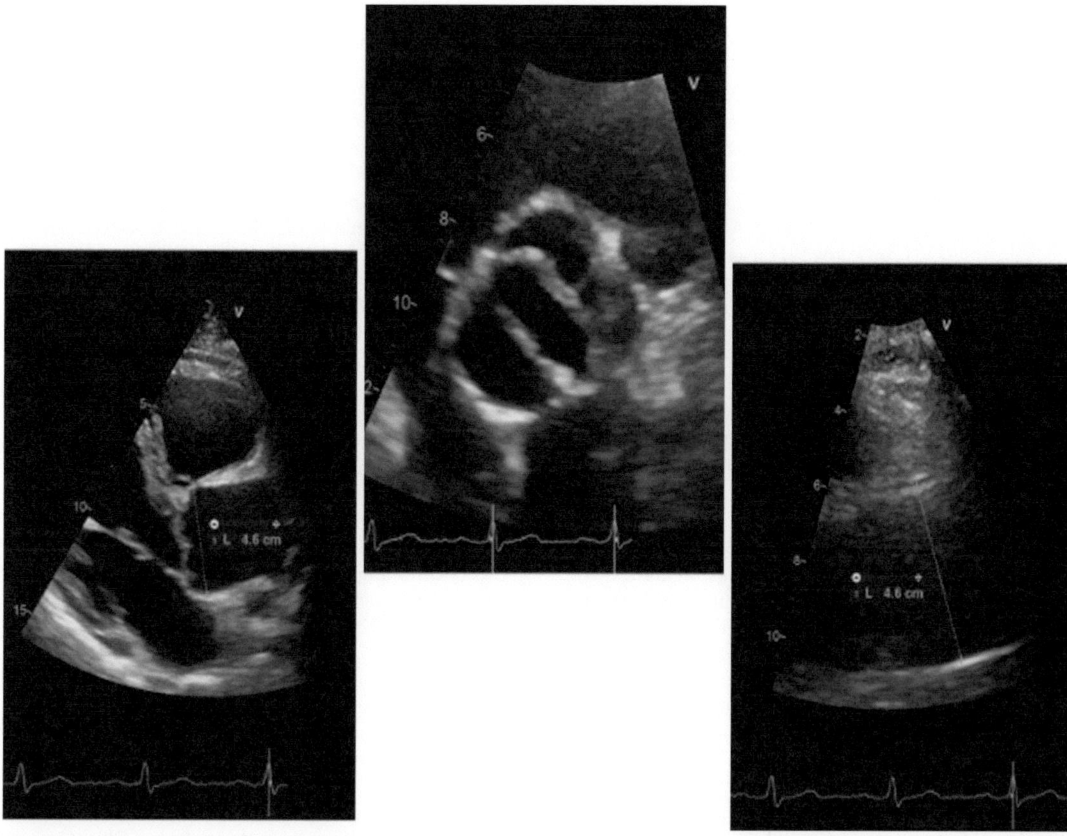

Fig. 10.6 Phenotyping thoracic aorta (TA). When measuring the TA, it should be taken into account that the complex Valsalva sinuses/sinotubular junction and the ascending aorta do not lie on the same plane. Thus, adjustments of the parasternal long-axis views to focus on the TA are needed. In this echocardiographic examination of a 67-year-old male with a bicuspid aortic valve *(middle panel),* TA was measured after angulating the probe to better align the walls of the aortic root *(left panel);* to visualize and measure the ascending aorta, the probe was first re-positioned in an upper intercostal space, than swept to improve alignment, and zoomed for better resolution *(right panel).* As both root and ascending aorta were dilated, this patient's aortopathy was classified as "extended"

of different genetic mutations on aortic development, with a potential impact on prognosis and better management of the diseases [59]. Further translational studies are thus needed to clarify these issues in BAV patients. TA dissection is a ominous, though rare, complication of BAV [4, 23, 26]. Since early diagnosis is still a major pitfall in managing these patients, a diagnostic multiparametric algorithm is proposed by guidelines [59] Although a dissection flap can be visualized sometimes by TTE at the emergency department, it is mostly recommended to assess pericardial effusion, wall motion abnormalities, aortic regurgitation, and aortic diameters. Otherwise, TEE is especially useful pre-, intra, and post-operatively to monitor changes in the anatomical configuration of TA dissection or for surgical complications [60].

References

1. Nistri S, Grande-Allen J, Noale M, Basso C, Siviero P, Maggi S, Crepaldi G, Thiene G. Aortic elasticity and size in bicuspid aortic valve syndrome. Eur Heart J. 2008;29:472–9.
2. Michelena HI, Prakash SK, Della Corte A, Bissell MM, Anavekar N, Mathieu P, et al. Bicuspid aortic valve: identifying knowledge gaps and rising

to the challenge from the International Bicuspid Aortic Valve Consortium (BAVCon). Circulation. 2014;129:2691–704.
3. Michelena HI, Suri RM, Katan O, Eleid MF, Clavel MA, Maurer MJ, et al. Sex differences and survival in adults with bicuspid aortic valves: verification in 3 contemporary echocardiographic cohorts. J Am Heart Assoc. 2016;5:e004211.
4. Michelena HI, Della Corte A, Evangelista A, Maleszewski JJ, Edwards WD, Roman MJ, Devereux RB, et al. International consensus statement on nomenclature and classification of the congenital bicuspid aortic valve and its aortopathy, for clinical, surgical, interventional and research purposes. Eur J Cardiothorac Surg. 2021;60(3):448–76.
5. Pepe G, Nistri S, Giusti B, Sticchi E, Attanasio M, Porciani C, et al. Identification of fibrillin 1 gene mutations in patients with bicuspid aortic valve (BAV) without Marfan syndrome. BMC Med Genet. 2014;15:23.
6. Sticchi E, De Cario R, Magi A, Giglio S, Provenzano A, Nistri S, Pepe G, Giusti B. Bicuspid aortic valve: role of multiple gene variants in influencing the clinical phenotype. Biomed Res Int. 2018;2018:8386123.
7. Giusti B, Sticchi E, De Cario R, Magi A, Nistri S, Pepe G. Genetic bases of bicuspid aortic valve: the contribution of traditional and high-throughput sequencing approaches on research and diagnosis. Front Physiol. 2017;8:612.
8. Nistri S, Basso C, Marzari C, et al. Frequency of bicuspid aortic valve in young male conscripts by echocardiogram. Am J Cardiol. 2005;96:718–21.
9. Nistri S, Basso C, Thiene G. La valvola aortica bicuspide. G Ital Cardiol. 2012;13(1):25–37.
10. Hillebrand M, Koschyk D, Ter Hark P, Schüler H, Rybczynski M, Berger J, et al. Diagnostic accuracy study of routine echocardiography for bicuspid aortic valve: a retrospective study and meta-analysis. Cardiovasc Diagn Ther. 2017;7(4):367–79.
11. Szöcs K, Toprak B, Schön G, Rybczynski M, Brinken T, Mahlmann A, Girdauskas E, Blankenberg S, von Kodolitsch Y. Concomitant cardiovascular malformations in isolated bicuspid aortic valve disease: a retrospective cross-sectional study and meta-analysis. Cardiovasc Diagn Ther. 2022;12(4):400–14.
12. Kumar P, Bhatia M, Arora N. Computed tomographic evaluation of congenital left ventricular outflow obstruction. Curr Cardiol Rev. 2023;19(6):31–49.
13. Kim WK, Liebetrau C, Fischer-Rasokat U, Renker M, Rolf A, Doss M, Möllmann H, Nef H, Walther T, Hamm CW. Challenges of recognizing bicuspid aortic valve in elderly patients undergoing TAVR. Int J Cardiovasc Imaging. 2020;36(2):251–6.
14. You YJ, Ko SM. Computed tomography and magnetic resonance imaging findings of bicuspid aortic valve and related abnormalities of the heart and thoracic aorta. Korean J Radiol. 2023;24(10):960–73.
15. Kang JW, Song HG, Yang DH, Baek S, Kim DH, Song JM, et al. Association between bicuspid aortic valve phenotype and patterns of valvular dysfunction and bicuspid aortopathy: comprehensive evaluation using MDCT and echocardiography. JACC Cardiovasc Imaging. 2013;6:150–61.
16. Kern LM, Bynum JPW, Pincus HA. Care fragmentation, care continuity, and care coordination-how they differ and why it matters. JAMA Intern Med. 2024;184(3):236–7.
17. Ram D, Bouhout I, Karliova I, Schneider U, El-Hamamsy I, Schäfers HJ. Concepts of bicuspid aortic valve repair: a review. Ann Thorac Surg. 2020;109(4):999–1006.
18. Karwat P, Klimonda Z, Styczyński G, Szmigielski C, Litniewski J. Aortic root movement correlation with the function of the left ventricle. Sci Rep. 2021;11(1):4473.
19. Hahn RT, Abraham T, Adams MS, Bruce CJ, Glas KE, Lang RM, Reeves ST, Shanewise JS, Siu SC, Stewart W, Picard MH. Guidelines for performing a comprehensive transesophageal echocardiographic examination: recommendations from the American Society of Echocardiography and the Society of Cardiovascular Anesthesiologists. J Am Soc Echocardiogr. 2013;26(9):921–64.
20. Evangelista A, Gallego P, Calvo-Iglesias F, Bermejo J, Robledo-Carmona J, Sánchez V, et al. Anatomical and clinical predictors of valve dysfunction and aortic dilation in bicuspid aortic valve disease. Heart. 2018;104(7):566–73.
21. Mai Z, Guan L, Mu Y. Association between bicuspid aortic valve phenotype and patterns of valvular dysfunction: a meta-analysis. Clin Cardiol. 2021;44(12):1683–91.
22. Krieger EV, Hung J. Bicuspid aortic valve type: it takes two. Heart. 2018;104:544–5.
23. Tzemos N, Therrien J, Yip J, Thanassoulis G, Tremblay S, Jamorski MT, et al. Outcomes in adults with bicuspid aortic valves. JAMA. 2008;300:1317–25.
24. Roberts WC, Ko JM. Frequency by decades of unicuspid, bicuspid, and tricuspid aortic valves in adults having isolated aortic valve replacement for aortic stenosis, with or without associated aortic regurgitation. Circulation. 2005;111:920–5.
25. Roberts WC, Janning KG, Ko JM, Filardo G, Matter GJ. Frequency of congenitally bicuspid aortic valves in patients >_80 years of age undergoing aortic valve replacement for aortic stenosis (with or without aortic regurgitation) and implications for transcatheter aortic valve implantation. Am J Cardiol. 2012;109:1632–6.
26. Michelena HI, Desjardins VA, Avierinos JF, Russo A, Nkomo VT, Sundt TM, Pellikka PA, Tajik AJ, Enriquez-Sarano M. Natural history of asymptomatic patients with normally functioning or minimally dysfunctional bicuspid aortic valve in the community. Circulation. 2008;117(21):2776–84.
27. Yang LT, Benfari G, Eleid M, Scott CG, Nkomo VT, Pellikka PA, Anavekar NS, Enriquez-Sarano M, Michelena HI. Contemporary differences between bicuspid and tricuspid aortic valve in chronic aortic

regurgitation. Heart. 2021;107(11):916–24. https://doi.org/10.1136/heartjnl-2020-317466. Epub 2020 Oct 27
28. Detaint D, Messika-Zeitoun D, Maalouf J, et al. Quantitative echocardiographic determinants of clinical outcome in asymptomatic patients with aortic regurgitation: a prospective study. J Am Coll Cardiol Img. 2008;1(1):1–11.
29. Moore MK, Jones GT, Whalley G, Prendergast B, Williams MJA, Coffey S. Outcomes of patients with early calcific aortic valve disease detected by clinically indicated echocardiography. Eur Heart J Cardiovasc Imaging. 2024;25(3):356–64.
30. Coffey S, Cox B, Williams MJ. The prevalence, incidence, progression, and risks of aortic valve sclerosis: a systematic review and meta-analysis. J Am Coll Cardiol. 2014;63(25 Pt A):2852–61.
31. Nemchyna O, Soltani S, Solowjowa N, et al. Validity of visual assessment of aortic valve morphology in patients with aortic stenosis using two-dimensional echocardiography. Int J Cardiovasc Imaging. 2021;37:813–23.
32. Kraler S, Blaser MC, Aikawa E, Camici GG, Lüscher TF. Calcific aortic valve disease: from molecular and cellular mechanisms to medical therapy. Eur Heart J. 2022;43(7):683–97.
33. Gollmann-Tepeköylü C, Nägele F, Engler C, Stoessel L, Zellmer B, et al. Different calcification patterns of tricuspid and bicuspid aortic valves and their clinical impact. Interact Cardiovasc Thorac Surg. 2022;35(6):ivac274.
34. Beusch CM, Simonson OE, Wedin JO, et al. Analysis of local extracellular matrix identifies different aetiologies behind bicuspid and tricuspid aortic valve degeneration and suggests therapies. Cell Mol Life Sci. 2023;80:268.
35. Sticchi E, Giusti B, Cordisco A, Gori AM, Sereni A, Sofi F, Mori F, Colonna S, Fugazzaro MP, Pepe G, Nistri S, Marcucci R. Role of lipoprotein (a) and LPA KIV2 repeat polymorphism in bicuspid aortic valve stenosis and calcification: a proof of concept study. Intern Emerg Med. 2019;14(1):45–50.
36. Baumgartner H, Hung J, Bermejo J, Chambers JB, Edvardsen T, Goldstein S, Lancellotti P, LeFevre M, Miller F Jr, Otto CM. Recommendations on the echocardiographic assessment of aortic valve stenosis: a focused update from the European Association of Cardiovascular Imaging and the American Society of Echocardiography. J Am Soc Echocardiogr. 2017;30(4):372–92.
37. Zoghbi WA, Adams D, Bonow RO, Enriquez-Sarano M, Foster E, Grayburn PA, Hahn RT, Han Y, Hung J, Lang RM, Little SH, Shah DJ, Shernan S, Thavendiranathan P, Thomas JD, Weissman NJ. Recommendations for noninvasive evaluation of native valvular regurgitation: a report from the American Society of Echocardiography Developed in Collaboration with the Society for Cardiovascular Magnetic Resonance. J Am Soc Echocardiogr. 2017;30(4):303–71.
38. Lancellotti P, Pibarot P, Chambers J, La Canna G, Pepi M, Dulgheru R, Dweck M, Delgado V, Garbi M, Vannan MA, Montaigne D, Badano L, Maurovich-Horvat P, Pontone G, Vahanian A, Donal E, Cosyns B. Scientific Document Committee of the European Association of Cardiovascular Imaging. Multi-modality imaging assessment of native valvular regurgitation: an EACVI and ESC council of valvular heart disease position paper. Eur Heart J Cardiovasc Imaging. 2022;23(5):e171–232.
39. Benfari G, Gori AM, Rossi A, Papesso B, Vassanelli C, Zito GB, Nistri S. Feasibility and relevance of right parasternal view for assessing severity and rate of progression of aortic valve stenosis in primary care. Int J Cardiol. 2017;240:446–51.
40. Benfari G, Nistri S, Cerrito LF, Maritan L, Tafciu E, Setti M, Bursi F, Tadiello E, De Manna ND, Rossi A, Ribichini FL. Usefulness of the right parasternal echocardiographic view to improve the hemodynamic assessment after valve replacement for aortic stenosis. Am J Cardiol. 2021;142:103–8.
41. Yang L, Enriquez-Sarano M, Michelena H, et al. Predictors of progression in patients with stage B aortic regurgitation. JACC. 2019;74(20):2480–92.
42. Spilias N, Martyn T, Denby KJ, Harb SC, Popovic ZB, Kapadia SR. Left ventricular systolic dysfunction in aortic stenosis: pathophysiology, diagnosis, management, and future directions. Struct Heart. 2022;6(5):100089. https://doi.org/10.1016/j.shj.2022.100089. PMID: 37288060; PMCID: PMC10242576
43. Tsampasian V, Victor K, Bhattacharyya S, et al. Echocardiographic assessment of aortic regurgitation: a narrative review. Echo Res Pract. 2024;11:1.
44. Delgado V, Ajmone Marsan N, de Waha S, Bonaros N, Brida M, Burri H, et al. 2023 ESC guidelines for the management of endocarditis. Eur Heart J. 2023;44(39):3948–4042.
45. Nistri S, Sorbo MD, Marin M, Palisi M, Scognamiglio R, Thiene G. Aortic root dilatation in young men with normally functioning bicuspid aortivalves. Heart. 1999;82:19–22.
46. Sillesen AS, Vøgg O, Pihl C, Raja AA, Sundberg K, Vedel C, Zingenberg H, Jørgensen FS, Vejlstrup N, Iversen K, Bundgaard H. Prevalence of bicuspid aortic valve and associated aortopathy in newborns in Copenhagen, Denmark. JAMA. 2021;325(6):561–7.
47. Bray JJH, Freer R, Pitcher A, Kharbanda R. Family screening for bicuspid aortic valve and aortic dilatation: a meta-analysis. Eur Heart J. 2023;44(33):3152–64.
48. Frasconi P, Baracchi D, Giusti B, Kura A, Spaziani G, Cherubini A, Favilli S, Di Lenarda A, Pepe G, Nistri S. Two-dimensional aortic size normalcy: a novelty detection approach. Diagnostics (Basel). 2021;11(2):220.
49. Campens L, Demulier L, De Groote K, Campens L, Demulier L, De Groote K, Vandekerckhove K, De Wolf D, Roman MJ, Devereux RB, et al. Reference values for echocardiographic assessment of the diam-

eter of the aortic root and ascending aorta spanning all age categories. Am J Cardiol. 2014;114:914–20.
50. Asch FM, Yuriditsky E, Prakash SK, Roman MJ, Weinsaft JW, Weissman G, et al. The need for standardized methods for measuring the aorta: multimodality core lab experience from the GenTAC Registry. JACC Cardiovasc Imaging. 2016;9(3):219–26.
51. Tamborini G, Galli CA, Maltagliati A, Andreini D, Pontone G, Quaglia C, Ballerini G, Pepi M. Comparison of feasibility and accuracy of transthoracic echocardiography versus computed tomography in patients with known ascending aortic aneurysm. Am J Cardiol. 2006;98(7):966–9.
52. Park JY, Foley TA, Bonnichsen CR, Maurer MJ, Goergen KM, Nkomo VT, Enriquez-Sarano M, Williamson EE, Michelena HI. Transthoracic echocardiography versus computed tomography for ascending aortic measurements in patients with bicuspid aortic valve. J Am Soc Echocardiogr. 2017;30(7):625–35.
53. Milleron O, Masi P, Eliahou L, Paul JF, Arnoult F, Ouali NO, Lansac E, Raffoul R, Pellenc Q, Ou P, Jondeau G. Aortic root anatomy is related to the bicuspid aortic valve phenotype. J Am Soc Echocardiogr. 2022;35(3):278–86.
54. Della Corte A, Michelena HI, Citarella A, Votta E, Piatti F, Lo Presti F, Ashurov R, Cipollaro M, Forte A. Risk stratification in bicuspid aortic valve aortopathy: emerging evidence and future perspectives. Curr Probl Cardiol. 2019; https://doi.org/10.1016/j.cpcardiol.2019.06.002.
55. Pepe G, Nistri S, Giusti B, Sticchi E, Attanasio M, Porciani C, Abbate R, Bonow RO, Yacoub M, Gensini GF. Identification of fibrillin 1 gene mutations in patients with bicuspid aortic valve (BAV) without Marfan syndrome. BMC Med Genet. 2014;15:23–9.
56. Girdauskas E, Elefteriades JA. Importance of bicuspid aortic valve phenotype on aortopathy. Ann Cardiothorac Surg. 2022;11(4):470–2.
57. Borger MA, Fedak PWM, Stephens EH, Gleason TG, Girdauskas E, Ikonomidis JS, et al. The American Association for Thoracic Surgery consensus guidelines on bicuspid aortic valve-related aortopathy: full online-only version. J Thorac Cardiovasc Surg. 2018; https://doi.org/10.1016/j.jtcvs.2018.02.115.
58. Pepe G, Giusti B, Colonna S, Fugazzaro MP, Sticchi E, De Cario R, et al. When should a rare inherited connective tissue disorder be suspected in bicuspid aortic valve by primary-care internists and cardiologists? Proposal of a score. Intern Emerg Med. 2021;16(3):609–15.
59. Lovato L, Cefarelli M, Di Marco L, Arcioni D, Tortora G, Dormi A, et al. Marfan and Loeys-Dietz aortic phenotype: a potential tool for diagnosis and management. JTCVS Open. 2024;19:223–40.
60. Mazzolai L, Teixido-Tura G, Lanzi S, Boc V, Bossone E, Brodmann M, et al. 2024 ESC guidelines for the management of peripheral arterial and aortic diseases. Eur Heart J. 2024;45(36):3538–700.

11 Cardiac MRI

Martina Perazzolo Marra and Annagrazia Cecere

11.1 Introduction

Bicuspid aortic valve (BAV), characterized by a native bileaflet valve or as a result of two aortic leaflets fusion, is the most common congenital heart disease with a reported prevalence of 0.4–2.25% [1]. It's an inherited disease, transmitted as an autosomal dominant trait with incomplete penetrance and variable expressivity, responsible for different phenotypical expressions [2]. BAV is nowadays considered a *valvulo-aortopathy*, that frequently could be complicated by valve dysfunction and ascending aorta dilation (AAo) [3].

Aortic regurgitation (AR), frequently due to annular dilation and cups prolapse or retraction, is a frequent cause of valve dysfunction in BAV patients, requiring a valve replacement [4]. European Guidelines recommended an aortic valve replacement in symptomatic patients with severe AR (Class I, Level B) or asymptomatic ones with left ventricular (LV) dilation or systolic dysfunction [Class I, Level B) [5]. Conversely to elderly patients more likely to develop aortic stenosis, young patients with BAV are more likely to develop AR with a prevalence of valve dysfunction between 47% and 64% and moderate-to-severe grade in less than 30% [6, 7]. AAo is a progressive phenomenon with an estimated mean annual growth of the aortic root of 0.15–0.23 mm/year [8]. Aortic dissection is the most fearsome complication in BAV patients with an incidence of 3.1 cases per 10.000 patient-years (8 times higher than in the general population), increasing to 0.5% in patients with aortic diameters >45 mm [9]. European Guidelines recommended an ascending aortic surgery in presence of a maximal ascending aorta diameter ≥ 55 mm in all patients and an aortic diameter ≥50 mm in BAV patients with additional risk factors (family history of dissection, severe aortic or mitral regurgitation, desire for pregnancy, uncontrolled systemic arterial hypertension, and/or aortic size increase >3 mm/year) or coarctation (Class IIA, Level C) [5].

11.2 Cardiac Magnetic Resonance

A prompt diagnosis of BAV patients is crucial to avoid long-term complications and improve prognosis. In addition, considering the genetic role in this aortic valve anomaly, a screening of the first-degree relatives is highly recommended, especially in cases of aortic aneurysm or dissection family history [5].

Transthoracic echocardiography is able to identify the aortic valve anomaly and to evaluate

M. P. Marra (✉) · A. Cecere
Cardiology Clinic, Department of Cardiac, Thoracic, Vascular Sciences and Public Health, University of Padua, Padua, Italy
e-mail: martina.perazzolomarra@unipd.it; annagrazia.cecere@unipd.it

G. Thiene et al. (eds.), *Bicuspid Aortic Valve Syndrome*,
https://doi.org/10.1007/978-3-031-85589-4_11

valve dysfunction and aortic dimensions. Due to its high availability, quick execution, low cost, and absence of radiation exposure, echocardiography is the first-line examination in BAV patients. However, due to the well-known limitations (poor acoustic window, AR eccentric jet with reduced possibility of quantitative measurements), echocardiography alone isn't able to accurately evaluate BAV patients in terms of valve anatomy and associated complications. Therefore, the use of multimodality imaging is nowadays encouraged in order to better evaluate the aortic valve anatomy, quantify valve dysfunction, measure aortic dimensions, and select patients for valve replacement.

Cardiac magnetic resonance imaging (cardiac MRI) demonstrated to accurately quantify biventricular volumes and function, as well as a precise tissue characterization [10].

In the last years, cardiac MRI proved its role in the valve disease evaluation and in BAV patients, despite the high acquisition time, the relative lower temporal resolution, and the lesser availability in comparison to echocardiography [3]. In particular, European Guidelines highly recommended the use of cardiac MRI in patients with inadequate echocardiographic acoustic window or discrepant result, in order to assess the severity of valvular lesions, ventricular volumes and function, abnormalities of ascending aorta, and presence of myocardial fibrosis [5].

11.3 Valve Anatomy

Balanced steady-state free precession (SSFP) cine sequences allow an accurate assessment of the aortic valve morphology, due to the high blood-to-myocardium contrast and intrinsic signal-to-noise ratio [11]. The standard 3-chamber view and the orthogonal LVOT long axis are firstly acquired (Fig. 11.1a, b); subsequently, these sequences are used to obtain the perpendicular slice through the aortic valve, just below the aortic leaflet tips, allowing the valve morphology evaluation (Fig. 11.1c).

A normal aortic valve is characterized by three thin leaflets (left, right and non-coronary leaflets) that open symmetrically during systole (Fig. 11.2a). The more anterior leaflet is the right cuspid; conversely, the posterior leaflet near to interatrial septum is the non-coronary one. During the diastole, the leaflets close down forming a characteristic sign like a 3-pointed star (well known as the "Mercedes-Benz emblem") (Fig. 11.2b) [11]. Conversely, a stenotic aortic valve is characterized by thickened leaflets with an incomplete and asymmetrical opening, often associated with a turbulent flow, responsible for a signal void in aorta (black signal in the white cavity of the aorta) (Fig. 11.3a, b). The same sequences used for the aortic valve morphology assessment allow the evaluation of the planimetric area (Fig. 11.1c and 11.3a). In particular, after

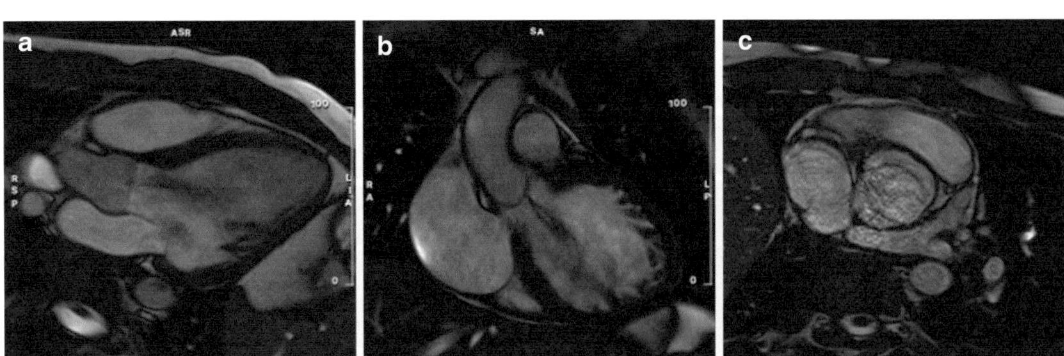

Fig. 11.1 Balanced steady-state free precession cine imaging acquired in standard 3-chamber view (**a**) and in the orthogonal LVOT long axis view (**b**). A perpendicular slice through the aortic valve is subsequently acquired, revealing a BAV (**c**)

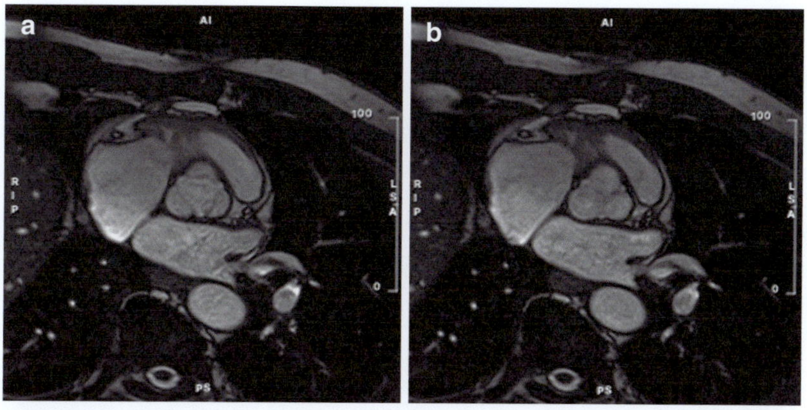

Fig. 11.2 The normal trileaflet aortic valve during systole (**a**) and diastole (**b**)

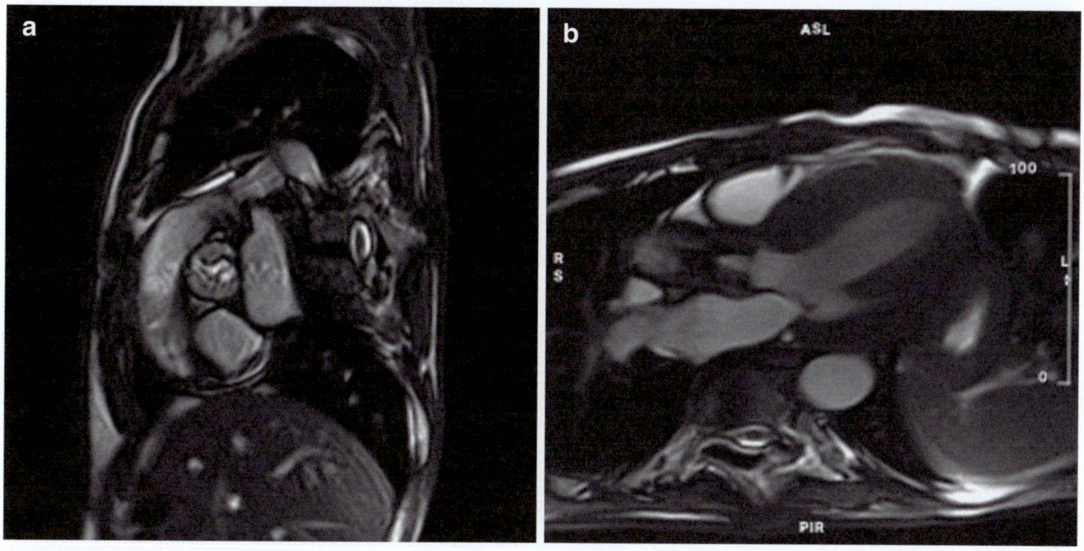

Fig. 11.3 A stenotic tricuspid aortic valve with a reduced systolic area (**a**). A turbulent systolic flow is visible in aorta (**b**), responsible for a typical signal void

acquiring several continuous slices perpendicular to the valve plane, the aortic valve orifice at the leaflet tips could be measured with a good agreement with aortic valve measurements on transesofageal echocardiography [12].

Patients with BAV could have a real bileaflet valve or be a result of a congenital fusion of two leaflets with a presence of rafe (Fig. 11.4a, b). So, the diagnosis of BAV is based on the visualization of the systolic valve opening in the balanced-SSFP cine imaging acquired perpendicularly to the valve. The opening of a BAV is quite similar to the open mouth of fish, so BAV is common called as a "*fish mouth*" valve. Also, the standard 3-chamber balanced SSFP-cine imaging may raise suspicion of BAV revealing an asymmetrical valve closure and a turbulent flow (Fig. 11.1a, b).

11.4 Morpho-Functional Evaluation of Left Ventricle

BAV is usually associated valve dysfunction, responsible for a progressive LV remodeling and adaptation (Fig. 11.5a–c). Cardiac MRI demonstrated to accurately evaluate LV volumes and systolic function with a good reproducibility [13]. However, a reduced LV ejection fraction

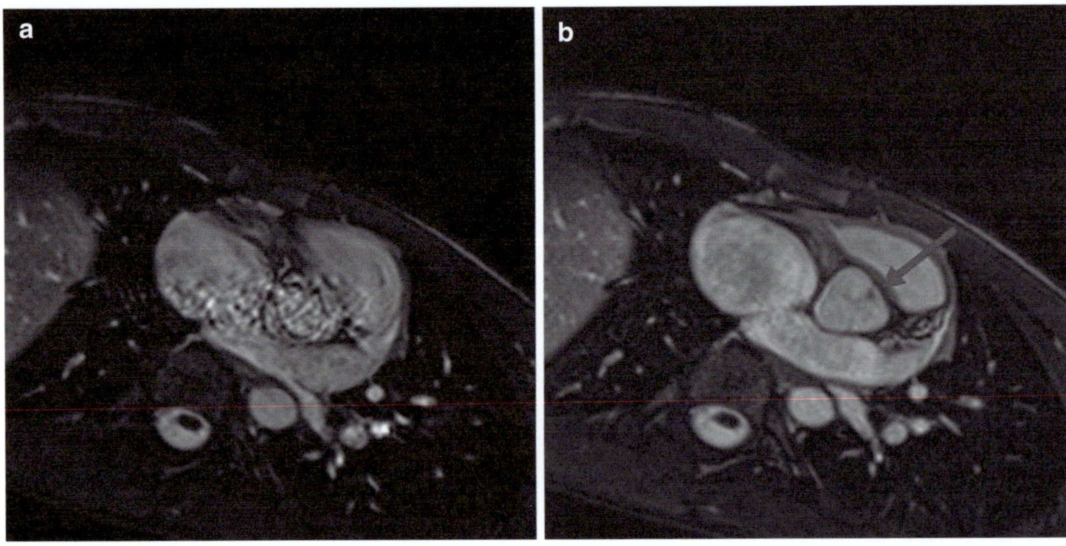

Fig. 11.4 A bicuspid aortic valve during systole (**a**) with a classical opening movement as a *"fish mouth"* and diastole (**b**). A raphe (red arrow) between the right and left aortic cusp is visible

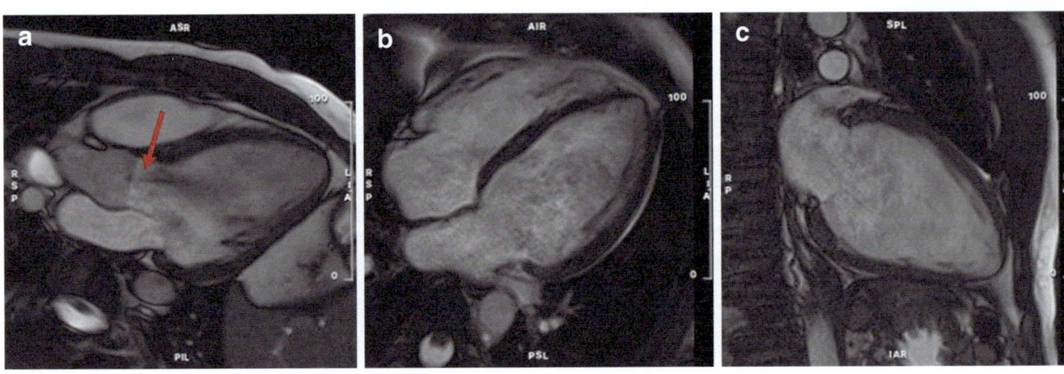

Fig. 11.5 A 27-year-old man with BAV performed CMR for morpho-functional evaluation. The 3-chamber view (**a**) evidenced an asymmetrical closure of the aortic valve and an eccentric, holosystolic aortic regurgitant flow in left ventricle (red arrow). Left ventricle resulted severely dilated (end-diastolic volume indexed 130 ml/mq) with mild reduction of ejection fraction (53%) (**a-c**)

(LVEF) means the presence of an overt LV dysfunction.

A precocious marker of subclinical dysfunction could be useful to early identify patients with a progressive LV negative remodeling. The evaluation of myocardial deformation with a dedicated cardiac MRI post-processing software demonstrated to identify a subclinical LV dysfunction, in presence of normal LVEF [14]. Asymptomatic BAV patients with normal LVEF presented an impaired circumferential strain, probably related to the increased mechanical stress secondary to volume overload [15]. Accordingly, a reduced global longitudinal strain demonstrated to be a marker of severity in patients with chronic AR, confirming the predicting role of an abnormal myocardial deformation the developing of an overt LV myocardial dysfunction [16].

Therefore, considering the pivotal role of LV dimensions and systolic function in the surgical management of asymptomatic patients with severe AR, cardiac MRI could be usefully considered in this setting.

11.5 Quantitative Assessment of Valve Dysfunction

The presence of a turbulent flow or a regurgitant diastolic flow could be visually identified in balanced SSFP-cine sequences (Figs. 11.1, 11.2 and 11.3). Cardiac MRI allows a quantitative assessment of valve dysfunction through the phase-contrast pulse sequences. Starting from balanced SSPF-cine sequences, the imaging plane could be positioned 10 mm distal to the aortic valve annulus, perpendicularly to the blood flow in order to evaluate the velocity of the blood passing through the imaging plane ("through plane" velocity mapping) or in the same direction of the flow ("in plane" velocity mapping). In clinical practice, through plane sequences are usually preferred to in plane ones for higher accuracy [11]. These sequences produce two sets of images: the first one is a magnitude image related to the chosen imaging plane; the second one is the phase map encoding the velocities within each voxel (Fig. 11.6a, b). The valve flow is derived by the through-plane velocity map with the integration of the pixel's velocity and its area during a single cardiac circle with a good accuracy and reproducibility [17]. In order to correctly estimate the peak and mean velocities in the aortic valve, it's crucial to choose the appropriate encoding velocity (VENC). A VENC lower than the real valve peak velocity produces aliasing (black spots inside the valve in the phase map figure); conversely, a higher VENC could increase the impact of noise on the velocity cine imaging, masking the true transvalvular peak velocity [11]. These phase contrast pulse sequences, acquired in through plane, provide a quantification of the AR jet, that is typically holodiastolic and eccentric, originating from the aortic valve and directed to the LVOT, as regurgitant volume (RV, ml) (Fig. 11.6c) [18]. Accordingly, the regurgitant fraction (RF, %) could be calculated as the percentage of the ratio between the RV and the forward volume, derived by the phase

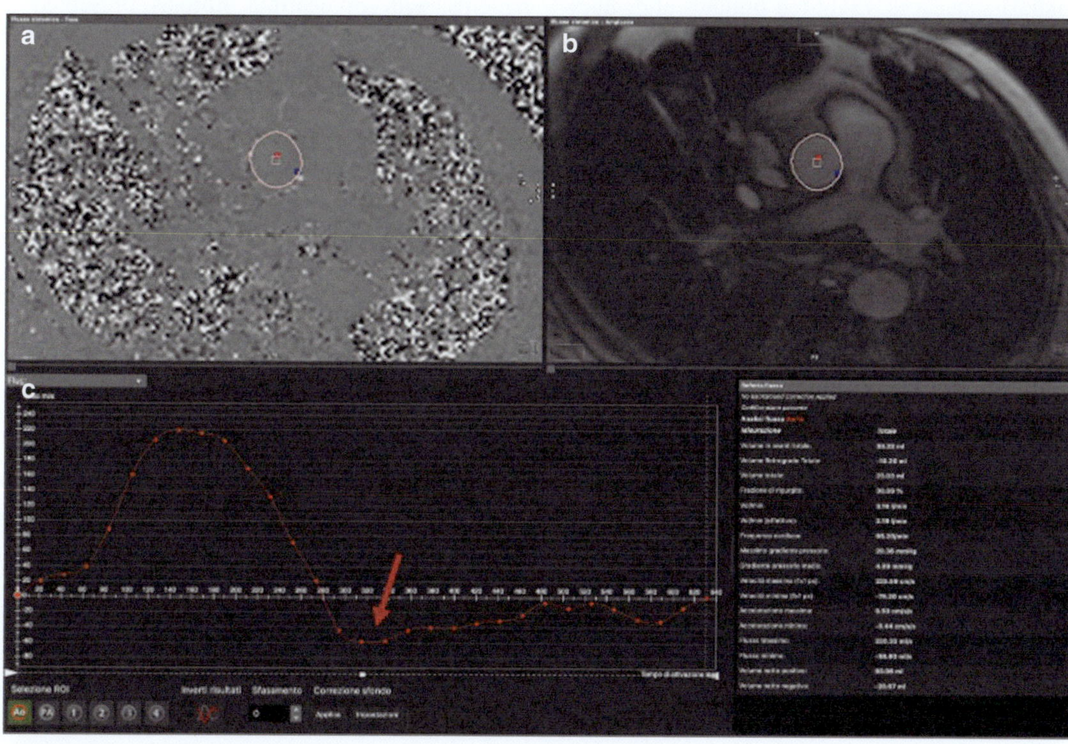

Fig. 11.6 A 76-year-old man with BAV complicated by valve dysfunction. The phase-contrast sequences (**a**, **b**) evidenced a moderate proto-mesodiastolic aortic regurgitation (red arrow, **c**)

contrast sequences [18]. However, due to the eccentricity of jet and the increased systolic aortic valve motion towards the apex, typically present in case of a significant AR, in BAV patients the quantification of the valve regurgitation with cardiac MRI could be underestimated [19].

In absence of measurable phase-contrast sequences and other valve disease, the RV could be also calculated with the difference between the left ventricular and right ventricular stroke volume, derived by the balanced SSFP-cine sequences [18].

11.6 Myocardial Tissue Characterization

The pressure and volume overload BAV-induced could increase the production of new sarcomeres in series, resulting in myocardial fibrosis [20, 21].

Cardiac MRI demonstrated to be able to assess the presence of myocardial fibrosis with post-contrast sequences. Eight-ten minutes after intravenous administration of gadolium-based contrast injection (0.2 mmol/kg), the phase sensitive inversion recovery sequences (PSIR) are acquired into 4-, 3-, and 2-chamber long axis as well as in the short axis. The myocardial fibrosis, identifiable as late gadolium enhancement (LGE), could be present as an ischemic pattern (subendocardial or transmural) or a non-ischemic one (midwall or subepicardial).

The negative prognostic role of myocardial fibrosis in patients with chronic AR has been widely documented. In chronic AR patients, myocardial fibrosis has been associated with a 2.5-fold increase in mortality in comparison with patients without LGE [22]. Accordingly, the presence of myocardial fibrosis in patients with severe AR could be considered as a marker of adverse LV remodeling and a worse prognosis after aortic valve replacement [21]. In addition, comparing patients with moderate or severe AR, the presence of BAV has been demonstrated as an independent factor for myocardial fibrosis, suggesting the increased mechanical stress in these patients (Fig. 11.7a, b) [23]. Accordingly, BAV patients with myocardial fibrosis presented a higher transvalvular gradients evaluated by echo-

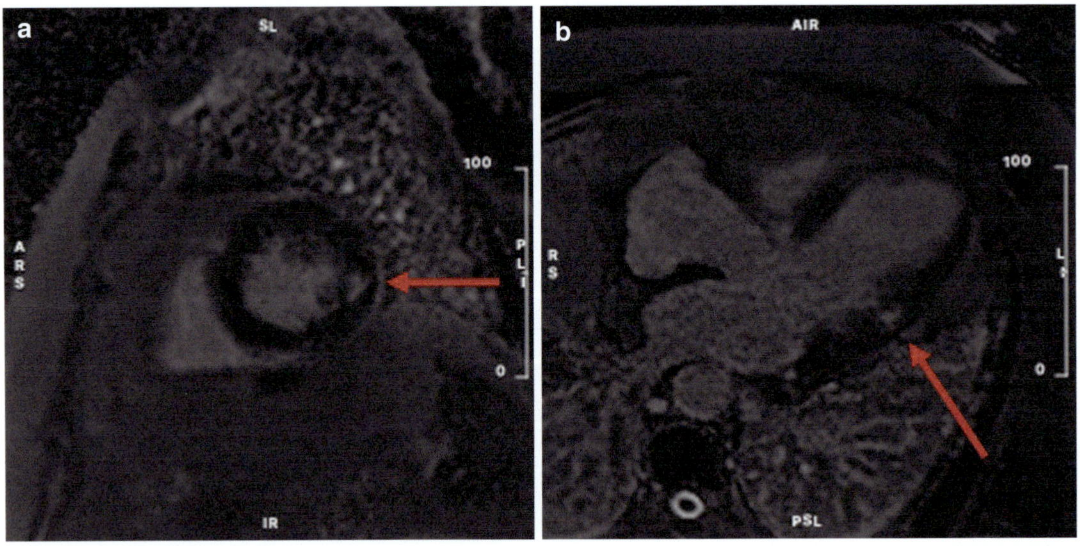

Fig. 11.7 A 62-year-old male patient with BAV and preserved left ventricular systolic function performed CMR for tissue characterization. Post-contrast sequences demonstrated an intramural LGE in the mid lateral wall in short (**a**) and long (**b**) axis view

cardiography as well as more LV hypertrophy. Subsequently, these BAV patients with LGE demonstrated to be 10 times more likely to need aortic valve replacement, suggesting that myocardial fibrosis is a marker of a more severe phenotype of the valve disease [24].

11.7 4D Flow and Future Perspectives

The presence of BAV significantly impacts on the valve motion during the cardiac cycle. The resulting asymmetric flow with a direction angle higher than normal to the center of aorta produces vortices with a relevant hemodynamic impact on the aortic wall [19]. BAV patients without aorta dilation revealed these flow abnormalities, suggesting that the only presence of an abnormal valve could produce flow alterations and a consequent vascular stress [19]. The 4D flow sequences are a particular type of phase-contrast sequences that uses blood flow encoded in three directions, allowing the visualization and quantification of hemodynamic and morphology alterations [25]. Based on the different anatomical BAV phenotype, the use of 4D Flow sequences demonstrated to evaluate the different hemodynamic impact of the asymmetrical flow on the vascular aortic wall [26]. In addition, considering the limitation of the classical phase contrast in the quantification of eccentric jets, the 4D flow sequences could be particularly useful in the AR quantification in BAV patients [19].

Finally, the progressive dilation of aorta could be estimated by the assessment of aortic stiffness, that represents the velocity of flow propagation. The stiffer the aorta is, the faster is the velocity of flow propagation. The thoracic aorta stiffness could be non-invasively assessed by measuring the flow velocity propagation (pulse wave velocity) or the distensibility, a marker of aortic elasticity, that is inversely correlated with aortic stiffness [19]. In BAV patients with AAo dilation, a lower aortic distensibility has been demonstrated [27]. Further studies will be necessary to better evaluate if the reduced aortic distensibility in BAV patients could be secondary to an increased aortic stiffness or it could be the result of an abnormal wall stress distribution dilation-related [28].

11.8 Limitations

Although the promising role of cardiac MRI in the BAV evaluation, some limitations should be taken into consideration. Firstly, the quality and the reproducibility of cardiac MRI sequences are closely related to the patient's heart rhythm and breathing compliance. So, the presence of frequents premature ventricular beats or irregular rhythm (for example atrial fibrillation with high variable RR interval) could significantly limit the quality of the images [29]. Secondly, the lower temporal resolution than echocardiography doesn't allow an accurate evaluation and tissue characterization of small size valvular masses, which are characterized by high mobility [30]. Finally, cardiac MRI can't be executed in patients hemodynamically unstable or with unsafe/condition cardiac devices, as well as in the emergency setting or catheterization laboratory in comparison to echocardiography that could be performed before, during and after the valvular interventions [11].

11.9 Conclusions

Cardiac MRI could provide an accurate evaluation of BAV patients regarding valve anatomy, dysfunction and dimension as well as LV morpho-functional analysis and tissue characterization. Considering the potential clinical impact of a comprehensive valve, aorta and LV evaluation, cardiac MRI should be considered integrant part of the multimodality imaging BAV patient's evaluation.

References

1. Hoffman JIE, Kaplan S. The incidence of congenital heart disease. J Am Coll Cardiol. 2002;39:1890–900. https://doi.org/10.1016/s0735-1097(02)01886-7.
2. Prakash SK, Bossé Y, Muehlschlegel JD, Michelena HI, Limongelli G, Della Corte A, Pluchinotta FR, Russo MG, Evangelista A, Woodrow Benson D, Body SC, Milewicz DM, BAVCon Investigators. A roadmap to investigate the genetic basis of bicuspid aortic valve and its complications: insights from the International BAVCon (Bicuspid Aortic Valve Consortium). J Am Coll Cardiol. 2014;64:832–9. https://doi.org/10.1016/j.jacc.2014.04.073.
3. Wassmuth R, von Knobelsdorff-Brenkenhoff F, Gruettner H, Utz W, Schulz-Menger J. Cardiac magnetic resonance imaging of congenital bicuspid aortic valves and associated aortic pathologies in adults. Eur Heart J Cardiovasc Imaging. 2014;15:673–9. https://doi.org/10.1093/ehjci/jet275.
4. Yang L, Michelena HI, Maleszewski JJ, Schaff HV, Pellikka PA. Contemporary etiologies, mechanisms, and surgical approaches in pure native aortic regurgitation. Mayo Clin Proc. 2019;94:1158–70. https://doi.org/10.1016/j.mayocp.2018.11.034.
5. Vahanian A, Beyersdorf F, Praz F, Milojevic M, Baldus S, Bauersachs J, Capodanno D, Conradi L, De Bonis M, De Paulis R, Delgado V, Freemantle N, Gilard M, Haugaa KH, Jeppsson A, Jüni P, Pierard L, Prendergast BD, Sádaba JR, Tribouilloy C, Wojakowski W, ESC/EACTS Scientific Document Group. 2021 ESC/EACTS guidelines for the management of valvular heart disease. Eur Heart J. 2022;(43):561–632. https://doi.org/10.1093/eurheartj/ehab395.
6. Tzemos N, Therrien J, Yip J, Thanassoulis G, Tremblay S, Jamorski MT, Webb GD, Siu SC. Outcomes in adults with bicuspid aortic valves. JAMA. 2008;300:1317–25. https://doi.org/10.1001/jama.300.11.1317.
7. Evangelista A, Gallego P, Calvo-Iglesias F, Bermejo J, Robledo-Carmona J, Sánchez V, Saura D, Arnold R, Carro A, Maldonado G, Sao-Avilés A, Teixidó G, Galian L, Rodríguez-Palomares J, García-Dorado D. Anatomical and clinical predictors of valve dysfunction and aortic dilation in bicuspid aortic valve disease. Heart. 2018;104:566–73. https://doi.org/10.1136/heartjnl-2017-311560.
8. Lopez A, Dentamaro I, Galian L, Calvo F, Alegret JM, Sanchez V, Citro R, Moreo A, Chirillo F, Colonna P, Carrero MC, Bossone E, Moral S, Sao-Aviles A, Gutiérrez L, Teixido-Tura G, Rodríguez-Palomares J, Evangelista A. Predictors of ascending aorta enlargement and valvular dysfunction progression in patients with bicuspid aortic valve. J Clin Med. 2021;10:5264. https://doi.org/10.3390/jcm10225264.
9. Evangelista Masip A, Galian-Gay L, Guala A, Lopez-Sainz A, Teixido-Turà G, Ruiz Muñoz A, Valente F, Gutierrez L, Fernandez-Galera R, Casas G, Panaro A, Marigliano A, Huguet M, González-Alujas T, Rodriguez-Palomares J. Unraveling bicuspid aortic valve enigmas by multimodality imaging: clinical implications. J Clin Med. 2022;11:456. https://doi.org/10.3390/jcm11020456.
10. Merlo M, Gagno G, Baritussio A, Bauce B, Biagini E, Canepa M, Cipriani A, Castelletti S, Dellegrottaglie S, Guaricci AI, Imazio M, Limongelli G, Musumeci MB, Parisi V, Pica S, Pontone G, Todiere G, Torlasco C, Basso C, Sinagra G, Perrone Filardi P, Indolfi C, Autore C, Barison A. Clinical application of CMR in cardiomyopathies: evolving concepts and techniques: a position paper of myocardial and pericardial diseases and cardiac magnetic resonance working groups of Italian society of cardiology. Heart Fail Rev. 2023;28:77–95. https://doi.org/10.1007/s10741-022-10235-9.
11. Looi J, Kerr AJ, Gabriel R. Morphology of congenital and acquired aortic valve disease by cardiovascular magnetic resonance imaging. Eur J Radiol. 2015;84:2144–54. https://doi.org/10.1016/j.ejrad.2015.07.022.
12. Reant P, Lederlin M, Lafitte S, Serri K, Montaudon M, Corneloup O, Roudaut R, Laurent F. Absolute assessment of aortic valve stenosis by planimetry using cardiovascular magnetic resonance imaging: comparison with transesophageal echocardiography, transthoracic echocardiography, and cardiac catheterization. Eur J Radiol. 2006;59:276–83. https://doi.org/10.1016/j.ejrad.2006.02.011.
13. Grothues F, Smith GC, Moon JCC, Bellenger NG, Collins P, Klein HU, Pennell DJ. Comparison of interstudy reproducibility of cardiovascular magnetic resonance with two-dimensional echocardiography in normal subjects and in patients with heart failure or left ventricular hypertrophy. Am J Cardiol. 2002;90:29–34. https://doi.org/10.1016/s0002-9149(02)02381-0.
14. Rajiah PS, Kalisz K, Broncano J, Goerne H, Collins JD, François CJ, Ibrahim ES, Agarwal PP. Myocardial strain evaluation with cardiovascular MRI: physics, principles, and clinical applications. Radiographics. 2022;42:968–90. https://doi.org/10.1148/rg.210174.
15. Burris NS, Lima APS, Hope MD, Ordovas KG. Feature tracking cardiac MRI reveals abnormalities in ventricular function in patients with bicuspid aortic valve and preserved ejection fraction. Tomography. 2018;4:26–32. https://doi.org/10.18383/j.tom.2018.00005.
16. Fernández-Golfín C, Hinojar-Baydes R, González-Gómez A, Monteagudo JM, Esteban A, Alonso-Salinas G, Fernández MA, García-Martín A, Santoro C, Pascual-Izco M, Jiménez-Nacher JJ, Zamorano JL. Prognostic implications of cardiac magnetic resonance feature tracking derived multidirectional strain in patients with chronic aortic regurgitation. Eur Radiol. 2021;31:5106–15. https://doi.org/10.1007/s00330-020-07651-6.
17. Chatzimavroudis GP, Oshinski JN, Franch RH, Walker PG, Yoganathan AP, Pettigrew RI. Evaluation

of the precision of magnetic resonance phase velocity mapping for blood flow measurements. J Cardiovasc Magn Reson. 2001;3:11–9. https://doi.org/10.1081/jcmr-100000142.
18. Herzog BA, Greenwood JP, Plein S. Cardiovascular magnetic resonance pocket guide. Second Edition 2017; 77–79.
19. La Mura L, Lembo M, Musella F, D'Amato M, D'Andrea A, Izzo R, Esposito G. Aortic regurgitation in bicuspid aortic valve: the role of multimodality imaging. J Clin Med. 2024;13:3924. https://doi.org/10.3390/jcm13133924.
20. Borer JS, Truter S, Herrold EM, Falcone DJ, Pena M, Carter JN, Dumlao TF, Lee JA, Supino PG. Myocardial fibrosis in chronic aortic regurgitation: molecular and cellular responses to volume overload. Circulation. 2002;105:1837–42. https://doi.org/10.1161/01.cir.0000014419.71706.85.
21. Azevedo CF, Nigri M, Higuchi ML, Pomerantzeff PM, Spina GS, Sampaio RO, Tarasoutchi F, Grinberg M, Rochitte CE. Prognostic significance of myocardial fibrosis quantification by histopathology and magnetic resonance imaging in patients with severe aortic valve disease. J Am Coll Cardiol. 2010;56:278–87. https://doi.org/10.1016/j.jacc.2009.12.074.
22. Malahfji M, Senapati A, Tayal B, Nguyen DT, Graviss EA, Nagueh SF, Reardon MJ, Quinones M, Zoghbi WA, Shah DJ. Myocardial scar and mortality in chronic aortic regurgitation. J Am Heart Assoc. 2020;9:e018731. https://doi.org/10.1161/JAHA.120.018731.
23. Zheng Y, Yang K, Chen X, Li R, Su G, Yin G, Wang K, Lu M, Zhao S. Prognostic significance of myocardial fibrosis and CMR characteristics in bicuspid aortic valve with moderate and severe aortic insufficiency. Eur Radiol. 2021;31(10):7262–72. https://doi.org/10.1007/s00330-021-07823-y.
24. Lluri G, Renella P, Finn JP, Vorobiof G, Aboulhosn J, Deb A. Prognostic significance of left ventricular fibrosis in patients with congenital bicuspid aortic valve. Am J Cardiol. 2017;120:1176–9. https://doi.org/10.1016/j.amjcard.2017.06.060.
25. Zhuang B, Sirajuddin A, Zhao S, Lu M. The role of 4D flow MRI for clinical applications in cardiovascular disease: current status and future perspectives. Quant Imaging Med Surg. 2021;11:4193–210. https://doi.org/10.21037/qims-20-1234.
26. Rodríguez-Palomares JF, Dux-Santoy L, Guala A, Kale R, Maldonado G, Teixidó-Turà G, Galian L, Huguet M, Valente F, Gutiérrez L, González-Alujas T, Johnson KM, Wieben O, García-Dorado D, Evangelista A. Aortic flow patterns and wall shear stress maps by 4D-flow cardiovascular magnetic resonance in the assessment of aortic dilatation in bicuspid aortic valve disease. J Cardiovasc Magn Reson. 2018;20:28. https://doi.org/10.1186/s12968-018-0451-1.
27. Guala A, Rodriguez-Palomares J, Dux-Santoy L, Teixido-Tura G, Maldonado G, Galian L, Huguet M, Valente F, Gutiérrez L, González-Alujas T, Johnson KM, Wieben O, Sao Avilés A, Garcia-Dorado D, Evangelista A. Influence of aortic dilation on the regional aortic stiffness of bicuspid aortic valve assessed by 4-dimensional flow cardiac magnetic resonance: comparison with Marfan syndrome and degenerative aortic aneurysm. JACC Cardiovasc Imaging. 2019;12:1020–9. https://doi.org/10.1016/j.jcmg.2018.03.017.
28. Galian-Gay L, Rodríguez-Palomares J, Guala A, Michelena HI, Evangelista A. Multimodality imaging in bicuspid aortic valve. Prog Cardiovasc Dis. 2020;63:442–51. https://doi.org/10.1016/j.pcad.2020.06.003.
29. Myerson SG, Francis JM, Neubauer S. Direct and indirect quantification of mitral regurgitation with cardiovascular magnetic resonance, and the effect of heart rate variability. MAGMA. 2010;23:243–9. https://doi.org/10.1007/s10334-010-0222-y.
30. Tyebally S, Chen D, Bhattacharyya S, Mughrabi A, Hussain Z, Manisty C, Westwood M, Ghosh AK, Guha A. Cardiac tumors: JACC CardioOncology state-of-the-art review. JACC CardioOncol. 2020;2:293–311. https://doi.org/10.1016/j.jaccao.2020.05.009.

Computed Tomography Angiography in Bicuspid Aortic Valve

12

Gianluca Pontone and Maria Elisabetta Mancini

12.1 Introduction

Bicuspid aortic valve (BAV) is the most common congenital heart disease, affecting approximately 0.5% to 2% of the adult population, with a male-to-female ratio of 3:1 [1].

It accounts for a significant proportion of surgical aortic valve replacements (AVR) in adults, with surgical options including valve replacement with or without ascending aorta replacement, composite aortic root replacement, the Ross procedure, and aortic valve repair procedures. Nonetheless, transcatheter aortic valve implantation (TAVI) may be considered as an alternative to surgery after evaluating patient-specific factors, including individual preference [2, 3].

The most common clinical presentations of BAV include progressive aortic valvular dysfunction, such as aortic stenosis (14–50% of cases), and aortic regurgitation (23–70%) as well as aortic dilatation, primarily due to an abnormal aortic wall substrate. This dilatation may involve the aortic sinuses, ascending aorta, or aortic arch, occurring in 10–40% of patients [1, 4, 5]. As aortic diameters increase, the risk of acute aortic dissection also rises. Although the incidence of dissection in BAV patients is eight times higher than in the general population, the absolute risk remains relatively low (31/100.000 patients-years) and is significantly lower than that seen in Marfan syndrome or other heritable thoracic aortic diseases [6].

BAV is also frequently associated with other congenital cardiac malformations, such as aortic coartaction, Shone complex, patent ductus arteriosus and ventricular septal defects. Furthermore, it shows a higher prevalence of coronary artery anomalies and carries an heightened risk of infective endocarditis, with active inflammatory lesions detected in nearly 5% of patients [5, 7–9].

The purpose of this chapter is to highlight the role of computed tomography angiography (CTA) in understanding the anatomical and pathological complexities of the BAV spectrum, with a focus on its essential contribution to both diagnosis and therapeutic planning.

G. Pontone (✉)
Department of Perioperative Cardiology and Cardiovascular Imaging, Centro Cardiologico Monzino, IRCCS, Milan, Italy

Department of Biomedical, Surgical and Dental Sciences, University of Milan, Milan, Italy
e-mail: gianluca.pontone@cardiologicomonzino.it

M. E. Mancini
Department of Perioperative Cardiology and Cardiovascular Imaging, Centro Cardiologico Monzino, IRCCS, Milan, Italy
e-mail: Maria.Mancini@cardiologicomonzino.it

G. Thiene et al. (eds.), *Bicuspid Aortic Valve Syndrome*,
https://doi.org/10.1007/978-3-031-85589-4_12

12.2 Overview of CT Acquisition Protocol

Multimodality imaging is fundamental for the early diagnosis of BAV disease, assessment of valvular dysfunction and aortopathy, and timely planning of surgical or transcatheter interventions. Transthoracic echocardiography (TTE) is commonly used in clinical practice as the first-line imaging modality. However, some technical limitations can hinder precise evaluation of the valve and aorta, such as suboptimal acoustic windows (particularly in obese patients and those with chronic obstructive pulmonary disease), the presence of a prominent raphe, severely calcified valve leaflets, and the inherent operator dependence of the technique [10, 11].

In the last decades, the continuous advancements in CT technology have transformed CTA into an essential tool for supplementing echocardiography in the evaluation of aortic valve pathology and in decision-making for interventions, becoming the standard of reference for the pre-procedural work-up for TAVI. This is particularly true for patients with bicuspid aortic valve disease [12].

In most published studies on CTA of the aortic valve, scanning protocols are similar to those used in coronary CT and vary depending on the scanner generation and vendor [13]. However, some key principles can be identified to ensure the acquisition of high quality, motion-free ECG-gated angiographic CT images of the aortic root and the thoracic aorta, which are basic for the accurate diagnosis of BAV (Table 12.1) [14].

A 64-slice CT scanner is recommended as the minimum technological requirement, with obvious advantages offered by last-generation dual-source and wide-detector CT scanners, allowing for the acquisition of the whole heart volume within a single heartbeat [15].

Data should be acquired at the smallest available collimation; this should be no more than 0.625 mm, to achieve high-spatial-resolution image datasets with isotropic voxels [16].

ECG-gating can be obtained using a retrospective or prospective technique. Retrospective ECG-gated scanning acquires continuous data throughout the entire cardiac cycle, allowing the reconstruction of the aortic valve in any phase of the R-R interval (0–100%, reconstructed at 5–10% intervals) and enabling qualitative assessment of valve leaflet motion in cine mode. Prospective ECG-gated scanning, or "step and shoot"/"axial" mode, keeps the table static while the X-ray tube rotates around the patient, advancing only after data acquisition is completed. This means that the X-ray tube is activated only during a pre-specified phase of the cardiac cycle (R-R interval). Therefore, the use of adequate ECG-padding or a wide pulsing window—effectively widening the acquisition window for a longer interval on each side of the R-R interval—is recommended to allow the analysis of both systolic and diastolic phases. For systems with whole-heart coverage (i.e., 16 cm detector coverage), an ECG-gated one beat acquisition should be employed.

A preliminary noncontrast CT scan, covering from approximately 1 cm below the carina to below the apex of the heart, may be used to assess the valvular calcium score. For the contrast injection strategy in the angiographic scan, the amount of contrast medium should be adjusted according to the capabilities of the CT system and the patient's body habitus, typically ranging between 50 and 100 mL. This should be administered as a bolus through an 18-gauge antecubital intravenous cannula at a rate of 5 mL/s, followed immediately by a 50–60 mL saline chaser. Bolus tracking or bolus timing software should be used to trigger contrast-enhanced acquisitions once the bolus reaches the ascending aorta. Scan parameters (kVp, mAs) should be tailored to each patient's size, ensuring that the radiation dose remains "as low as reasonably achievable" (ALARA) while maintaining adequate diagnostic quality [12, 15–18].

For TAVI diagnostic work-up—subject to specific consensus documents [12, 17, 18]—the scan range should extend from the subclavian arteries to the superficial femoral arteries at the level of the femoral head. There are two main approaches for combining, within the same intravenous con-

Table 12.1 CT scan protocol

Premedication	Nitroglycerine and beta-blockers are usually not used
Scanner type	≥64-slice
ECG gating	• Retrospective or prospective • The coverage of the entire cardiac cycle is recommended (0–100% R-R; reconstruction at 5–10% intervals)
Slice thickness	≤1 mm
Tube voltage and current	ALARA principles according on individual patient's size: • BMI ≤ 30 kg/m²: 100 kV • BMI > 30 kg/m²: 120 kV • mAs adjusted to the lowest setting that guarantees acceptable image noise
Contrast medium injection strategy	• Unenhanced scan for evaluation of aortic valve calcium score • 50–100 ml of contrast material at 5 mL/s followed by saline chaser (50–60 mL at same rate)
Triggering of acquisition	• Bolus tracking method (ROI placed in ascending or descending aorta with opacification threshold of 100–150 HU for initiating scan acquisition) • Bolus test technique (small bolus of contrast material administered to measure contrast material arrival and travel time)
Scan range	• Depends on the indication; usually from 2 cm above carina to the diaphragm • In cases of TAVI, from subclavian arteries to the superficial femoral arteries at the level of the femoral heads
Post-processing	• Aortic valve calcium score (Agatson score) • Double oblique multiplanar reconstructions orthogonal to the aortic valve, aortic root and thoracic aorta at different levels • 3D volume rendering reconstructions • Cine 4D reconstructions • Semiautomatic dedicated softwares for TAVI planning
Key parameters to assess	• Aortic valve morphology and function (number of cusps, pattern of fusion and simmetry, presence of raphe, degree of calcification, planimetry, leaflet motion in cine-mode) • Aortic root configuration and dimensions (LVOT, annulus, sinuses of Valsalva, sinotubular junction) • Ascending aorta dimensions • Exclusion of aortic coartaction • Coronary artery anatomy (origin, ostial height) and CAD • Thoraco-abdominal aorta and ileo-femoral vessels (course, burden of atherosclerosis, aneurysm, dissection, calcifications, minimal artery diameters—In case of TAVI planning) • Signs of aortic dissections or endocarditis (if high clinical suspect)

trast injection, the ECG-gated acquisition of the aortic root structures and a subsequent faster non-ECG gated CTA of the aorta-ileo-femoral vascular axis, to simultaneously assess the device access route:

- ECG-gated (spiral or sequential) of the chest, followed by an ungated spiral scan of the abdomen and pelvis.
- ECG-gated (spiral or sequential) of the heart and aortic root, followed by an ungated spiral scan of the thorax, abdomen, and pelvis.

Nitroglycerine and beta-blockers are contraindicated in severe aortic stenosis and should not be administered prior to scanning [12, 17, 18].

Image datasets should be reconstructed in 10% increments, from 0% to 100% of the R-R interval, using iterative or deep learning algorithms if available, with a standard soft tissue kernel at a maximum slice thickness of <1 mm (preferably 0.5–0.625 mm) with moderate overlap [16]. A small reconstruction field of view centered on the aortic root and a 512x512 matrix is recommended to optimize spatial resolution [18].

Post-processing includes the evaluation of both static and cine images, employing 3D multiplanar reconstructions (MPR) to generate double-oblique short-axis and long-axis planes, aligned perpendicular to the aortic valve and parallel to the aortic root. These images should be analyzed in a cranio-caudal direction from the sinotubular junction to the left ventricular outflow tract, allowing for detailed assessment of the valve morphology and anatomical relationships.

The application of semi-automatic softwares with a specific aortic module, capable of extracting the aortic centerline and generating images orthogonal to the aortic root, is particularly useful for TAVI planning [12].

The main objectives of CT analysis can be summarized as follows: (1) BAV diagnosis and phenotype description (number of cusps, pattern of fusion, presence of raphe, degree of calcification, planimetry of the aortic valve orifice, leaflet motion); (2) Evaluation of the aortic root and ascending aorta; (3) Exclusion of aortic coartaction; (4) Evaluation of coronary arteries anatomy and rule-out of significant coronary artery disease (CAD); (5) Evaluation of the thoraco-abdominal aorta and peripheric vasculature (for TAVI planning); (6) Exclusion of severe complications, such as aortic dissection and endocarditis (in specific clinical contexts) [11, 13].

12.3 Essential Anatomical Concepts

Before proceeding with the evaluation of CT diagnosis of bicuspid aortic valve, it is important to briefly clarify the anatomical concepts that will be used in the subsequent descriptions (Table 12.2).

12.4 Diagnosis of BAV and Phenotype Description

BAV is commonly diagnosed through short-axis imaging of the aortic valve at the base of the heart, which reveals only two commissures defining two functional valve cusps [21].

Several studies have demonstrated that CTA offers superior diagnostic accuracy compared to TTE for distinguishing BAVs from TAVs, with surgical findings used as the reference standard. CTA has shown excellent sensitivity (92–97%) and specificity (95–100%), as well as high intra- and inter-observer reproducibility. This diagnostic advantage is particularly evident in challenging anatomical substrates, such as severe valvular calcifications, which cause significant acoustic shadowing on echocardiography [22–25].

A recent study by Takagi et al., involving 212 consecutive patients with severe aortic stenosis, including 78 patients (37%) with BAV, reported the incremental clinical value of adding CTA findings to a model that also included age, aortic diameter index, and TTE results. Incorporating CTA significantly improved the ability to diagnose BAV at surgery (AUC 0.91 vs 0.97, $P = 0.003$), as well as valve morphological classification (net reclassification improvement = 1.79, $P < 0.001$) [24]. These findings emphasize the role of CTA as a complementary modality to TTE for detecting BAV, particularly in patients with inconclusive echocardiographic results or those being considered for surgery. Therefore, all patients with echocardiographically suspected BAV should undergo a CTA to confirm the diagnosis and guide treatment.

Regarding the phenotype description, BAVs differ in their morphologic, positional, and functional characteristics, all of which are significant for selecting the appropriate treatment.

Over time, multiple classification systems have been proposed (Fig. 12.1), based on pathological, surgical, and imaging features, reflecting the heterogeneity of this condition. The most widely adopted to date is the Sievers-Schmidtke system, which categorizes BAVs into three types based on the number of raphes:

- **Type 0**: no raphe, with two usually simmetrical cusps.
- **Type 1**: one raphe, shared by two smaller fused cusps and one larger cusp (usually the non-coronary).
- **Type 2**: two raphes.

Table 12.2 Anatomical terminology [adapted from [19, 20]]

Aortic root	The aortic root is the segment of aorta that extends from the basal attachment of the aortic valve leaflets within the left ventricle to their peripheral attachment at the level of the sinotubular junction. It is composed of five key components: The ventriculo-arterial junction, the annulus, the leaflets, the aortic sinuses, and the sinotubular junction
Left ventricle outflow tract	The left ventricular outflow (LVOT) tract is the region of the left ventricle that lies between the anterior cusp of the mitral valve and the ventricular septum
Ventriculo-arterial junction	The ventriculo-arterial junction is the border between the left ventricular myocardium and the fibroelastic walls of the aortic sinuses
Aortic annulus	The aortic annulus is not an anatomically discrete structure, but represents the virtual basal ring, formed by joining the three most caudal connection points of the aortic valvar leaflets (nadir) to the wall of the LVOT tract slightly below the ventriculo-arterial junction
Aortic sinuses or sinuses of Valsalva	The aortic sinuses (SoV) are pouch-like structures of the aortic root, composed of collagen and elastic tissue, located behind each valve leaflet and extending distally to the sinotubular junction. The coronary ostia are usually located in the upper third of the SoV. Normally, there are three sinuses, named according to the origin of the coronary arteries: Left-coronary (LC), right-coronary (RC), and non-coronary (NC) sinus, with the NCS being the largest and the LCS the smallest
Sinotubular junction	The sinotubular junction is a circular fibrous-elastic ring that defines the border between the aortic sinuses and the tubular segment of the ascending aorta, typically appearing as a mild waist
Ascending aorta	The ascending aorta is the segment of the aorta that extends between the sinotubular junction and the origin of the brachiocephalic trunk
Cusps/leaflets	The "cusps" or "leaflets" are the moving parts of the aortic root, composed of fibrous tissue covered on both sides by an endothelial layer. Each of them has several anatomic regions: (a) the body; (b) the lunula, which represents the appositional surface between the leaflets; (c) the free margin, in the middle of which there is a thickened nodule, called the nodule of Arantius; (d) the insertion or hinge line, which is characterized by a "crown-shaped" or "parabolic" configuration
Commissures	The commissures are the most superior points where the insertions of two adjacent aortic valve leaflets on the aortic wall join each other, at the level of sinotubular junction
Interleaflet triangles	The interleaflet triangles are three triangle-shaped fibrous structures, considerable as extensions of the left ventricular outflow tract, located between the aortic valve leaflets, rising from the plane of aortic annulus and reaching the commissures
Raphe	A raphe is a fibrous ridge located at the site of fusion between two undeveloped conjoint or fused adjacent aortic valve leaflets, forming a pseudocommissure with a hypoplastic interleaflet triangle. It can be solid or fenestrated and may become calcified over time. The raphe is characterized by a variable length of fusion between the leaflets and a variable height of insertion on the aortic root wall. Notably, its attachment to the aortic wall is typically lower than the height of the true commissures

Further subclassifications are based on the spatial orientation of the cusp or raphes relative to the sinuses of Valsalva and the valve's functional status, including predominant insufficiency, stenosis, a balance of both, or neither [26].

In 2016, Jilaihawi et al. proposed a CTA-based classification following a multicenter study involving 130 consecutive patients with severe native aortic stenosis and BAV undergoing TAVI. This classification identified three distinct morphologies, emphasizing the interaction

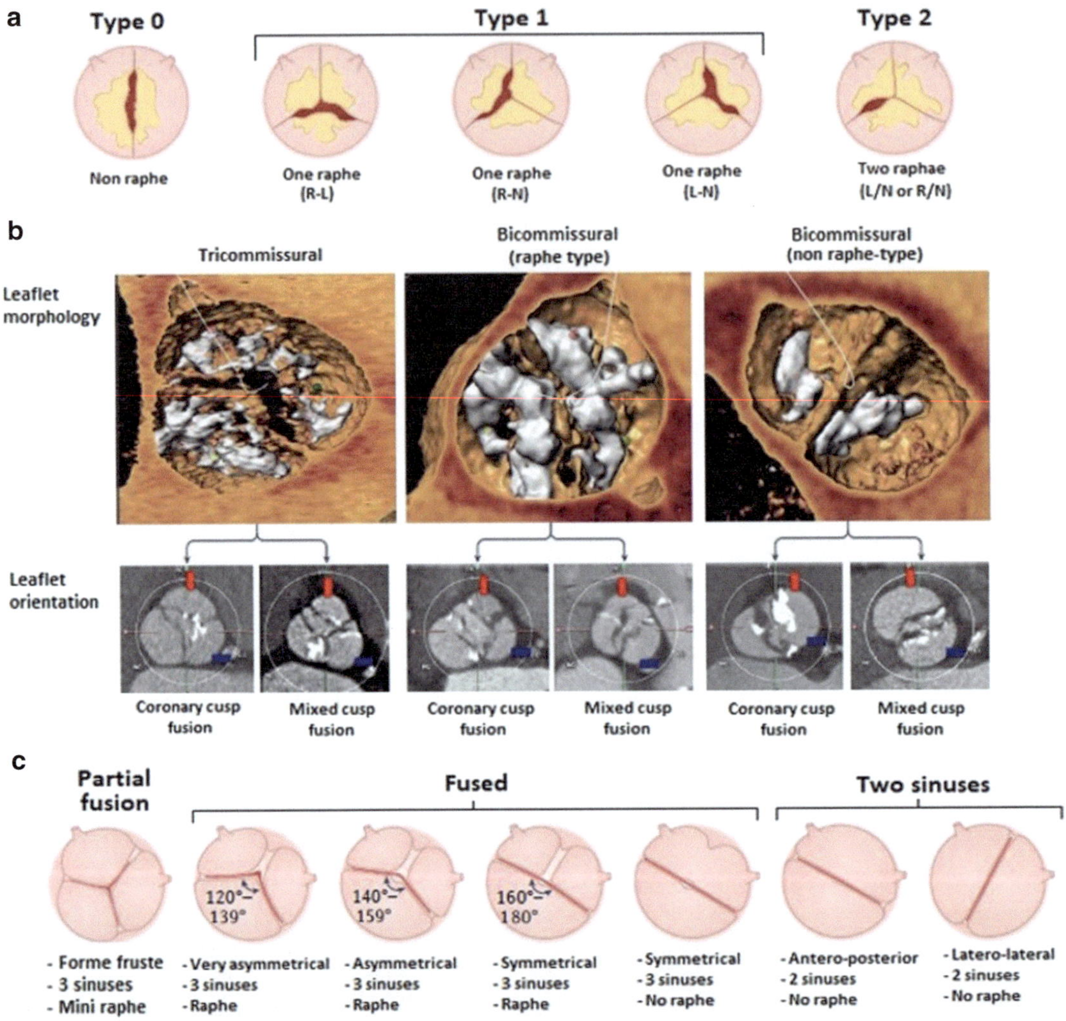

Fig. 12.1 Classification systems for BAV. (**a**) Sievers classification system; (**b**) Jilaihawi et al. classification system; (**c**) Classification from the International Consensus Statement on Nomenclature and Classification of the Congenital Bicuspid Aortic Valve by Michelena et al. Adapted from [46], with permission

between the implanted valve and the aortic valve complex at both the basal leaflet plane (presence or absence of a raphe) and the commissural level (two or three commissures) [27]:

- **Tricommissural** (18.5%): two cusps fused at or near the commissural level, often referred to as "functional" or "acquired" BAV (arising from rheumatic or degenerative processes).
- **Bicommissural raphe type** (56.9%): two cusps fused by a fibrous or calcified ridge of various vertical heights, located at or proximal to the basal third of the sinus, without reaching the commissure.
- **Bicommissural non-raphe type** (16.2%): two cusps fused from their basal origin, with no evidence of a raphe or third commissure.

In 2021, Michelena et al. aimed to establish a common language for describing all BAV phenotypes, viewed as a continuous anatomical spectrum of "bicuspidity" related to the severity of its embryological mechanisms. Their consensus document outlines three main BAV types [28]:

- **Fused bicuspid aortic valve (90–95%)**: two of the three cusps are fused within three distinguishable aortic sinuses, resulting in two functional cusps, usually different in size and shape. A congenital fibrous ridge (raphe) is present in 70–90% of cases. Subtypes include BAV-RL (right-left cusp fusion, 70–80%), BAV-RN (right-non coronary cusp fusion, 20–25%), and BAV-LN (left-non coronary cusp fusion, ~ 2%).
- **Two-sinus bicuspid aortic valve (5–7%)**: two nearly equal-sized cusps are located within two aortic sinuses, each occupying 180° of the annular circumference. Subtypes include BAV-AP (anterior-posterior, where the left and right coronary cusps are fused), BAV-LA (latero-lateral, where the left or right coronary cusp is fused with the non coronary cusp). In BAV-AP, both coronary arteries arise from a single cusp, whereas in BAV-LA, each cusp gives rise to one coronary artery.
- **Partial-fusion bicuspid aortic valve (or forme fruste BAV, prevalence unknown)**: resembles a tricuspid aortic valve, but with less than 50% cusp fusion at the base of a commissure, forming a "mini-raphe."
- To effectively apply this classification systems in clinical practice using CTA, the following features should be evaluated [24, 28] (Fig. 12.2):

1. Number of distinguishable cusps and commissures.
2. Number of aortic sinuses.
3. Presence and the pattern of cusp fusion (L-R, R-NC, L-NC).
4. Visibility, axial length, and height of the raphe.
5. Burden and location of calcification (base/tips of leaflets, raphe, LVOT).
6. Relative cusp size (different vs roughly equal).
7. Cusp spatial orientation within the short-axis (antero-posterior vs latero-lateral).
8. Commissural angle of the non-fused cusp (160°-180°: symmetric, 140–159°: asymmetric, 120–139°: very asymmetric).
9. Valve opening pattern of the valve and systolic orifice shape (oval or "fish-mouth"shape).

Notably, the degree of asymmetry is particularly important for guiding surgical repair and

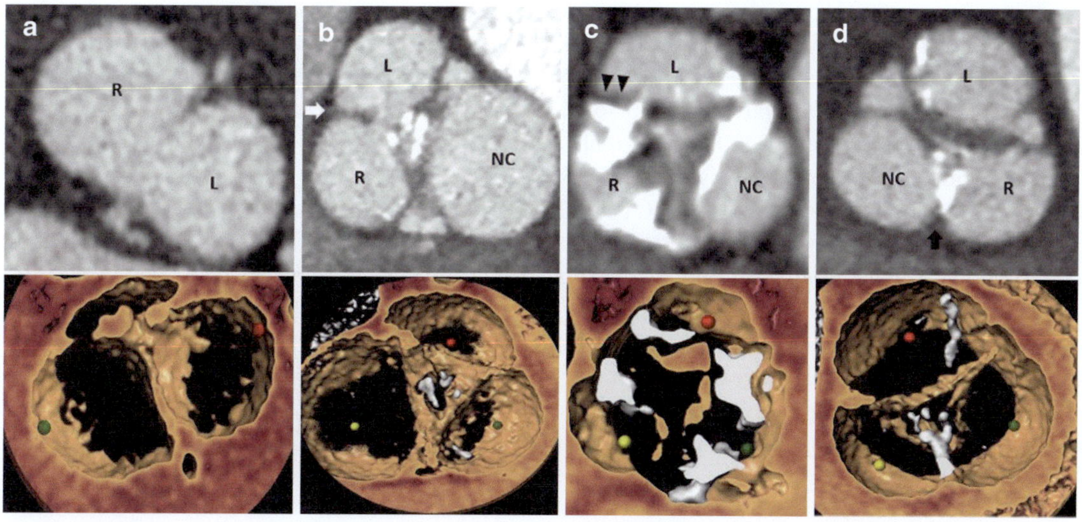

Fig. 12.2 Examples of BAV anatomy with CTA. (**a**) Two sinuses latero-lateral BAV; (**b**) Fused BAV with fibrotic raphe (white arrow) between right and left coronary cusps; (**c**) Fused BAV with calcific raphe between right and left coronary cusps (black arrowheads); (**d**) Fused BAV with calcific raphe between right and noncoronary cusp (black arrow)

predicting long-term durability. The closer the phenotype resembles a 2-sinus configuration with a symmetrical non-fused cusp commissural angle, the more feasible repair becomes. In cases of significant asymmetry, the surgeon may treat the valve as though it were tricuspid [28, 29].

12.5 Functional Evaluation of BAV

TTE is the primary tool for diagnosing and quantifying BAV-associated dysfunctions, such as stenosis or regurgitation, though CTA also provides valuable additional diagnostic information.

Valve calcification, a complex and dynamic process, influenced by genetic, inflammatory, and hemodynamic factors, typically begins at the leaflets bases and gradually extends to the margins. In BAV, the primary site of calcification is often the raphe, leading to a reduction in the aortic valve area (AVA), restricted leaflet movement, and stenosis, which occurs earlier and more prominently than in TAV. However, the degree of calcification does not correlate with specific BAV phenotypes [10, 30].

CTA is the reference modality for assessing aortic valve calcification. The aortic valve calcium score (CT-AVC), expressed as an Agatson score, can be measured similarly to coronary calcium score on non-contrast ECG-gated CT scans and correlates with stenosis severity. Pawade et al. suggest that an aortic valve calcium score above 1300 AU in females or 2000 AU in males effectively identifies severe stenosis and predicts adverse clinical outcomes. As in TAVs, CT-AVC is particularly useful for patients with borderline or discordant echocardiographic findings and in low-flow states [31].

CTA also aids in assessing aortic stenosis by measuring the aortic valve area (AVA), using images obtained at multiple levels orthogonal to the valve plane, during peak systolic valve opening.

CTA-derived AVA measurements correlate well with those obtained via transesophageal echocardiography (TEE) and outperform TEE when calcified leaflets cause acoustic shadowing, impending the visualization of the valve orifice.

Recent studies have shown that AVA measurements from CTA tend to be larger than those from echocardiography, both by direct planimetry and CT-derived LVOT area calculation, likely due to more accurate LVOT measurements in CTA. As a result, a cutoff of 1.2 cm^2 (rather than 1 cm^2) is suggested for defining severe aortic stenosis when using CTA, with good sensitivity values [10, 32]. Incorporating CTA-derived LVOT area into the continuity equation can lead to changes in stenosis gradients, potentially downgrading severity in up to 30% of patients, particularly in those with low transvalvular gradients [33].

Aortic regurgitation (AR) occurs when inadequate leaflet coaptation during diastole leads to blood backflow from the ascending aorta into the left ventricle [30]. Although CTA has a limited role in assessing regurgitation and lacks a validated method for measuring flow velocity or volume [34], it still provides useful diagnostic insights. Leaflet malcoaptation can be visualized in coronal, sagittal, and transverse CT images during the end-diastolic phase. Images reconstructed around 70% of the R-R interval allow a qualitative assessment of AR severity, with good sensitivity and specificity for moderate to severe AR [22, 30].

CTA quantification of the anatomic regurgitant orifice area (ARO) is feasible, particularly when echocardiography or CMR are unavailable. Based on Jeon et al., a CTA-based grading scheme for AR severity classifies mild AR as ARO < 0.3 cm^2, moderate AR as 0.3 cm^2 to <0.5 cm^2, and severe AR as ≥ 0.5 cm^2 [35]. Finally, CTA allows for visualization of related features such as compensatory left ventricular hypertrophy and dilatation, which can result from chronic AR.

12.6 BAV-Related Aortopathy

Bicuspid aortopathy is a complex pathological process influenced by genetic, biomechanical, and hemodinamic factors, including altered aortic flow from asymmetrical valve opening, flow eccentricity, and wall shear stress. These mechanisms lead to elastin and collagen fibers disruption, smooth muscle cell loss, metalloproteinase release, and loss of fibrillin 1 microfibrils, ultimately causing fragmentation of the medial elastic layer in thoracic aorta [36, 37]. Clinically, BAV-associated aortopathy manifests as vessel dilatation, progressing faster than in degenerative aneurysms related to TAVs, with an annual growth rate of 0.2–2 mm (mean 0.3–0.8 mm) [37].

Based on the most affected segments, three BAV-related aortopathy patterns can be identified: the most common (70%) involves preferential dilatation of the tubular ascending aorta, particularly at the anterior wall. The second (20%) affects the aortic root (sinuses of Valsalva), with possible involvement of the ventriculo-arterial junction. The third pattern includes extended or combined phenotypes, where dilatation of both the aortic root and ascending aorta, or the ascending aorta and arch, is the most frequently observed [28].

Patients with fused right-noncoronary cusps have a higher incidence of ascending aortic dilatation, whereas right-left cusp fusion often presents with aortic root involvement, sparing the arch [10]. The aortic root phenotype is typically associated with an earlier age of diagnosis and may represent a more aggressive form, characterized by faster growth rates and higher risk of adverse aortic events [37].

Aortic dissection (AD) is the most serious complication of BAV-related aortopathy. A recent meta-analysis of 4330 patients reported a cumulative incidence of 0.6% for BAV-related AD, with good long-term survival rates (97.2%, 96.7%, 92.45%, and 81.1% freedom from mortality at 1, 3, 5, and 10 years, respectively) [38]. While the overall incidence is low, the risk of AD is 8.4 times higher than in the general population, rising to 0.5% when aortic diameter exceeds 45 mm [11] (Fig. 12.3).

Current guidelines recommend aortic surgery for BAV patients when the aortic diameter is ≥55 mm. However, in cases with coarctation or additional risk factors (family history of aortic dissection or unexplained early cardiac death, spontaneous vascular dissection, desire for pregnancy, uncontrolled hypertension, severe AR, or aortic growth >3 mm/year), surgery should be considered when the diameter reaches ≥50 mm [39].

TTE remains the first-line imaging modality for evaluating the thoracic aorta. However, CTA offers more accurate and reproducible visualization of aortic anatomy and is considered the reference standard for measuring aortic dimensions. CTA is recommended for a baseline or confirmatory examination to complement echocardiographic measurements [10, 37].

CTA allows for the measurement of thoracic aorta diameters at multiple levels, enabling the assessment of dilatation, identification of aortopathy phenotypes, and evaluation of additional aortic conditions such as coarctation (which affects 7–10% of adults with BAV, compared to BAV's prevalence of 50–85% in patients with coarctation), collateral branching patterns, and signs of aortic wall destabilization, intramural hematomas, or dissections [40].

Specific measurement sites include the aortic annulus, sinuses of Valsalva, sinotubular junction, and the most dilated segment of the ascending aorta. Other locations include the aortic arch (between the brachiocephalic and left subclavian arteries), the isthmus (below the left subclavian artery), and the proximal descending aorta at the level of the pulmonary trunk. Maximum orthogonal diameters (anteroposterior and laterolateral) should be measured in a double-oblique imaging plane, perpendicular to the aorta, at end-diastole, using inner-to-inner edges. This overcomes the limitations of 2D imaging, which can underestimate dimensions, especially in cases of aortic root asymmetry, common in BAV due to dilation opposite the commissural fusion [10, 37, 41]. The sinuses of Valsalva can be measured from the commissure to the opposite sinus or using cusp-to-cusp diameters. CTA-derived sinus-to-sinus diameters are generally 2–2.5 mm larger than sinus-to-commissure measurements. Therefore, a recent consensus suggests using a double-

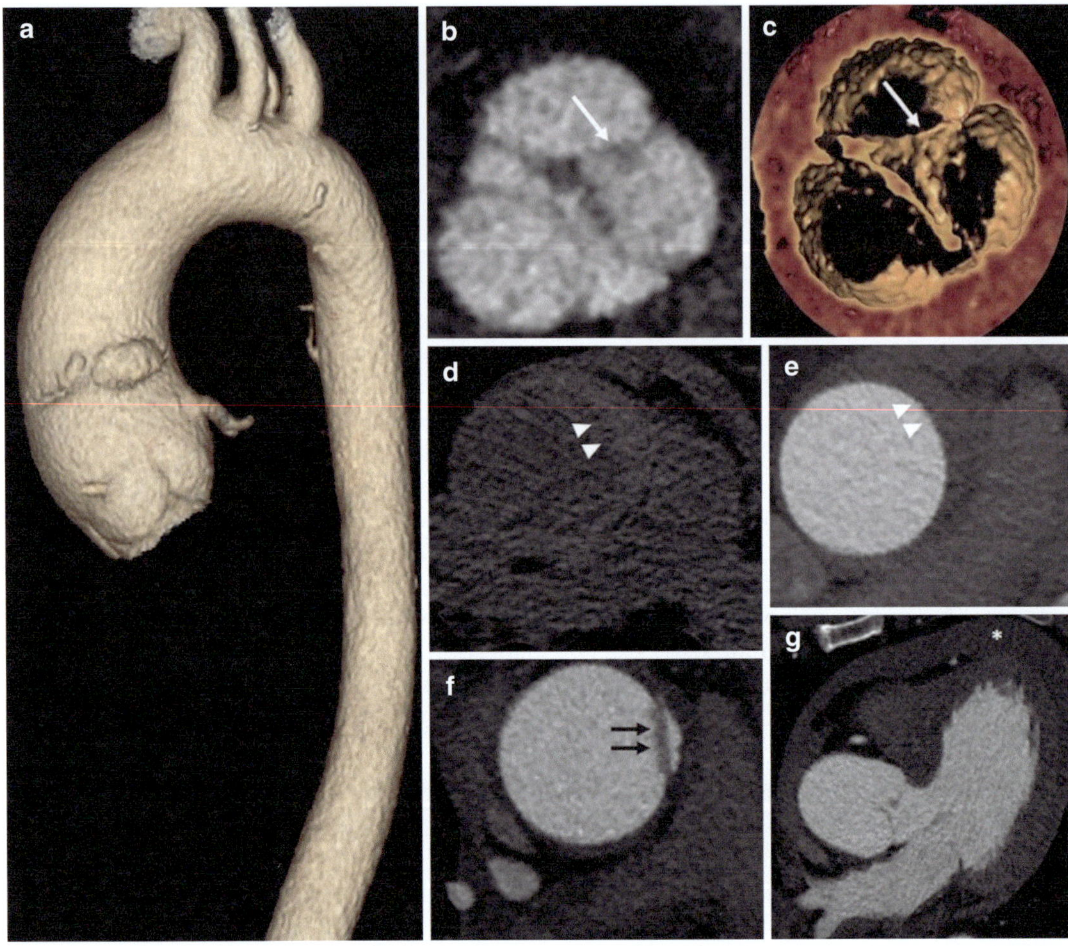

Fig. 12.3 Acute aortic syndrome in BAV. (**a**) Volume-rendering reconstruction showing combined dilatation of aortic root and ascending thoracic aorta, with a bulging of the anterior aortic wall; (**b**, **c**) Axial and 3-D views of the aortic valve, showing a BAV morphology with fibrotic raphe between right and left coronary cusps (white arrow); (**d**, **e**) Axial views showing crescent-shaped thickening of the aortic wall (white arrowheads), appearing hyperdense on the pre-contrast scan and hypodense with a smooth, regular interface with the aortic lumen on the arterial scan, suggestive of intramural hematoma; (**f**) Axial view showing an intimal dissection flap on the left wall of the middle ascending aortic wall (black arrows); (**g**) Severe circumferential pericardial effusion (white asterisk)

oblique, end-diastole, inner-to-inner edge measurement from sinus-to-sinus. In BAV cases with two sinuses, both longitudinal and transverse orthogonal diameters should be recorded [41].

Finally, the ability of CTA to simultaneously assess coronary arteries—evaluating their origin, course, and excluding coronary artery disease—reinforces its role as the gold standard for preoperative surgical evaluation of BAV-related valvulo-aortopathy, often eliminating the need for coronary angiography in many younger BAV patients [42].

12.7 Specific Considerations for TAVI in BAV

Historically, BAV anatomy, with its distinct and often unfavorable structural features, was considered as a relative contraindication for TAVI. Patients with BAV have been excluded from landmark randomized clinical trials comparing TAVI and surgical aortic valve replacement (SAVR). Early case series involving first-generation transcatheter heart valves (THVs) in high-risk BAV patients showed comparable mortality rates with TAVs, but higher rates of moderate-to-severe paravalvular leak (PVL), new permanent pacemaker implantation, prosthesis embolization, aortic injury, and conversion to open surgery.

Subsequent registries and observational studies have reported improved procedural outcomes, attributed to new-generation THVs, better imaging-based procedural planning, and increased operator experience. These advancements have brought outcomes for BAV patients closer to those seen in TAV stenosis. Over time, accumulating data have supported the safety and feasibility of TAVI in carefully selected BAV patients with aortic stenosis [43, 44].

Notably, 20–25% of severe BAV stenoses occur in octogenarians, and the proportion of BAV patients undergoing TAVI in Western populations ranges from 2% to 10%, according to single-centre and multicentre registries. As TAVI expands into younger and lower-risk populations, the number of BAV patients considered for percutaneous treatment is expected to increase [45, 46].

Currently, surgery remains the treatment of choice for BAV patients, particularly those younger than 70 years or with aortic root dilatation (≥45 mm and aged <75 years), in the absence of surgical controindications. TAVI is a reasonable alternative for patients over 70 years with intermediate or high surgical risk, provided no aortopathy requiring standalone surgery or other anatomical contraindications exist. The final decision should be made by the Heart Team, considering all relevant clinical, anatomical, and procedural factors [46].

CTA is fundamental for preprocedural TAVI planning, especially in BAV patients. A careful evaluation of valve anatomy and associated aortic and coronary anomalies, which may pose challenges during the procedure, is essential for selecting the appropriate prosthesis type and size, as well as minimizing complications [44]. Several distinct anatomical features of BAV, differing from TAV, can be assessed with CTA, each playing a critical role in TAVI planning. These include [32, 43–46] (Fig. 12.4):

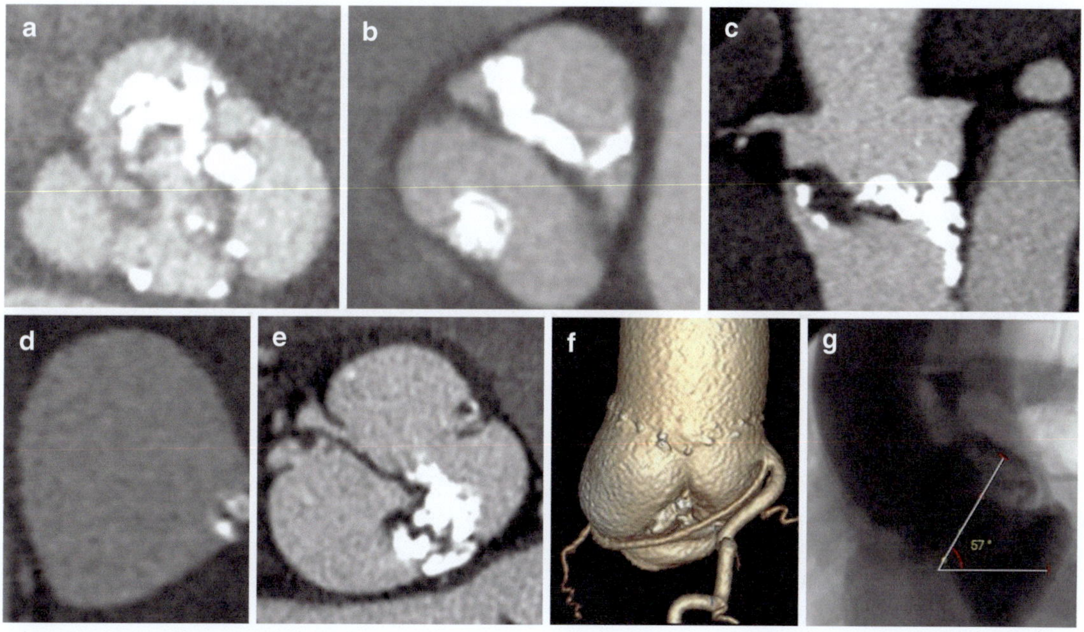

Fig. 12.4 Anatomic features of BAV challenging for TAVI procedure. (**a**) Elongated and severely calcific valve leaflets; (**b**) Severely calcific raphe; (**c**) Extensive calcifications of LVOT; (**d**) Larger and more elliptical annulus; (**e**) Larger sinuses of Valsalva; (**f**) Higher prevalence of coronary artery anomalies (anomalous left circumflex artery from the right sinus with a retroaortic course); (**g**) Horizontal aorta

- Asymmetrical and elongated valve leaflets.
- Presence of raphe.
- Larger annular dimensions, sometimes exceeding those recommended for commercial THVs.
- More elliptical configurations of the annular and supra-annular portions of the aortic valve complex, potentially resulting in an annular and supra-annular sizing mismatch, with the narrowest dimensions often located above the annular plane.
- Larger sinuses of Valsalva.
- Larger ascending aorta dimensions and/or horizontal ascending aorta (defined as an angle >49 degrees between virtual basal ring and horizontal reference line), increasing the difficulties of accurately positioning the THV within a vertical annulus.
- Increased prevalence of coronary arteries anomalies, such as left-dominant coronary circulation, separate left coronary ostia, a short left main coronary lenght (<5 mm), a high coronary take-off above the sino-tubular junction, and a two- or threefold increased risk of anomalous coronary arteries originating from the opposite sinus, with the most common being an anomalous left circumflex artery from the right sinus.
- More eccentric and heavier calcification burden at the tips and body of the leaflets, LVOT, and raphe, which can restrict the stent frame expansion and increase the likelihood of PVL.
- Shorter membranous interventricular septum, which could predispose to conduction disturbances after TAVI.

The optimal method for THV sizing remains controversial, with no single approach currently accepted for BAV patients undergoing TAVI.

Two sizing approaches using CTA are typically considered: annular and supra-annular sizing. In fused BAVs, with three distinguishable aortic sinuses, the hinge points of the three cusps can be used to defined the annulus plane as in TAVs. In non-fused BAVs, the two hinge points of the existing cusps are used, while the third point is derived from a plane perpendicular to the prosthesis landing zone. Annular measurements should include both area and perimeter (with respective derived mean diameters) for standard prosthesis sizing, as well as major and minor diameters for annular eccentricity assessment [32, 43, 47].

Supra-annular sizing is important because, in a significant proportion of BAV patients, the narrowest part of the aortic root may be located above the annulus [48]. Several supra-annular sizing methods using CTA have been proposed and are summarized in Table 12.3 and Fig. 12.5; however, further validation in prospective registries is required.

Although initial results from the prospective BIVOLUT-X registry suggest that combined annular and supra-annular sizing offers similar 30-day outcomes to conventional annular sizing [52], a recent consensus of 23 interventional cardiologists and surgeons with extensive TAVI experience recommends incorporating supra-annular structures into THV sizing algorithms for BAV patients, particularly when self-expanding valves (SEVs) are considered, resulting in a change in the selected valve size in 10–25% of cases [46].

Finally, THV selection should be individualized, considering device-specific features, patient anatomy, local availability, and operator experience. While there is no strong preference between the two major devices currently available (BEVs and SEVs) for tricuspid and bicuspid valve morphology, the aforementioned consensus recommends SEVs for BAV patients with severe leaflet calcification, BEVs for those likely to need post-TAVI coronary interventions, or in cases of a horizontal aorta, with no specific preference for device type in other anatomical scenarios [46].

Table 12.3 Supra-annular sizing methods

Intercommissural distance [48]	Proposed by Tchetche et al. in the BAVARD registry, this method involves measuring the intercommissural distance (ICD) from the midpoint of one commissure to the midpoint of the opposite one, 4 mm above the annulus, to determine the prosthesis landing zone configuration: (1) flared (ICD larger than the mean perimeter-derived diameter of the annulus); (2) tubular (ICD approximately equal to the annulus); and (3) tapered (ICD smaller than the annulus). In flared and tubular configurations, which account for 88% of patients, the annulus diameter can be used for sizing, while in tapered configurations, the ICD should be considered, as an annulus-based sizing alone could result in selecting a THV that is too large for the patient
Lira (level of implantation at the raphe) method [49]	Proposed by Iannopollo et al., this method evaluates the perimeter at both the annular level and the point of maximum raphe protrusion, with THV sizing based on the smallest of these measurements, assuming that the raphe, often stiff and heavily calcified, can interfere with proper THV expansion
Casper (Calcium Algorithm Sizing for bicusPid Evaluation with Raphe) alghoritm [50]	Proposed by Petronio et al., this multiparametric algorithm focuses on three key valvular features: (1) the valvular calcium score (greater or less than 300 mm^3); (2) the ratio between raphe length and area/perimeter-derived mean annulus diameter (greater or less than 50%); and (3) the localization of calcium relative to the raphe (predominantly at the raphe site or not), suggesting specific indications for THV downsizing relative to annular dimensions. It is based on the assumption that both calcium burden and distribution, as well as raphe length, are linked to incomplete THV expansion
Circle method [51]	Proposed by Blackman et al. for balloon-expandable valves (BEVs), this method involves drawing circles with a diameter identical to the proposed BEV, at 3 mm intervals from the annular plane up to the sinotubular junction. The goal is to assess the prosthesis' interactions with surrounding structures (leaflets, commissures, raphe, calcifications, coronary ostia) to predict sealing, expansion, and potential complications such as rupture and coronary obstruction. Three scenarios can occur: (1) if the circle is too large and extends beyond the commissures, there is a risk of annular rupture; (2) if the circle is large enough to contact the commissures, effective sealing is expected; (3) if the circle is undersized and does not reach the commissures, there is the risk of PVL or valve embolization

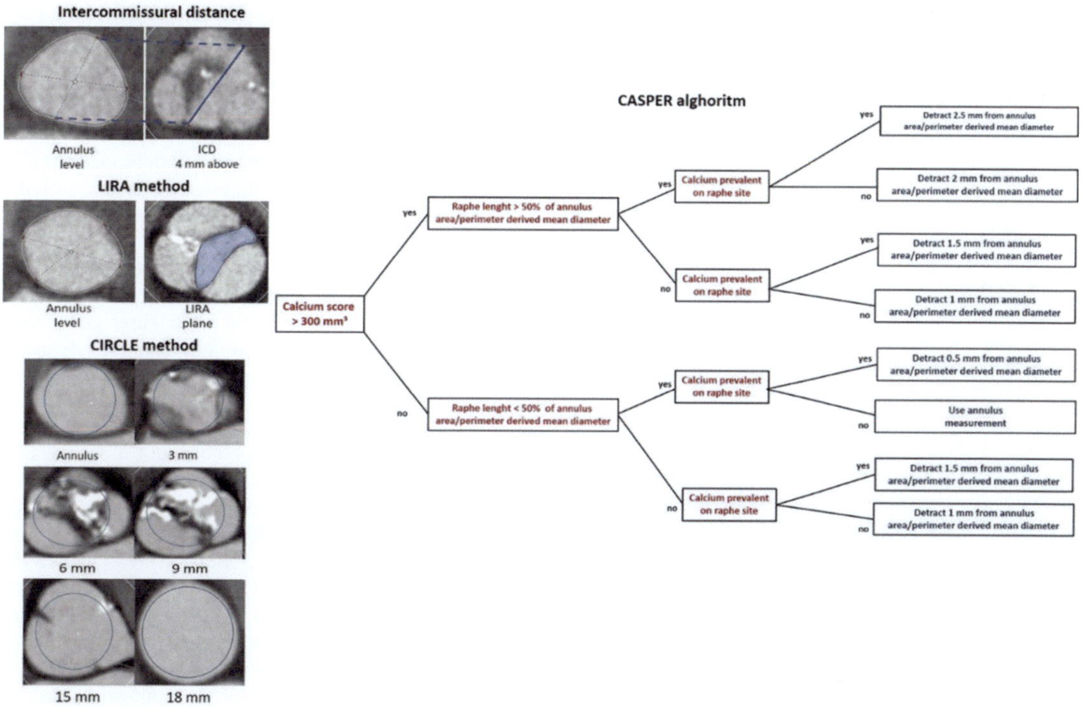

Fig. 12.5 Supra-annular sizing method for TAVI procedure in BAV

References

1. Otto CM, Nishimura RA, Bonow RO, Carabello BA, Erwin JP, Gentile F, et al. 2020 ACC/AHA guideline for the management of patients with valvular heart disease: a report of the American College of Cardiology/American Heart Association Joint Committee on Clinical Practice Guidelines. Circulation. 2021;143(5):e72–e227.
2. Michelena HI, Vallabhajosyula S, Prakash SK. Nosology spectrum of the bicuspid aortic valve condition: complex-presentation valvulo-aortopathy. Circulation. 2020;142(3):294–9.
3. Arnaoutakis GJ, Sultan I, Siki M, Bavaria JE. Bicuspid aortic valve repair: systematic review on long-term outcomes. Ann Cardiothorac Surg. 2019;8(3):302–12.
4. La Mura L, Lembo M, Musella F, D'Amato M, D'Andrea A, Izzo R, et al. Aortic regurgitation in bicuspid aortic valve: the role of multimodality imaging. J Clin Med. 2024;13:13.
5. Machowiec P, Przybylski P, Czekajska-Chehab E, Drop A. Patients with a bicuspid aortic valve (BAV) diagnosed with ECG-gated cardiac multislice computed tomography-analysis of the reasons for referral, classification of morphological phenotypes, co-occurring cardiovascular abnormalities, and coronary artery stenosis. J Clin Med. 2024;13:13.
6. Baumgartner H, De Backer J, Babu-Narayan SV, Budts W, Chessa M, Diller GP, et al. 2020 ESC guidelines for the management of adult congenital heart disease. Eur Heart J. 2021;42(6):563–645.
7. You YJ, Ko SM. Computed tomography and magnetic resonance imaging findings of bicuspid aortic valve and related abnormalities of the heart and thoracic aorta. Korean J Radiol. 2023;24(10):960–73.
8. DeFaria YD. Coronary artery variants and bicuspid aortic valve disease: gaining insight into genetic underpinnings. Heart. 2018;104(5):363–4.
9. Vallabhajosyula S. Anomalous coronary artery origin from the opposite sinus in patients with bicuspid aortic valve: comparison with tricuspid aortic valve. In: Fuchs Margaret YL-T, Medina IJ, Tajouri Tanya H, Enriquez SM, Phillips Sabrina D, Rajiv G, Klarich Kyle W, Hector M, editors. . Open Heart: BMJ; 2021.
10. Ahmad Y, Agarwal V, Williams ML, Wang DD, Reardon MJ, Cavalcante JL, et al. Imaging, treatment options, patient selection, and outcome considerations for patients with bicuspid aortic valve disease. J Soc Cardiovasc Angiogr Interv. 2022;1(6):100506.
11. Evangelista Masip A, Galian-Gay L, Guala A, Lopez-Sainz A, Teixido-Turà G, Ruiz Muñoz A, et al. Unraveling bicuspid aortic valve enigmas by multimodality imaging: clinical implications. J Clin Med. 2022;11:2.
12. Marano R, Pontone G, Agricola E, Alushi B, Bartorelli A, Cameli M, et al. Recommendations in pre-procedural imaging assessment for TAVI intervention: SIC-SIRM position paper part 2 (CT and MR angiography, standard medical reporting, future perspectives). Radiol Med. 2022;127(3):277–93.
13. Bennett CJ, Maleszewski JJ, Araoz PA. CT and MR imaging of the aortic valve: radiologic-pathologic correlation. Radiographics. 2012;32(5):1399–420.
14. Hell MM, Emrich T, Lurz P, von Bardeleben RS, Schmermund A. Cardiac CT beyond coronaries: focus on structural heart disease. Curr Heart Fail Rep. 2023;20(6):484–92.
15. Pontone G, Rossi A, Guglielmo M, Dweck MR, Gaemperli O, Nieman K, et al. Clinical applications of cardiac computed tomography: a consensus paper of the European Association of Cardiovascular Imaging-part I. Eur Heart J Cardiovasc Imaging. 2022;23(3):299–314.
16. Pulerwitz TC, Khalique OK, Leb J, Hahn RT, Nazif TM, Leon MB, et al. Optimizing cardiac CT protocols for comprehensive acquisition prior to percutaneous MV and TV repair/replacement. JACC Cardiovasc Imaging. 2020;13(3):836–50.
17. Francone M, Budde RPJ, Bremerich J, Dacher JN, Loewe C, Wolf F, et al. CT and MR imaging prior to transcatheter aortic valve implantation: standardisation of scanning protocols, measurements and reporting-a consensus document by the European Society of Cardiovascular Radiology (ESCR). Eur Radiol. 2020;30(5):2627–50.
18. Blanke P, Weir-McCall JR, Achenbach S, Delgado V, Hausleiter J, Jilaihawi H, et al. Computed tomography imaging in the context of transcatheter aortic valve implantation (TAVI)/transcatheter aortic valve replacement (TAVR): an expert consensus document of the Society of Cardiovascular Computed Tomography. J Cardiovasc Comput Tomogr. 2019;13(1):1–20.
19. Faletra FF, Ferrari E, Pedrazzini G, Ho SY, Narula J. The aortic root. In: Faletra F, Narula J, Ho S, editors. Atlas of non-invasive imaging in cardiac anatomy, vol. 2020. Switzerland: Springer-Nature; 2020. p. 21–32.
20. Paiocchi VL, Faletra FF, Ferrari E, Schlossbauer SA, Leo LA, Maisano F. Multimodality imaging of the anatomy of the aortic root. J Cardiovasc Dev Dis. 2021;8:5.
21. Michelena HI, Prakash SK, Della Corte A, Bissell MM, Anavekar N, Mathieu P, et al. Bicuspid aortic valve: identifying knowledge gaps and rising to the challenge from the International Bicuspid Aortic Valve Consortium (BAVCon). Circulation. 2014;129:2691–704.
22. Alkadhi H, Leschka S, Trindade PT, Feuchtner G, Stolzmann P, Plass A, et al. Cardiac CT for the differentiation of bicuspid and tricuspid aortic valves: comparison with echocardiography and surgery. AJR Am J Roentgenol. 2010;195(4):900–8.
23. Tanaka R, Yoshioka K, Niinuma H, Ohsawa S, Okabayashi H, Ehara S. Diagnostic value of cardiac CT in the evaluation of bicuspid aortic stenosis: comparison with echocardiography and operative findings. AJR Am J Roentgenol. 2010;195(4):895–9.
24. Takagi H, Yoshizawa M, Orii M, Kumagai A, Tashiro A, Chiba T, et al. Additive value of CT to age, aor-

tic diameter, and echocardiography in diagnosis and classification of bicuspid aortic valve in patients with severe aortic stenosis. Radiol Cardiothorac Imaging. 2021;3(1):e200423.

25. Kim J, Kim SM, Ahn J, Choe YH. Diagnostic performance of cardiac CT and transthoracic echocardiography for detection of surgically confirmed bicuspid aortic valve: effect of calcium extent and valve subtypes. J Korean Soc Radiol. 2023;84(6):1324–36.
26. Sievers HH, Schmidtke C. A classification system for the bicuspid aortic valve from 304 surgical specimens. J Thorac Cardiovasc Surg. 2007;133(5):1226–33.
27. Jilaihawi H, Chen M, Webb J, Himbert D, Ruiz CE, Rodés-Cabau J, et al. A bicuspid aortic valve imaging classification for the TAVR era. JACC Cardiovasc Imaging. 2016;9(10):1145–58.
28. Michelena HI, Della Corte A, Evangelista A, Maleszewski JJ, Edwards WD, Roman MJ, et al. International consensus statement on nomenclature and classification of the congenital bicuspid aortic valve and its aortopathy, for clinical, surgical, interventional and research purposes. J Thorac Cardiovasc Surg. 2021;162(3):e383–414.
29. Tretter JT, Spicer DE, Franklin RCG, Béland MJ, Aiello VD, Cook AC, et al. Expert consensus statement: anatomy, imaging, and nomenclature of congenital aortic root malformations. Ann Thorac Surg. 2023;116(1):6–16.
30. Hoey ET, Ganeshan A. Multi-detector CT angiography of the aortic valve-part 2: disease specific findings. Quant Imaging Med Surg. 2014;4(4):273–81.
31. Pawade T, Clavel MA, Tribouilloy C, Dreyfus J, Mathieu T, Tastet L, et al. Computed tomography aortic valve calcium scoring in patients with aortic stenosis. Circ Cardiovasc Imaging. 2018;11(3):e007146.
32. Kong WKF, Delgado V, Bax JJ. Bicuspid aortic valve: what to image in patients considered for Transcatheter aortic valve replacement? Circ Cardiovasc Imaging. 2017;10:9.
33. Weber M, Jaenisch M, Spilker M, Pingel S, Schueler R, Stundl A, et al. TAVR outcome after reclassification of aortic valve stenosis by using a hybrid continuity equation that combines computed tomography and echocardiography data. Catheter Cardiovasc Interv. 2020;96(4):958–67.
34. Lancellotti P, Pibarot P, Chambers J, La Canna G, Pepi M, Dulgheru R, et al. Multi-modality imaging assessment of native valvular regurgitation: an EACVI and ESC council of valvular heart disease position paper. Eur Heart J Cardiovasc Imaging. 2022;23(5):e171–232.
35. Jeon MH, Choe YH, Cho SJ, Park SW, Park PW, Oh JK. Planimetric measurement of the regurgitant orifice area using multidetector CT for aortic regurgitation: a comparison with the use of echocardiography. Korean J Radiol. 2010;11(2):169–77.
36. Michelena HI. Bicuspid aortopathy: toward individualized risk assessment with imaging biomarkers. JACC Cardiovasc Imaging. 2022;15(1):57–9.
37. Rodríguez-Palomares JF, Dux-Santoy L, Guala A, Galian-Gay L, Evangelista A. Mechanisms of aortic dilation in patients with bicuspid aortic valve: JACC state-of-the-art review. J Am Coll Cardiol. 2023;82(5):448–64.
38. Wilson-Smith AR, Eranki A, Muston B, Kamalanathan H, Yung A, Williams ML, et al. Incidence of bicuspid valve related aortic dissection: a systematic review and meta-analysis. Ann Cardiothorac Surg. 2022;11(4):363–8.
39. Vahanian A, Beyersdorf F, Praz F, Milojevic M, Baldus S, Bauersachs J, et al. 2021 ESC/EACTS guidelines for the management of valvular heart disease. Eur Heart J. 2022;43(7):561–632.
40. Galian-Gay L, Rodríguez-Palomares J, Guala A, Michelena HI, Evangelista A. Multimodality imaging in bicuspid aortic valve. Prog Cardiovasc Dis. 2020;63(4):442–51.
41. Evangelista A, Sitges M, Jondeau G, Nijveldt R, Pepi M, Cuellar H, et al. Multimodality imaging in thoracic aortic diseases: a clinical consensus statement from the European Association of Cardiovascular Imaging and the European Society of Cardiology working group on aorta and peripheral vascular diseases. Eur Heart J Cardiovasc Imaging. 2023;24(5):e65–85.
42. Nappi F, Giacinto O, Lusini M, Garo M, Caponio C, Nenna A, et al. Patients with bicuspid aortopathy and aortic dilatation. J Clin Med. 2022;11:20.
43. Tarantini G, Fabris T. Transcatheter aortic valve replacement for bicuspid aortic valve stenosis: a practical operative overview. Circ Cardiovasc Interv. 2021;14(7):e009827.
44. Gutierrez L, Boiago M, De Biase C, Oliva O, Laforgia P, Feliachi S, et al. Transcatheter aortic valve implantation for bicuspid aortic valve disease: procedural planning and clinical outcomes. J Clin Med. 2023;12:22.
45. Vincent F, Ternacle J, Denimal T, Shen M, Redfors B, Delhaye C, et al. Transcatheter aortic valve replacement in bicuspid aortic valve stenosis. Circulation. 2021;143(10):1043–61.
46. Xiong TY, Ali WB, Feng Y, Hayashida K, Jilaihawi H, Latib A, et al. Transcatheter aortic valve implantation in patients with bicuspid valve morphology: a roadmap towards standardization. Nat Rev Cardiol. 2023;20(1):52–67.
47. Costa G, Angelillis M, Petronio AS. Bicuspid valve sizing for Transcatheter aortic valve implantation: the missing link. Front Cardiovasc Med. 2021;8:770924.
48. Tchetche D, de Biase C, van Gils L, Parma R, Ochala A, Lefevre T, et al. Bicuspid aortic valve anatomy and relationship with devices: the BAVARD multicenter registry. Circ Cardiovasc Interv. 2019;12(1):e007107.
49. Iannopollo G, Romano V, Buzzatti N, Ancona M, Ferri L, Russo F, et al. Supra-annular sizing of transcatheter aortic valve prostheses in raphe-type bicuspid aortic valve disease: the LIRA method. Int J Cardiol. 2020;317:144–51.

50. Petronio AS, Angelillis M, De Backer O, Giannini C, Costa G, Fiorina C, et al. Bicuspid aortic valve sizing for transcatheter aortic valve implantation: development and validation of an algorithm based on multi-slice computed tomography. J Cardiovasc Comput Tomogr. 2020;14(5):452–61.
51. Blackman D, Gabbieri D, Del Blanco BG, Kempfert J, Laine M, Mascherbauer J, et al. Expert consensus on sizing and positioning of SAPIEN 3/ultra in bicuspid aortic valves. Cardiol Ther. 2021;10(2):277–88.
52. Tchétché D, Ziviello F, De Biase C, De Backer O, Hovasse T, Leroux L, et al. Transcatheter aortic valve implantation with the Evolut platform for bicuspid aortic valve stenosis: the international, multicentre, prospective BIVOLUTX registry. EuroIntervention. 2023;19(6):502–11.

Part IV

Treatment

Surgical Treatment of Bicuspid Valve Syndrome

13

Uberto Bortolotti and Aldo Domenico Milano

13.1 Introduction

The bicuspid aortic valve (BAV) is the most frequent, albeit substantially rare, isolated congenital valvular disease, occurring in approximately 2% of the general population and affecting mostly males [1, 2]. Although the steady use of echocardiographic screening presently allows an early diagnosis of BAV in childhood, it is currently believed that prevalence of BAV in the general population may be underestimated since many BAV patients are long asymptomatic; symptoms are generally absent early in life and appear later when BAV-related complications become clinically evident. Indeed, a broad spectrum of complications associated with BAV has been described and the natural history of BAV indicates that valve function may be preserved for many years but structural deterioration of aortic cusps eventually occurs leading to either valve incompetence or prevalent stenosis, the latter being considered as the most common form of symptomatic BAV-related disease, with possible calcification of the valve components [3].

Since besides valvular involvement in BAV subjects there is an increased prevalence of aortic abnormalities, ranging from reduced aortic elasticity to aortic dilatation, aneurysm, and dissection, BAV is generally considered as a syndrome [4]. It has also been demonstrated that BAV is a genetically transmitted disease with various genetic mutations being involved [5, 6].

In the past, most patients with symptomatic BAV dysfunction were treated by prosthetic aortic valve replacement (AVR); however, a better knowledge of BAV anatomy and pathophysiology has stimulated an increasing interest to develop conservative techniques to repair incompetent or stenotic BAVs, thus avoiding AVR and all possible prosthesis-related complications.

This chapter focuses of the surgical treatment of BAV reviewing the most frequent operative strategies with particular emphasis on those utilized for aortic valve repair.

13.2 Background

BAV was most likely identified and described for the first time by James Paget in 1844 [7]. Later, William C. Roberts, analyzing 85 autopsy cases, classified BAV according to valve function: functionally normal, with pure regurgitation and with prevalent stenosis [8]; he also recognized from the anatomical viewpoint two basic types based

U. Bortolotti (✉)
Cardio-Thoracic and Vascular Department, University Hospital, Pisa, Italy

A. D. Milano
Division of Cardiac Surgery, University Hospital, Bari, Italy
e-mail: aldo.milano@uniba.it

G. Thiene et al. (eds.), *Bicuspid Aortic Valve Syndrome*,
https://doi.org/10.1007/978-3-031-85589-4_13

on the presence or absence of a *raphe*, each one with various possible orientations of cusps and commissures.

Subsequently, other classifications were proposed the most popular one being that formulated by Sievers and Schmidtke in 2007, based on the observations on a large number of surgical specimens [9]. They evaluated three characteristics of BAV, namely the number of *raphes*, the spatial position of cusps or *raphes*, and the functional status of the valve. Accordingly, three major BAV types were identified: *type 0* with no *raphe*, *type 1* with one *raphe*, and *type 2* with two *raphes* (Fig. 13.1). Most frequently a type 1 BAV was found, with the *raphe* positioned between the left and right coronary sinuses and with a hemodynamic predominant stenosis.

Recently, a consensus statement on BAV nomenclature and classification endorsed by several International Societies proposed a new classification of BAV considered more useful than the previous one for clinical, surgical, interventional, and research purposes [10]. This new classification considers three BAV types: the *fused BAV*, the *2-sinus BAV,* and the *partial-fusion BAV*, each of them with a specific phenotype. The *fused BAV* is the most common type found in up to 95% of cases and is characterized by fusion of 2 of the 3 cusps within 3 aortic sinuses but resulting in 2 functional cusps generally of different sizes. In this category, three different phenotypes are recognized: right-left cusp fusion, right-non coronary cusp fusion, and left-non coronary cusp fusion, the first one being the most common. The *2-sinus BAV* is characterized by 2 cusps almost equal in size and shape which occupy half of the circumference of the aortic annulus with 180° commissural angles; this type, without any *raphe*, is the less frequent one. Finally, the *partial-fusion BAV,* also termed *forme fruste BAV,* resembles a typical tricuspid aortic valve with three symmetrical cusps with a systolic triangular opening and commissural angles of 120°; cusp fusion of less than 50% is noted at the base of a commissure forming a small "mini-*raphe*."

All proposed classifications are intended mostly for surgical purposes, to facilitate selection and undertaking of the optimal reparative approach.

13.3 Surgical Options in BAV

13.3.1 Aortic Valve Replacement

A severely calcified stenotic BAV needs to be replaced since diffuse cusp calcification precludes a durable repair. Occasionally, either a stenotic or incompetent BAV with calcific deposits may be repaired by aortic annulus decalcification

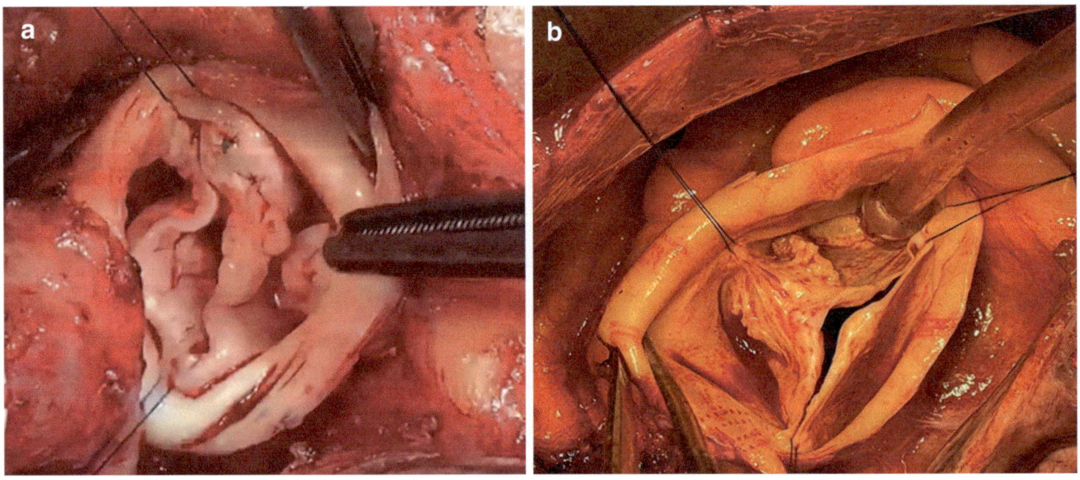

Fig. 13.1 Intraoperative view of bicuspid aortic valves, type 0 (**a**) and type 1 (**b**) according to Sievers and Schmidtke [9], based on the presence of a *raphe*

associated to excision of part of the calcified cusp which is then replaced with a tailored patch of pericardium [11]. However, this type of repair is generally quite complex and although the early results are encouraging long-term data are lacking.

In most cases of calcified BAV, prosthetic AVR is performed using standard surgical strategies. The aortic valve can be approached through a median sternotomy, a mini-sternotomy, or with a minimally invasive thoracoscopic approach, according to each surgeon preference. In any case, operation is undertaken under cardiopulmonary bypass, generally with moderate hypothermia and cardioplegic cardiac arrest. The aortic valve is excised and the annulus is cleared from all calcific infiltrates. Prostheses are implanted using interrupted stitches, preferably reinforced by subannular felts, and secured to the annulus. In the presence of two commissures with variable orientations, when using a stented bioprosthesis, care is taken to position the prosthetic struts so that the coronary ostia are not obstructed. Alternatively, a sutureless or a rapid deployment bioprosthesis, once considered contraindicated for the BAV anatomy, can be used since current experience shows that they can be easily implanted also when replacing a BAV [12]. Stentless bioprostheses have also been used for AVR in patients with BAV; however, compared to AVR for tricuspid aortic valves, BAV patients had a higher reoperation rate suggesting caution in using such devices in this setting [13].

13.3.2 Isolated Aortic Valve Repair

Aortic valve repair for aortic insufficiency (AI) has provided excellent early and medium-term results in the presence of a tricuspid aortic valve [14]. In most cases AI is caused by cusp prolapse which is considered as the ideal pathological substrate to achieve an optimal repair and this pertains also to BAV. In repairing a BAV, techniques similar to those used for tricuspid AI are employed. In case of cusp prolapse, central plication of the prolapsing cusp using 5/0 or 6/0 polypropylene sutures is performed (Fig. 13.2). The aortic attachment of the raphe can be cut to achieve improved cusp mobility or the raphe can be cut and the two halves of the cusp reattached with a continuous polypropylene suture (Fig. 13.3). Alternatively, cusp resuspension and reinforcement may be obtained with 6/0 or 7/0 Gore-Tex running sutures (*W. L. Gore and Associates, Flagstaff, AZ*) passed along the entire length of the free margin [15], a technique which appears to be particularly useful also to close cusp fenestrations without adversely affecting valve durability [16] (Fig. 13.4). The free margin of the cusp is shortened by applying tension on both Gore-Tex suture arms, which are locked outside the commissures after appropriate correction is reached. In case of severe prolapse, the excess tissue can be resected after plication and when part of a cusp has to be excised it can be replaced by a tailored pericardial patch. In the presence of mild calcific deposits, either cusp or *raphe* decalcification can be performed if cusp pliability is not compromised; triangular cusp resection can also be required and a pericardial patch is used for cusp augmentation when inadequate tissue remains. Subcommissural plication can be added which reduces the width of the interleaflet triangle, improves cusp coaptation, and stabilizes the ventriculo-aortic junction [17]. Others suggest systematic plication of the sino-tubular junction in patients with a normal aortic root, with pledgeted 4-0 polypropylene sutures at the level of the commissures, to maintain an ideal relationship between the annulus and the sinotubular junction [15].

Techniques to stabilize the aortic annulus have been proposed which include internal and external annuloplasty. A circular annuloplasty technique uses an internal continuous mattress suture that passes through the aortic wall at the leaflet attachment site around the entire valvular circumference and, once tied, imbricates the aortic wall. The external aortic annuloplasty is obtained with a flexible ring or a Dacron band to reduce the ventriculoaortic junction dilatation. Recently, a geometric annuloplasty ring has been developed to be used in BAV repair in addition to previously described maneuvers to prevent future annular dilatation [18]. Early results are

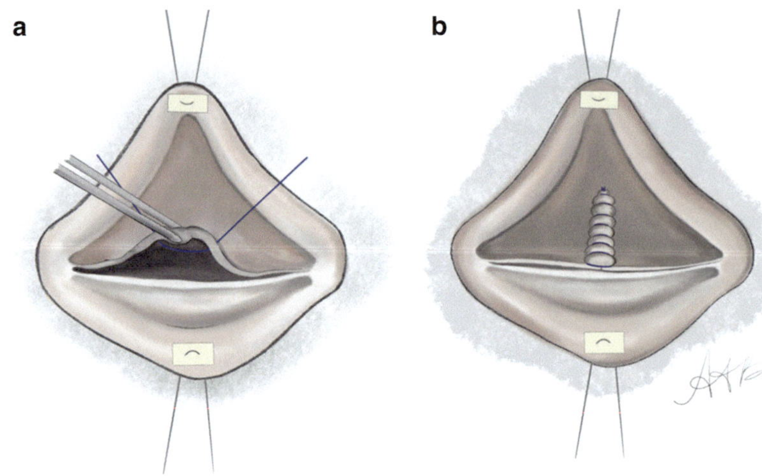

Fig. 13.2 Plication of a prolapsing cusp. (**a**) After the amount of excessive cusp tissue is identified, central plication is performed with a polypropylene suture (**b**). *Courtesy of A.T. Bortolotti*

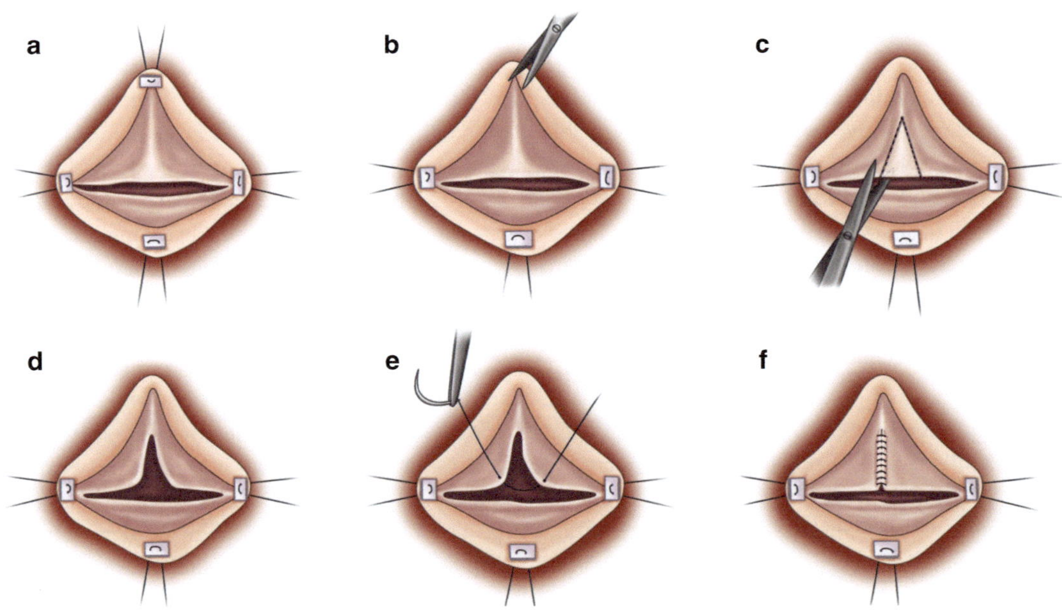

Fig. 13.3 A bicuspid aortic valve is exposed using stay stitches (**a**). The *raphe* can be detached from the aortic wall insertion to gain more cusp mobility (**b**) or it can be cut eliminating excessive cusp tissue (**c**, **d**) and the two halves of the conjoined cusp approximated with a continuous suture line (**e**). (**f**) The bicuspid aortic valve at the end of the procedure. *Courtesy of A.T. Bortolotti*

encouraging indicating that annular diameter was effectively reduced which allowed to recruit more central leaflet tissue, obtaining a better cusp coaptation without the need for pericardial augmentation. In any case, current experience indicates that BAV repair is an appealing technique for restoring valve competency avoiding the risks and complications associated to prosthetic AVR and with adequate long-term results and repair stability up to 15 years postoperatively [19–21].

13.3.3 Management of BAV and Aortopathy

Over 50% of patients with either a normal, stenotic or incompetent BAV often develop dilatation of the aortic root, the ascending aorta or both and are considered at risk for aortic rupture or dissections [10, 22]. According to the presence of aortopathy in BAV patients, three specific phenotypes are recognized based on the

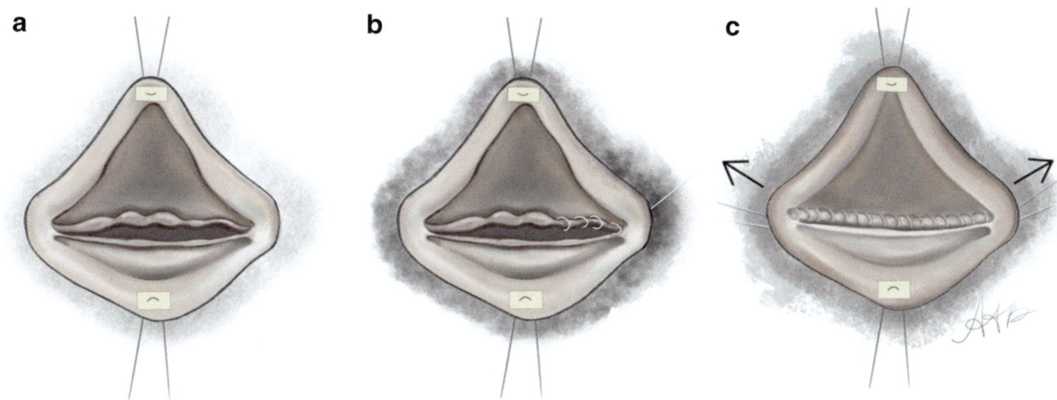

Fig. 13.4 This caption should be modified as follows: (**a**) Bicuspid aortic valve with a prolapsing leaflet: (**b**) Reinforcement of the free margin of the prolapsing cusp using expanded polytertrafluoroethylene sutures; (**c**) At the end the sutures are pulled outside and knotted applying tension (*arrows*), so that the free margin of the cusp matches the length of the opposite one

site and extension of aortic dilatation: the ascending (aorta) phenotype (70% of cases), the root phenotype (20% of cases), and the extended (root and ascending aorta) phenotype (not >10% of cases). This classification should aid in the surgical indications and type of procedure according to recent guidelines [23].

An aortic dilatation, root or ascending aorta, of ≥50 mm in a patient with BAV and AI or a root phenotype should represent an indication for combined treatment of the aortic valve and aorta; indeed, the root phenotype has been associated with greater rates of acute aortic dissection especially in patients with BAV who had undergone simple AVR compared to the ascending phenotype [24]. In BAV patients with the most common aortopathy phenotype (ascending dilatation) with a normal or only mildly dilated root/arch, replacement of the tubular portion alone will suffice and further extension of repair to the root or distal aorta and arch are deemed not necessary [23]. This issue, however, still represents a matter of debate among supporters of a limited surgical procedure and those who advocate a more aggressive strategy. It has been shown that in patients with BAV and dilatation of the ascending aorta, AVR and supracoronary replacement of the aorta provided results similar to those of a Bentall procedure, with stability of the repair and low risk of aortic-related events in the long-term [25]. Alternatively, the wrapping technique, with bovine pericardium or fabric material, has also been considered a valid option for treating a moderately dilated ascending aorta in selected patients with BAV undergoing AVR [26, 27] (Fig. 13.5). Others have shown that in the presence of a normal sized aorta (≤40 mm) isolated AVR has a protective effect against late aortic events [28].

It is also well known how BAV aortopathy is based on genetic and hemodynamic factors and progressive aortic dilatation can occur in BAV patients with normal valvular function or after AVR, all which suggest inherent lesions of the arterial wall of genetic etiology [29]. Additional studies have demonstrated that BAV aortopathy is associated to specific morphologic types of BAV and to different hemodynamic patterns [29, 30]. A recent report by Roberts et al. compared the structure of the aortic valve and of the operatively excised portion of aneurysmal ascending aorta in patients with aortic stenosis versus those with pure regurgitation showing that in the aortic wall of those with AI there was a significantly higher loss of aortic medial elastic fibers [31]. All these findings indicate that a patient with a stenotic BAV and moderate dilatation of the ascending aorta should be treated more conservatively compared to the same patient with pure BAV regurgitation; indeed, by eliminating the hemodynamic effect of an eccentric jet of the ste-

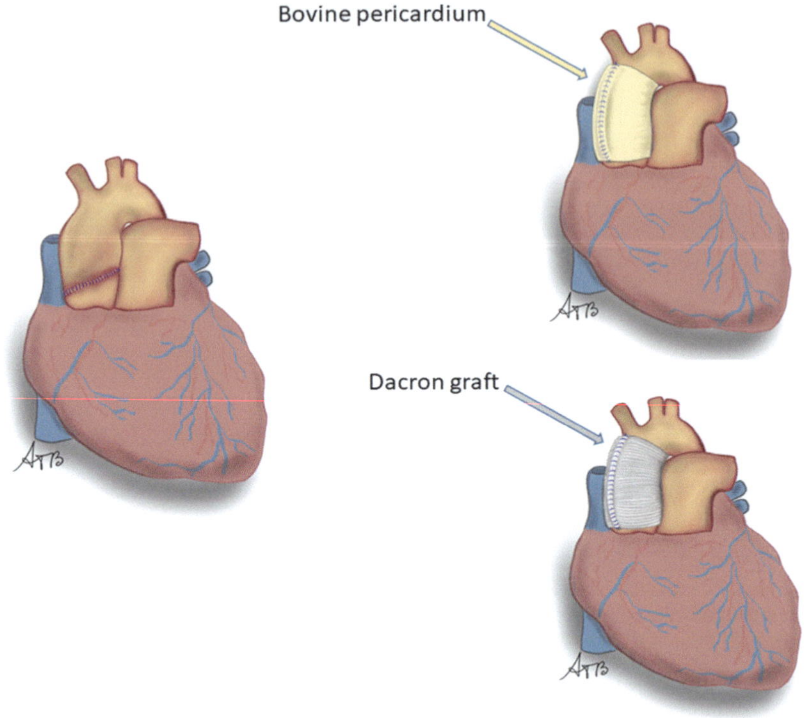

Fig. 13.5 Technique used for wrapping of a moderately dilated ascending aorta after aortic valve replacement. Further dilatation of the aorta can be prevented by using either a sheath of bovine pericardium or a segment of a tubular Dacron graft. *Courtesy of A.T. Bortolotti*

notic BAV, the risk of further aortic dilatation and aortic events can be reduced [25].

Patients with a stenotic or incompetent BAV deemed not reparable with aortic dilatation may undergo combined AVR and aortic root/ascending aorta replacement. This procedure is classically performed with the modified Bentall operation which consists in excision of the aortic valve and dilated aorta with detachment and adequate mobilization of two aortic wall buttons around the coronary ostia [32, 33]. A fabric valved-conduit, containing either a mechanical or a biological prosthesis, is than inserted on the aortic annulus followed by separate reattachment of the coronary ostia to the graft. The distal end of the graft is then sutured to the ascending aorta below the aortic cross-clamp to reestablish aortic continuity.

Originally, in the technique first described by Bentall and De Bono, the coronary ostia were directly sutured to the fabric graft and not detached from the dilated aortic wall which, at the end of the procedure, was wrapped around the neoaorta for hemostatic purposes [34]. This caused in many cases undue tension on the coronary anastomoses, eventually leading to suture dehiscence with pseudoaneurysm formation. This complication has been virtually eliminated by modification of the original technique with wide mobilization of the coronary buttons and their suture to the graft [35].

As most operations which require multiple sutures also the Bentall procedure has the inherent risk of postoperative bleeding. To reduce such risk, we have added some technical modifications, which have proved effective in improving hemostasis, and include (1) a supplemental suture between the aortic wall remnants and the prosthetic sewing ring and (2) isolation of the coronary ostia surrounded by large aortic wall buttons which allow their suture to the graft in a double layer fashion [36] (Figs. 13.6 and 13.7).

The same technique may be used when a not reparable BAV is associated to acute type A aortic dissection. In this case an open distal anastomosis is recommended after reconstruction of the aortic layers during a brief period of circulatory arrest under continuous cerebral perfusion; when also the aortic arch has to be replaced, a quadrifurcated graft is generally employed, to recon-

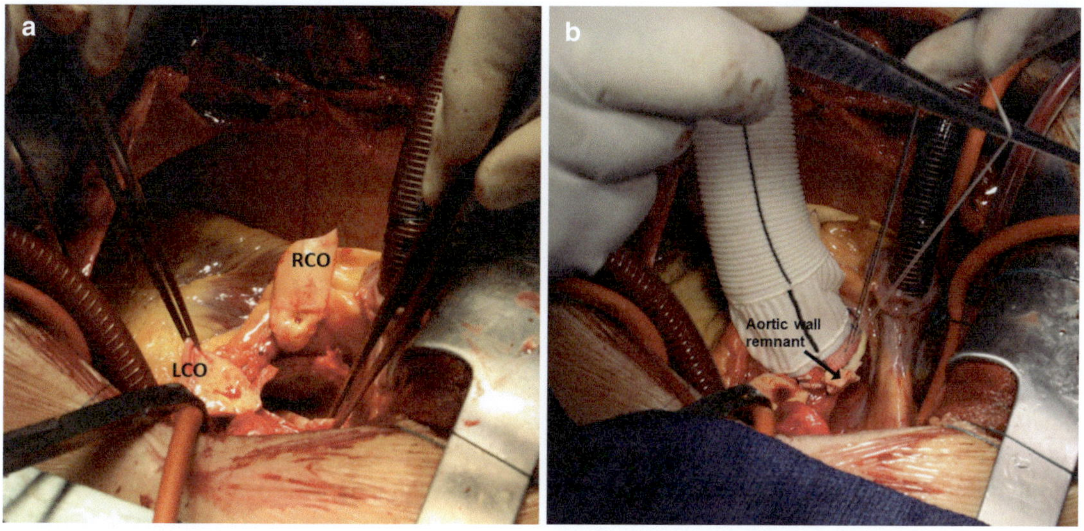

Fig. 13.6 Technique of the modified Bentall procedure. (**a**) The aortic valve is explanted and the aortic root and ascending aorta are excised; the right (RCO) and left (LCO) coronary ostia are isolated surrounded by adequate aortic tissue. (**b**) A valved conduit, used to replace both the valve and ascending aorta, is parachuted on the aortic annulus and the sutures tied

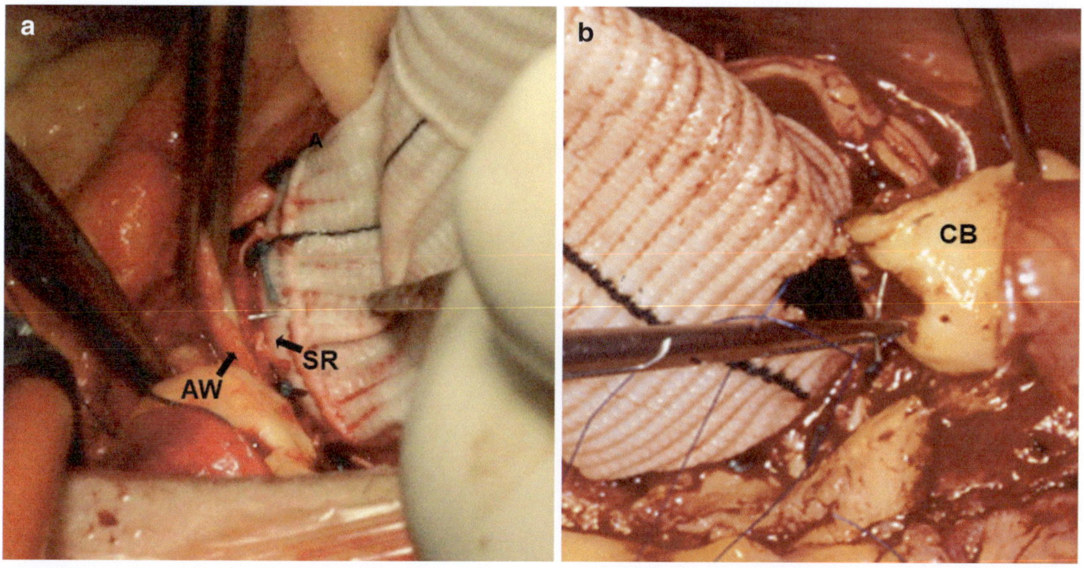

Fig. 13.7 (**a**) After the conduit is sutured an additional continuous suture joins the aortic wall (AW) remnant to the prosthetic sewing ring (SR) for hemostatic purposes. (**b**) The coronary button (CB) is then reattached using a double layer suture to a hole made in the graft using electrocautery

nect all the epiaortic vessels, and preferably using a frozen elephant trunk technique.

The modified Bentall procedure may be performed with a mechanichal or biological valved-conduit. The choice of the type of conduit much depends on the patient's age, since in younger subjects, mechanical prostheses are generally preferred having shown an extended durability over 20 years with excellent survival and freedom from major valve-related complications

[37]. Biological conduits are generally reserved to elderly patients but they currently may represent an attractive solution also in young subjects to avoid chronic anticoagulation; indeed, in case of future conduit valve failure, a valve-in-valve procedure can be planned or the use of sutureless bioprostheses can be considered having demonstrated to simplify a complex and hazardous reoperation [38].

A possible attractive alternative to the classic modified Bentall operation may be represented in this setting by the Ross procedure, where the dysfunctioning aortic valve is replaced by the patient's pulmonary valve and the pulmonary valve replaced with a pulmonary homograft obtained from a healthy donor [39]. Nevertheless, there is generally hesitation towards this operation particularly because of its complexity and the risk of perioperative complications; furthermore, the use of the pulmonary autograft in BAV patients has been questioned by concerns over inherent structural wall abnormalities in the pulmonary trunk, potentially leading to dilatation and an increased failure rate of the autograft [40]. However, excellent results of the Ross operation have been recently reported with a 20-year freedom from reoperation and from significant residual AI of 85% in adults with BAV presenting with pure AI by using specific surgical modifications to provide adequate support of the neo-aortic root to prevent late dilatation [41, 42].

In patients with an incompetent BAV and root dilatation, a valve sparing operation or a remodeling procedure may be indicated if the aortic valve is reparable, both in an elective or acute setting. This can be performed either with a David or a Yacoub procedure which have emerged as the preferred approaches in aortic surgery, preserving the native aortic valve and thereby conferring significant prognostic benefits to the patient (Fig. 13.8).

The technique originally described by David and Feindel in 1992 [43] was intended to repair an anatomically normal but incompetent tricuspid aortic valve due to aortic root dilatation. The David technique includes excision of the aortic sinuses, detachment of the coronary ostia and reimplantation of the aortic valve inside a Dacron graft which originally was a straight tube more recently replaced by a commercially manufactured graft reproducing the aortic sinuses [44]. Alternatively, Sarsam and Yacoub [45] proposed the remodeling procedure as aortic valve-sparing operation, where the aortic sinuses are also excised and a tubular Dacron graft is tailored to recreate the aortic sinuses, and an external annuloplasty can be added to correct or prevent annular dilatation. The current trend in valvular surgery, to repair instead of replace, has been increasingly applied also in BAV patients showing promising results also in this setting.

Since both valve sparing techniques are quite complex operations, preoperative assessment of valve morphology with 2D echocardiography is critical but, most importantly, a careful intraoperative evaluation of the aortic valve and precise understanding of the mechanism of valve dysfunction are of paramount importance to dictate the surgical strategy being prerequisites for a successful repair.

The congenitally fused cusp of a bicuspid valve is smaller than two cusps of a normal valve; moreover, its insertion is higher than the insertion of the larger noncoronary cusp and the rudimentary commissure between the two parts of the fused cusp is lower than the two commissures of the noncoronary cusp [46]. In addition, the depth of the sinuses differs significantly, the noncoronary sinus being unusually deeper, making appropriate realignment a complex task [47]. Another important feature is related to commissural orientation which, according to each specific BAV phenotype, may vary from symmetrical to very asymmetrical [48]. While cases of symmetrical commissural orientation may be corrected by simple cusp plication, more complex phenotypes may require additional maneuvers such as sinus plication, closure of the non-fused part of the raphe, or commissural reconstruction with pericardial patches [48].

Once the valve is repaired, the aortic root is replaced with a vascular graft. To obtain this a remodeling technique is suggested to better preserve the root asymmetry [46]; others, however, support a reimplantation approach which is considered to be superior since the entire aortic valve

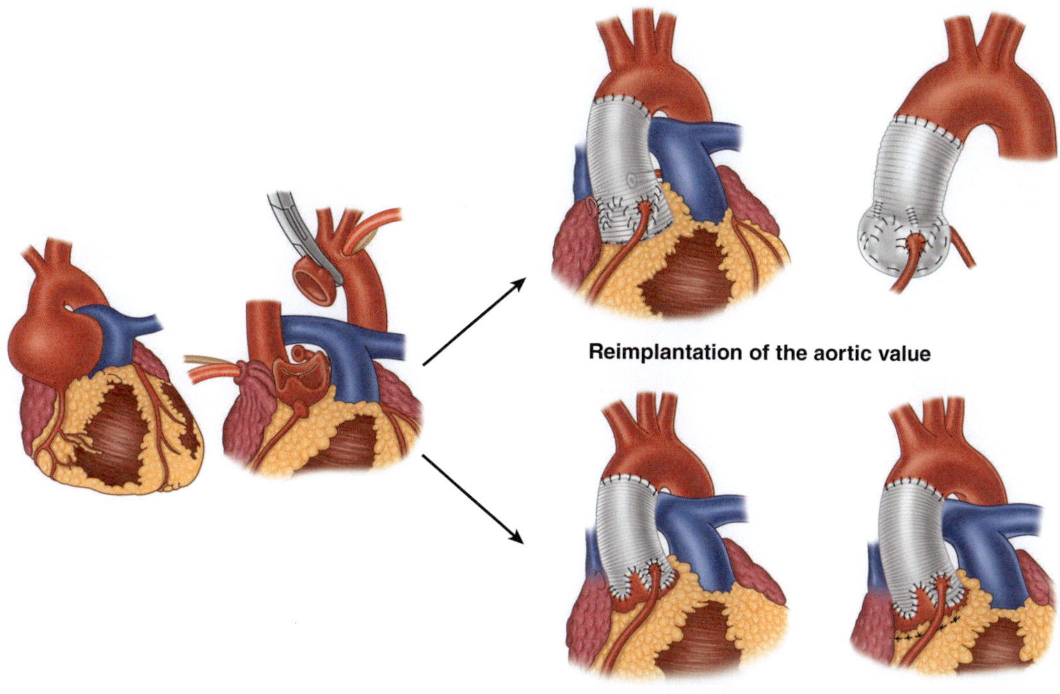

Fig. 13.8 Aortic valve sparing procedure using the David (*upper right*) or the Yacoub (*lower right*) procedures. The aortic sinuses are excised. In the reimplantation (David) procedure, the aortic valve is implanted inside a straight Dacron graft. Neoaortic sinuses can be created by plication of intercommissural spaces or, alternatively, a commercially available Valsalva graft incorporating aortic sinuses (*Terumo Aortic, Renfrewshire, UK*) can be used. In the remodeling procedure (Yacoub), a tubular Dacron graft is tailored to recreate the aortic sinuses, and an external annuloplasty can be added to correct or prevent annular dilatation. (*Reproduced with permission by Elsevier from: David TE. Aortic valve-sparing operations. Ann Thorac Surg. 2024; 117:45-53*)

apparatus is supported within the prosthetic graft [49, 50] (Fig. 13.9). Occasionally, when dilatation of only one aortic sinus is present, a simpler procedure may be indicated. In such a case replacing only the dilated sinus using a patch obtained by the same graft used to replace the ascending aorta may be quite effective also in the long-term [51] (Fig. 13.10).

Regardless of the technique used, at the end of the procedure, careful evaluation of the aortic valve is performed with transesophageal echography to assess valve continence and cusp coaptation height and length. In case of unsatisfactory repair, a second pump run is required for an optimal surgical result.

A recent review has demonstrated durability and safety of both David and Yacoub valve-sparing procedures during a long-term follow-up in BAV patients [47]. However, it has also evidenced, in patients with BAV aortopathy, a clear preference for the resuspension technique since it is considered superior to the remodeling approach mostly because it provides a better annular stabilization and includes all the aortic tissue within the graft without leaving unsupported residual external aortic root tissue which may be prone to further dilatation [47]. Indeed, recent modifications have been included also in the remodeling technique with the addition of an external aortic ring annuloplasty [52].

Long-term results for both remodeling and resuspension valve sparing techniques are available for a considerable number of patients having repair of an incompetent tricuspid valve while data on freedom from reoperation are reported for a more limited number of patients with BAV

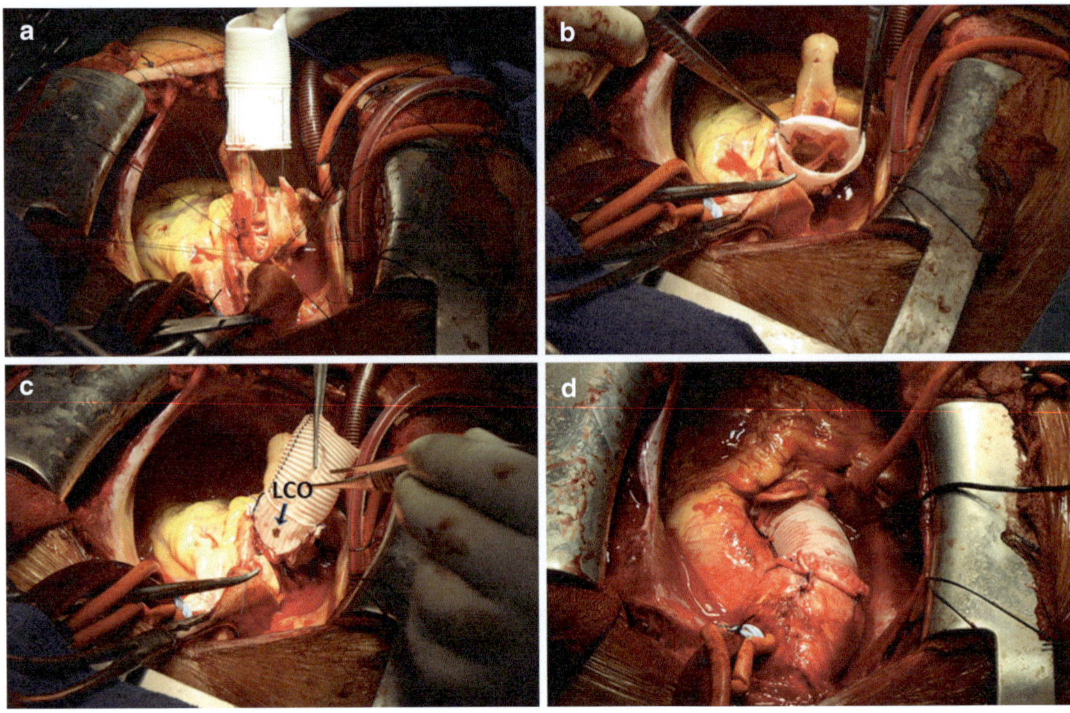

Fig. 13.9 Sequential steps of a valve-sparing repair of a bicuspid valve with aortopathy using the resuspension technique. (**a**) After the aortic valve has been repaired, the vascular graft is parachuted towards and sutured externally to the aortic annulus. The aortic valve is then resuspended by suturing it inside the graft wall (**b**). The coronary ostia are separately reattached to the graft sinuses after adequate mobilization (**c**). Finally, the graft is sutured to the distal aortic stump to recreate the aortic continuity (**d**). *LCO* Left coronary ostium

repair. Moreover, the results of such operations are generally reported cumulatively without separating BAV patients from those with a normal aortic valve. Nevertheless, recent experiences indicate a 91% freedom from reoperation and 97% freedom from moderate AI at 12 years with the David procedure for BAV repair [50] and a 100% freedom from reoperation at 10 years with the remodeling technique [51].

13.4 Conclusions

BAV and associated aortopathy represent an important issue from the clinical, diagnostic and surgical point of view. The true incidence of BAV in the general population may be underestimated and this can have important implications considering the reported increased risk of aortic complications in these patients [1]. Since most of patients with BAV and AI are young with long life expectancy, early diagnosis is vital and adequate repair aimed to provide long-term freedom from reoperations and life-long anticoagulation.

In most subjects presenting with isolated BAV or BAV aortopathy, the often surgical complexity of the repair demands that such patients should be referred only to operators experienced in the management of BAV aortopathy in high-volume centers to minimize as much as possible the likelihood of future reoperation.

Among the various possible technical solutions summarized in this chapter, a BAV-repair and reimplantation oriented approach appears to be currently the preferred surgical strategy in this challenging subset of patients. In any case, the variety of techniques adopted and validated in BAV repair indicates that an individualized patient approach should be considered whenever repairing such valves.

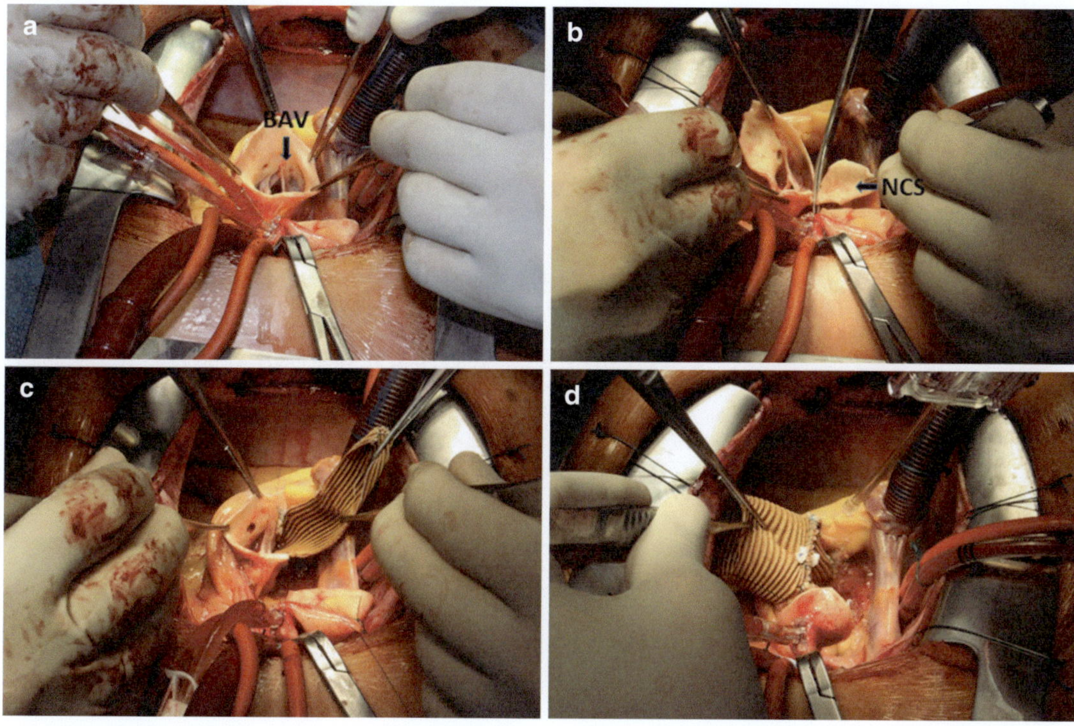

Fig. 13.10 Technique of replacement of the noncoronary sinus. The dilated sinus (**a**) is cut leaving few millimeters of supraannular aortic tissue (**b**). A drop-shaped fabric patch, used to replace the sinus, is sutured to the annulus to reach the superior border of the aortic root over the coronary ostia (**c**). A vascular graft is used to reconstruct the sino-tubular junction and to replace the ascending aorta (**d**). BAV = Bicuspid aortic valve; NCS = Noncoronary sinus

Acknowledgement We wish to thank Anna Teresa Bortolotti for the technical assistance and artwork.

References

1. Girdauskas E, Borger MA. Bicuspid aortic valve and associated aortopathy: an update. Semin Thorac Cardiovasc Surg. 2013;25:310–6. https://doi.org/10.1053/j.semtcvs.2014.01.004.
2. Martín M, Lorca R, Rozado J, Alvarez-Cabo R, Calvo J, Pascual I, et al. Bicuspid aortic valve syndrome: a multidisciplinary approach for a complex entity. J Thorac Dis. 2017;(Suppl 6):S454–64. https://doi.org/10.21037/jtd.2017.05.
3. Kusner JJ, Brown JY, Gleason TG, Edelman ER. The natural history of bicuspid aortic valve disease. Struct Heart. 2022;7:100119. https://doi.org/10.1016/j.shj.2022.100119.
4. Nistri S, Giusti B, Pepe G, Cademartiri F. Another piece in the puzzle of bicuspid aortic valve syndrome. Eur Heart J Cardiovasc Imaging. 2016;17:1248–9. https://doi.org/10.1093/ehjci/jew169.
5. Liu T, Xie M, Lv Q, Li Y, Fang L, Zhang L, et al. Bicuspid aortic valve: an update in morphology, genetics, biomarker, complications, imaging diagnosis and treatment. Front Physiol. 2019;9:1921. https://doi.org/10.3389/fphys.2018.01921.
6. Rashed ER, Dembar A, Riasat M, Zaidi AN. Bicuspid aortic valves: an up-to-date review on genetics, natural history, and management. Curr Cardiol Rep. 2022;24:1021–30. https://doi.org/10.1007/s11886-022-01716-2.
7. Paget J. On obstructions of the branches of the pulmonary artery. Med Chir Trans. 1844;27:162–89. https://doi.org/10.1177/095952874402700115.
8. Roberts WC. The congenitally bicuspid aortic valve. A study of 85 autopsy cases. Am J Cardiol. 1970;26:72–83. https://doi.org/10.1016/0002-9149(70)90761-7.
9. Sievers HH, Schmidtke C. A classification system for the bicuspid aortic valve from 304 surgical specimens. J Thorac Cardiovasc Surg. 2007;133:1226–33. https://doi.org/10.1016/j.jtcvs.2007.01.039.
10. Michelena HI, Della Corte A, Evangelista A, Maleszewski JJ, Edwards WD, Roman MJ, et al. International consensus statement on nomenclature

and classification of the congenital bicuspid aortic valve and its aortopathy, for clinical, surgical, interventional and research purposes. Eur J Cardiothorac Surg. 2021;60:448–76. https://doi.org/10.1093/ejcts/ezab038.
11. Urbanski PP. Repair of the calcified bicuspid aortic valve. Eur J Cardiothorac Surg. 2011;39:603–4. https://doi.org/10.1016/j.ejcts.2010.07.008.
12. Miceli A, Beretta P, Fiore A, Andreas M, Solinas M, Santarpino G, et al. Sutureless and rapid deployment implantation in bicuspid aortic valve: results from the sutureless and rapid-deployment aortic valve replacement international registry. Ann Cardiothorac Surg. 2020;9:298–304. https://doi.org/10.21037/acs-2020-surd-33.
13. Brown B, Le T, Naeem A, Malik A, Norton EL, Wu X, et al. Stentless valves for bicuspid and tricuspid aortic valve disease. JTCVS Open. 2021;8:177–88. https://doi.org/10.1016/j.xjon.2021.09.033.
14. de Kerchove L, Boodhwani M, Glineur D, Poncelet A, Rubay J, Watremez C, et al. Cusp prolapse repair in trileaflet aortic valves: free margin plication and free margin resuspension techniques. Ann Thorac Surg. 2009;88:455–61. https://doi.org/10.1016/j.athoracsur.2009.04.064.
15. El Khoury G, Vanoverschelde JL, Glineur D, Pierard F, Verhelst RR, Rubay J, et al. Repair of bicuspid aortic valves in patients with aortic regurgitation. Circulation. 2006;114(1 Suppl):I610–06. https://doi.org/10.1161/CIRCULATIONAHA.105.001594.
16. David TE, Armstrong S. Aortic cusp repair with Gore-Tex sutures during aortic valve-sparing operations. J Thorac Cardiovasc Surg. 2010;139(5):1340–2. https://doi.org/10.1016/j.jtcvs.2009.06.010.
17. Zhu Y, Woo YJ. Cusp repair techniques in bicuspid and tricuspid aortic valves. JTCVS Tech. 2021;7:109–16. https://doi.org/10.1016/j.xjtc.2021.01.029.
18. Rankin JS, Mazzitelli D, Fischlein TJM, Choi YH, Aicher D, Wei LM, et al. Bicuspid aortic valve repair using geometric ring annuloplasty: a first-in-humans pilot trial. JTCVS Tech. 2020;1:18–25. https://doi.org/10.1016/j.xjtc.2019.12.005.
19. Arnaoutakis GJ, Sultan I, Siki M, Bavaria JE. Bicuspid aortic valve repair. Systematic review on long-term outcomes. Ann Cardiothorac Surg. 2019;8:302–12. https://doi.org/10.21037/acs.2019.05.08.
20. Jasinski MJ, Kosiorowska K, Gocol R, Jasinski J, Nowicki R, Bielicki G, et al. Bicuspid aortic valve repair: outcomes after 17 years of experience. Eur J Cardiothorac Surg. 2021;60:1053–61. https://doi.org/10.1093/ejcts/ezab176.
21. Tamer S, Mastrobuoni S, Vancraeynest D, Lemaire G, Navarra E, Khoury GE, et al. Late results of aortic valve repair for isolated severe aortic regurgitation. J Thorac Cardiovasc Surg. 2023;165:995–1006.e3. https://doi.org/10.1016/j.jtcvs.2021.04.011.
22. Verma R, Cohen G, Colbert J, Fedak PWM. Bicuspid aortic valve associated aortopathy: 2022 guideline update. Curr Opin Cardiol. 2023;38:61–7. https://doi.org/10.1097/HCO.0000000000001020.
23. Borger MA, Fedak PWM, Stephens EH, Gleason TG, Girdauskas E, Ikonomidis JS, et al. The American Association for Thoracic Surgery consensus guidelines on bicuspid aortic valve-related aortopathy: executive summary. J Thorac Cardiovasc Surg. 2018;156:473–80. https://doi.org/10.1016/j.jtcvs.2017.10.161.
24. Girdauskas E, Disha K, Rouman M, Espinoza A, Borger MA, Kuntze T. Aortic events after isolated aortic valve replacement for bicuspid aortic valve root phenotype: echocardiographic follow-up study. Eur J Cardiothorac Surg. 2015;48:e71–6. https://doi.org/10.1093/ejcts/ezv259.
25. Vendramin I, Meneguzzi M, Sponga S, Deroma L, Cimarosti R, Lutman C, et al. Bicuspid aortic valve disease and ascending aortic aneurysm: should an aortic root replacement be mandatory? Eur J Cardiothorac Surg. 2016;49:103–9.
26. Choi MS, Jeong DS, Lee HY, Sung K, Kim WS, Lee YT, et al. Aortic wrapping for a dilated ascending aorta in bicuspid aortic stenosis. Circ J. 2015;79:778–84. https://doi.org/10.1253/circj.CJ-14-0933.
27. De Martino A, Milano AD, Bortolotti U. The use of pericardium in acquired heart diseases. A comprehensive review. Thorac Cardiovasc Surg. 2021;69:83–9. https://doi.org/10.1055/s-0039-1697918.
28. Disha K, Rouman M, Secknus M-A, Kuntze T, Girdauskas E. Are normal-sized ascending aortas at risk of late aortic events after aortic valve replacement for bicuspid aortic valve disease? Interact Cardiovasc Thorac Surg. 2016;22:465–71.
29. Wang J, Deng W, Lv Q, Li Y, Liu T, Xie M. Aortic dilatation in patients with bicuspid aortic valve. Front Physiol. 2021;12:615175. https://doi.org/10.3389/fphys.2021.615175.
30. Lopez A, Dentamaro I, Galian L, Calvo F, Alegret JM, Sanchez V, et al. Predictors of ascending aorta enlargement and valvular dysfunction progression in patients with bicuspid aortic valve. J Clin Med. 2021;10:5264. https://doi.org/10.3390/jcm10225264.
31. Roberts WC, Vowels TJ, Ko JM, Filardo G, Hebeler RF Jr, Henry AC, et al. Comparison of the structure of the aortic valve and ascending aorta in adults having aortic valve replacement for aortic stenosis versus for pure aortic regurgitation and resection of the ascending aorta for aneurysm. Circulation. 2011;123:896–903. https://doi.org/10.1161/CIRCULATIONAHA.110.972406.
32. Celiento M, Ravenni G, Margaryan R, Ferrari G, Blasi S, Pratali S, et al. The modified Bentall procedure: a single-institution experience in 249 patients with a maximum follow-p of 21.5 years. J Heart Valve Dis. 2016;25:448–55.
33. Vendramin I, Bortolotti U, De Manna DN, Lechiancole A, Sponga S, Livi U. Combined replacement of aortic valve and ascending aorta-A 70-year evolution of surgical techniques. Aorta (Stamford). 2021;9:118–23. https://doi.org/10.1055/s-0041-1729913.
34. Bentall H, De Bono A. A technique for complete replacement of the ascending aorta. Thorax. 1968;23:338–9. https://doi.org/10.1136/thx.23.4.338.

35. Milano AD, Pratali S, Mecozzi G, Boraschi P, Braccini G, Magagnini E, et al. Fate of coronary ostial anastomoses after the modified Bentall procedure. Ann Thorac Surg. 2003;75:1797–801. https://doi.org/10.1016/s0003-4975(03)00015-8.
36. Pratali S, Milano A, Codecasa R, De Carlo M, Borzoni G, Bortolotti U. Improving hemostasis during replacement of the ascending aorta and aortic valve with a composite graft. Tex Heart Inst J. 2000;27:246–9.
37. Lechiancole A, Vendramin I, Sponga S, Piani D, Benedetti G, Meneguzzi M, et al. Bentall procedure with the CarboSeal™ and CarboSeal Valsalva™ composite conduits: long-term outcomes. Interact Cardiovasc Thorac Surg. 2021;33:93–100. https://doi.org/10.1093/icvts/ivab045.
38. Vendramin I, Lechiancole A, Piani NG, Benedetti G, Sponga S, et al. Use of sutureless and rapid deployment prostheses in challenging reoperations. Review of the current evidence. J Cardiovasc Develop Dis. 2021;8:74. https://doi.org/10.3390/jcdd8070074.
39. Ross DN. Replacement of aortic and mitral valves with a pulmonary autograft. Lancet. 1967;2(7523):956–8. https://doi.org/10.1016/s0140-6736(67)90794-5.
40. Luciani GB, Barozzi L, Tomezzoli A, Casali G, Mazzucco A. Bicuspid aortic valve disease and pulmonary autograft root dilatation after the Ross procedure: a clinicopathologic study. J Thorac Cardiovasc Surg. 2001;122:74–9. https://doi.org/10.1067/mtc.2001.114638.
41. Poh CL, Buratto E, Larobina M, Wynne R, O'Keefe M, Goldblatt J, et al. The Ross procedure in adults presenting with bicuspid aortic valve and pure aortic regurgitation: 85% freedom from reoperation at 20 years. Eur J Cardiothorac Surg. 2018;54:420–6. https://doi.org/10.1093/ejcts/ezy073.
42. Mitchell ME, Woods RK, Geoffrion TR, Sow M. Late pulmonary autograft dilation: can we make a good operation great? The supported Ross. Semin Thorac Cardiovasc Surg Pediatr Card Surg Annu. 2024;27:47–51. https://doi.org/10.1053/j.pcsu.2024.01.007.
43. David TE, Feindel CM. An aortic valve-sparing operation for patients with aortic incompetence and aneurysm of the ascending aorta. J Thorac Cardiovasc Surg. 1992;103:617–21.
44. Pacini D, Settepani F, De Paulis R, Loforte A, Nardella S, Ornaghi D, et al. Early results of valve-sparing reimplantation procedure using the Valsalva conduit: a multicenter study. Ann Thorac Surg. 2006;82:865–71. https://doi.org/10.1016/j.athoracsur.2006.04.018.
45. Sarsam MA, Yacoub M. Remodeling of the aortic valve anulus. J Thorac Cardiovasc Surg. 1993;105:435–8.
46. Aicher D, Langer F, Kissinger A, Lausberg H, Fries R, Schäfers HJ. Valve-sparing aortic root replacement in bicuspid aortic valves: a reasonable option? J Thorac Cardiovasc Surg. 2004;128:662–8. https://doi.org/10.1016/j.jtcvs.2004.06.035.
47. Wilson-Smith AR, Wilson-Smith CJ, Strode Smith J, Ng D, Muston BT, Eranki A, et al. The outcomes of three decades of the David and Yacoub procedures in bicuspid aortic valve patients-a systematic review and meta-analysis. Ann Cardiothorac Surg. 2023;12:286–94. https://doi.org/10.21037/acs-2023-avs2-19.
48. de Kerchove L, Mastrobuoni S, Froede L, Tamer S, Boodhwani M, van Dyck M, et al. Variability of repairable bicuspid aortic valve phenotypes: towards an anatomical and repair-oriented classification. Eur J Cardiothorac Surg. 2019;56:351–9. https://doi.org/10.1093/ejcts/ezz033.
49. Patel PM, Chen EP. Valve sparing root replacement in bicuspid aortic valve disease. Op Tech Thorac Cardiovasc Surg. 2023;28:93–109. https://doi.org/10.1053/j.optechstcvs.2022.09.004.
50. Jahanyar J, El Khoury G, de Kerchove L. Reimplantation should be the gold standard to treat the regurgitant bicuspid aortic valve. JTCVS Tech. 2022;13:42–3. https://doi.org/10.1016/j.xjtc.2022.02.038.
51. Ugur M, Schaff HV, Suri RM, Dearani JA, Joyce LD, Greason KL, et al. Late outcome of noncoronary sinus replacement in patients with bicuspid aortic valves and aortopathy. Ann Thorac Surg. 2014;97:1242–6. https://doi.org/10.1016/j.athoracsur.2013.12.022.
52. Lansac E, Di Centa I, Sleilaty G, Lejeune S, Berrebi A, Zacek P, et al. Remodeling root repair with an external aortic ring annuloplasty. J Thorac Cardiovasc Surg. 2017;153:1033–42. https://doi.org/10.1016/j.jtcvs.2016.12.031.

14 Transcatheter Aortic Valve Implantation

Giuseppe Tarantini, Tommaso Fabris, Chiara Fraccaro, and Cristina Basso

14.1 Introduction

Bicuspid aortic valve (BAV) disease is the most common congenital valvular anomaly, affecting 0.5–2% of the population [1, 2]. It is characterized by the presence of two leaflets instead of three, which results in accelerated valve degeneration, calcification, and an increased risk of developing aortic stenosis (AS) at an earlier age compared to individuals with tricuspid aortic valves (TAV) [1, 2]. While surgical aortic valve replacement (SAVR) has long been the mainstay of treatment for BAV-related AS [3, 4], the advent of transcatheter aortic valve implantation (TAVI) has revolutionized treatment options, particularly for higher risk patients [5, 6] As the indications for TAVI progressively expand to include younger and lower-risk patients [7–9], the number of BAV patients being considered for percutaneous treatment is anticipated to rise. However, due to the unique anatomical challenges of BAV, the application of TAVI in this population requires careful planning and technical expertise.

This chapter provides an in-depth analysis of the technical challenges and procedural planning required for successful TAVI in BAV patients, with a focus on patient selection, imaging planning, valve sizing, and implantation techniques.

14.2 Challenges of BAV Anatomy in TAVI

The morphology of BAV poses unique challenges for TAVI. Unlike the classical TAV, BAV morphology is highly variable, often featuring asymmetric cusps, elliptical annuli, and calcified raphes [1, 2]. These factors create technical difficulties in valve implantation, increasing the risk of complications such as paravalvular leak (PVL), incomplete valve expansion, and aortic injury [10].

BAV is commonly classified according to the Sievers classification system [11], which categorizes valves based on the number of raphes:

G. Tarantini (✉) · T. Fabris · C. Fraccaro
Cardiology Clinic, Department of Cardiac, Thoracic, Vascular Sciences and Public Health, University of Padua, Padua, Italy
e-mail: giuseppe.tarantini.1@unipd.it; tommaso.fabris@aopd.veneto.it; chiara.fraccaro@unipd.it

C. Basso
Cardiovascular Pathology Unit, Department of Cardiac, Thoracic, Vascular Sciences and Public Health, University of Padua, Padua, Italy
e-mail: cristina.basso@unipd.it

G. Thiene et al. (eds.), *Bicuspid Aortic Valve Syndrome*,
https://doi.org/10.1007/978-3-031-85589-4_14

- Type 0: No raphe, characterized by symmetrical cusps.
- Type 1: One raphe, typically causing fusion between two cusps.
- Type 2: Two raphes, resulting in two fused cusps.

While Sievers' classification is valuable, it has limitations for TAVI planning because it does not fully account for variations in commissures, raphe calcification, or supra-annular dimensions that directly impact transcatheter heart valve (THV) expansion and positioning. More recent classifications, tailored specifically for TAVI, differentiate between bicommissural valves with or without raphe and tricommissural configurations [12] (Fig. 14.1). The latter system emphasizes anatomical features critical to procedural success. In particular, the distinction between the bicommissural raphe type and tricommissural BAV has important operative functional implications, as a congenital raphe may oppose a higher resistance to adequate prosthesis expansion than an acquired commissural fusion.

14.3 Imaging for Procedural Planning

Imaging plays a pivotal role in TAVI procedural planning for BAV patients. Multi-slice computed tomography (MSCT) is the gold standard for assessing BAV anatomy and identifying key anatomical features that will affect the TAVI procedure, such as annular size, supra-annular dimensions, calcification burden, and coronary ostia position [13].

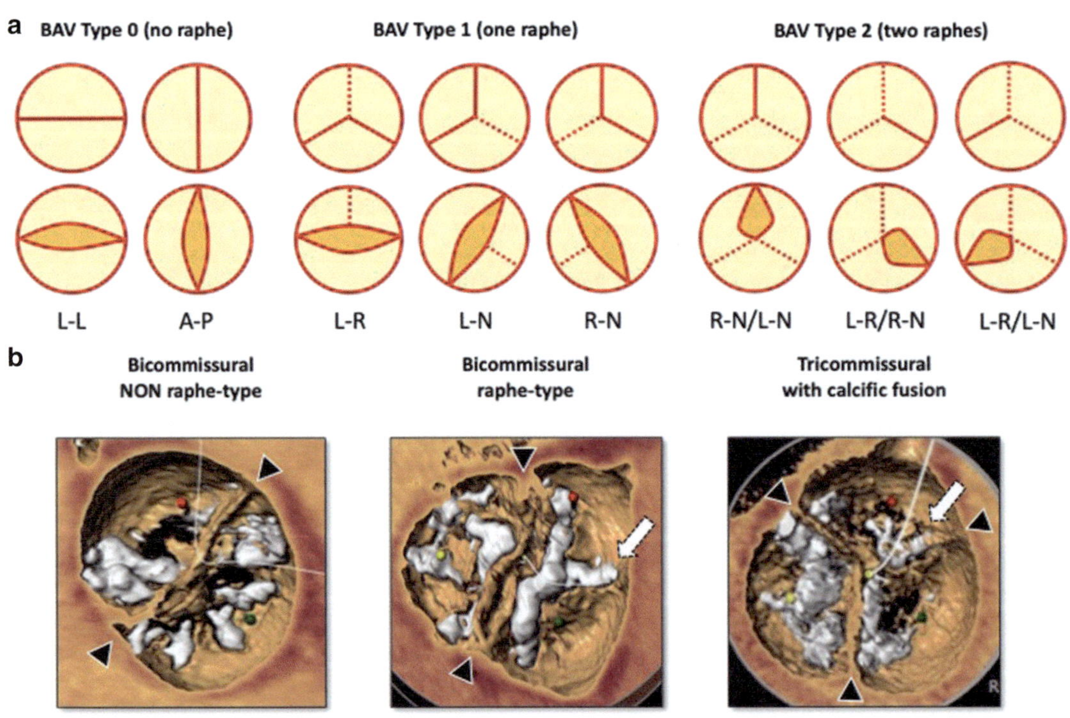

Fig. 14.1 Bicuspid aortic valve (BAV) classification systems. (Panel **a**) Sievers and Schmidtke (surgical-derived) classification. Dotted line(s) denote(s) the raphe(s). A indicates anterior; P, posterior; L, lateral (in BAV type 0) or left (in BAV type 1 or 2); and R, right. (Panel **b**) Jilaihawi et al. (CT-derived) classification. Head arrow indicates the commissure. Arrow indicates the raphe or commissural fixation. CT indicates computed tomography. (*From: Tarantini G, Fabris T. Transcatheter Aortic Valve Replacement for Bicuspid Aortic Valve Stenosis: A Practical Operative Overview. Circ Cardiovasc Interv. 2021 Jul;14(7):e009827. doi: 10.1161/CIRCINTERVENTIONS.120.009827. Epub 2021 Jun 16. PMID: 34130478; PMCID: PMC10332651*)

CT imaging should be performed across the entire cardiac cycle to obtain dynamic assessments of valve and aortic root anatomy (Fig. 14.2). Detailed measurements, including annular area, perimeter, and diameters at various levels of the aortic root, are essential for selecting the appropriate valve size and predicting areas of resistance during deployment [10]. MSCT assessment should always provide comprehensive data on:

- Annular size and shape: Unlike TAVs, BAVs tend to have elliptical, rather than circular, annuli, requiring careful evaluation of both major and minor diameters. In Type 0 BAV (bicommissural valve without raphe), the annular plane is often difficult to define because of the presence of only 2 anatomical hinge points [14].
- Supra-annular dimensions: BAV anatomy often features a "volcanic" shape at the supra-annular level, which can impact valve positioning and expansion.
- Raphe characteristics: Calcification as well location of raphe(s) can significantly impact valve expansion and occurrence of procedural complications.
- Coronary ostia height and position: In BAV patients, the coronary ostia can be more closely aligned with the commissures, creating a higher risk of coronary obstruction during valve deployment. Anomalies in coronary artery origin may also be present.
- Aorta: BAV is often associated with a markedly horizontal aorta and frequently accompanied by aortic dilation (aortopathy), which increases the challenges during valve deployment and elevates the risk of aortic injury.

14.4 Valve Sizing Strategies

Accurate evaluation of the annulus, supra-annular dimensions, and calcified raphes helps guide the appropriate selection of valve size and type (Fig. 14.3). In fact, THV sizing is one of the most challenging aspects of TAVI in BAV due to the variability in the anatomical structure of the aortic root and valve annulus. Typically, in TAV, the valve prosthesis is anchored at the virtual basal ring (VBR), which corresponds to the tightest part of the aortic annulus [15]. To address this, the virtual raphe ring (VRR) has been proposed as an alternative sizing strategy in cases where the supra-annular dimensions, specifically at the raphe, represent the narrowest point for valve anchoring [10]. Different sizing methods were proposed during the last years, providing specific measures to size the supra-annular structure in BAV [16–19]. The choice of whether to use an annular or supra-annular sizing strategy is influenced by the specific BAV anatomy encountered [10] (Fig. 14.4):

1. Annular (VBR) Sizing: Used in patients with annular-dominant or co-dominant configurations. Here, the prosthesis size is chosen based on the annular measurements from the CT scan.
2. Supra-annular (VRR) Sizing: In cases where the supra-annular dimensions at the raphe are significantly smaller than the annular size, operators may opt to size the prosthesis based on the supra-annular ring to avoid under-expansion and malpositioning of the valve.

The concept of supra-annular sizing remains a topic of debate and lacks standardization, with considerable variability among operators [16–20]. Recent findings from the prospective BIVOLUT-X registry [21], which compared annular sizing with combined (annular + supra-annular) sizing in patients with BAV treated using SEV, indicate that combined sizing achieves similar clinical outcomes and device success rates as conventional annular sizing. The multicenter retrospective AD HOC registry recently evaluated the safety and efficacy of modern-generation THVs in tapered type 1 BAV stenosis, focusing on the influence of annular and supra-annular sizing strategies on short- and mid-term outcomes [22]. After propensity matching, both groups demonstrated comparable rates of technical success, 30-day device success, and 30-day early safety [22]. However, the supra-annular

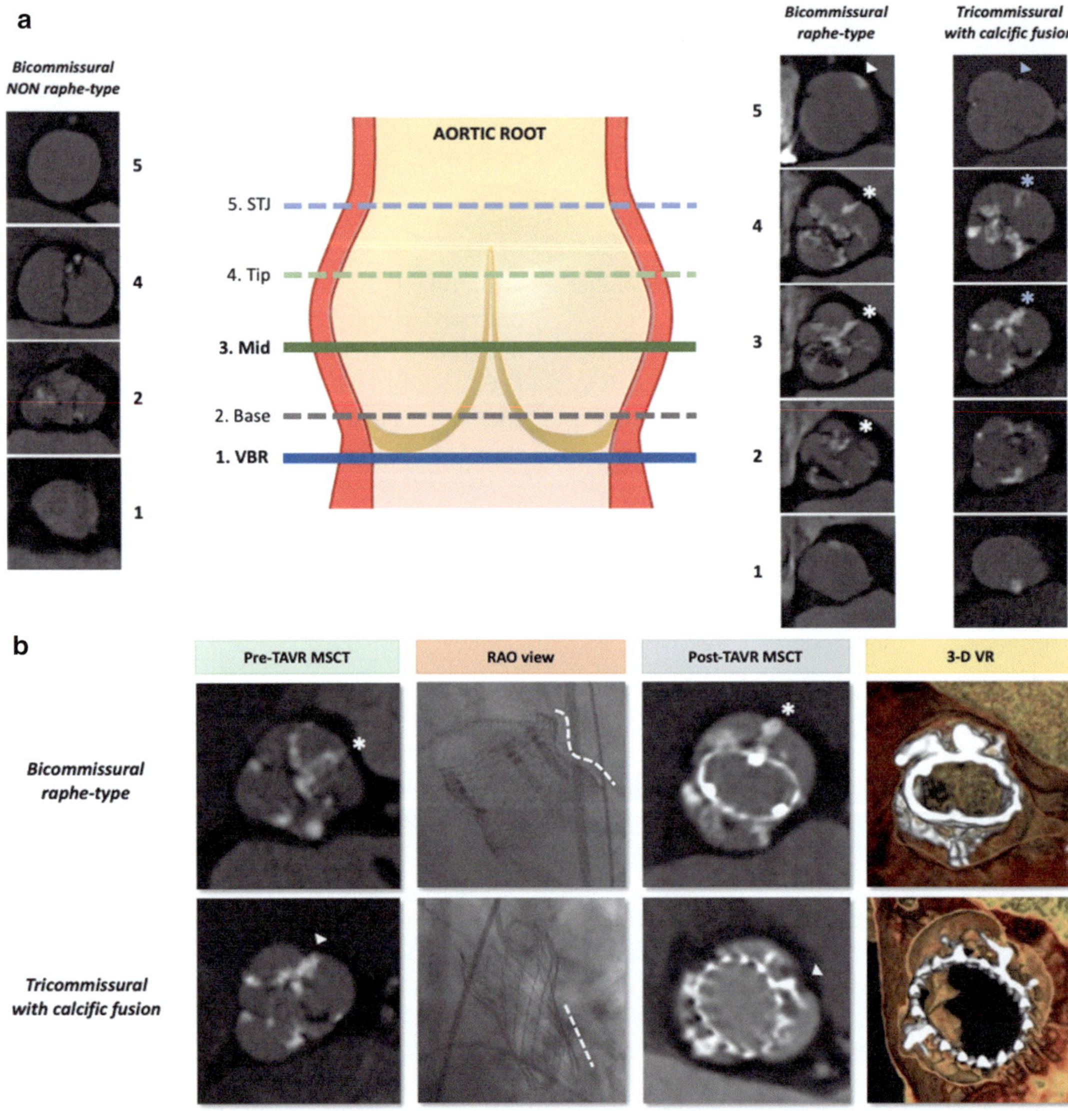

Fig. 14.2 MSCT imaging of BAV anatomy. (Panel **a**) MSCT analysis of the 3 TAVI-specific BAV morphologies at multiple cross-sectional levels: annular level (i.e., VBR) (1), base of sinus (2), mid-sinus (3), upper margin of cusps (4), and STJ (5). In bicommissural raphe-type, the raphe (white asterisks) extents from the basal part of sinus for almost the entire cusp height, and the "non-opening" commissure does not reach the STJ (white head arrow); on the contrary, in tricommissural BAV type, the acquired commissural fusion (blue asterisks) usually involves the upper portion of cusps, while the "non-opening" commissure reaches the STJ (blue head arrow). (Panel **b**) TAVI in bicommissural raphe-type (top) and in tricommissural BAV (bottom), with pre-procedural MSCT, procedural fluoroscopic RAO view, and post-TAVI MSCT images. To note, in tricommissural BAV with acquired commissural fusion, the prosthesis shows better expansion with respect to bicommissural with congenital severe calcified raphe. Asterisk indicates the raphe. Head arrow indicates the acquired calcific commissural fixation. *BAV* bicuspid aortic valve, *MSCT* multi-slice computed tomography, *TAVI* transcatheter aortic valve implantantion, *VBR* virtual basal ring, *STJ* sino-tubular junction, *RAO* right anterior oblique. (*From: Tarantini G, Fabris T. Transcatheter Aortic Valve Replacement for Bicuspid Aortic Valve Stenosis: A Practical Operative Overview. Circ Cardiovasc Interv. 2021 Jul;14(7):e009827. doi: 10.1161/CIRCINTERVENTIONS.120.009827. Epub 2021 Jun 16. PMID: 34130478; PMCID: PMC10332651*)

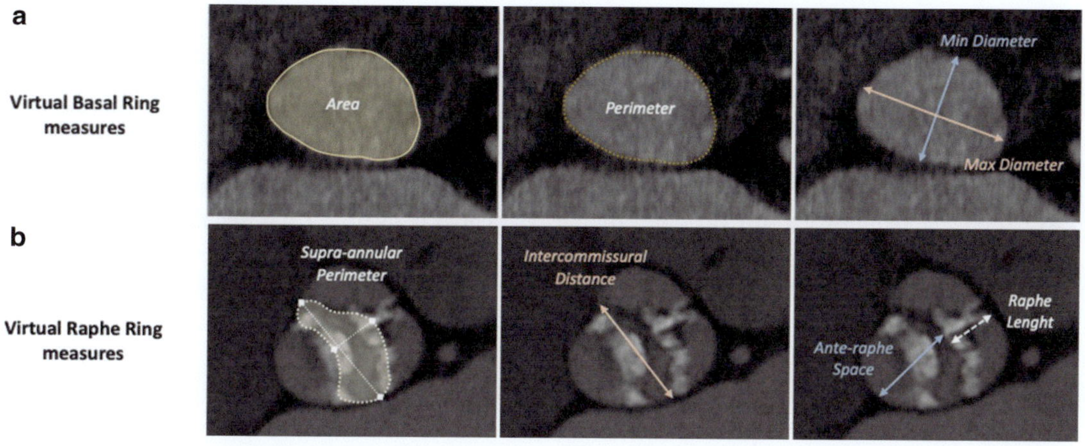

Fig. 14.3 Sizing strategies in bicuspid aortic valve. (Panel **a**) Standard CT-based measurements at VBR. (Panel **b**) Proposed CT-based measurements at VRR. CT indicates computed tomography; VBR, virtual basal ring; and VRR, virtual raphe ring. *From: Tarantini G, Fabris T. Transcatheter Aortic Valve Replacement for Bicuspid Aortic Valve Stenosis: A Practical Operative Overview. Circ Cardiovasc Interv. 2021 Jul;14(7):e009827. doi: 10.1161/CIRCINTERVENTIONS.120.009827. Epub 2021 Jun 16. PMID: 34130478; PMCID: PMC10332651*

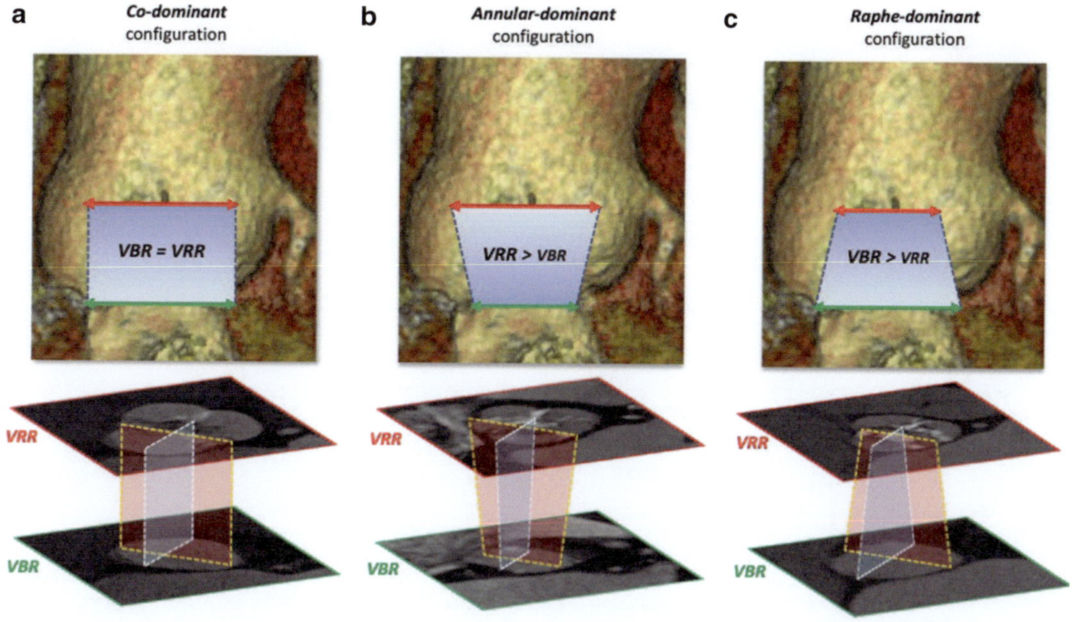

Fig. 14.4 Transcatheter heart valve (THV) sizing in bicuspid aortic valve (BAV). Different patterns of landing zone in BAV. Co-dominant (**a**) and Annular-dominant (**b**) with indicated standard THV annular-sizing strategy. Raphe-dominant (**c**) with advised THV supra-annular sizing strategy. (*From: Tarantini G, Fabris T. Transcatheter Aortic Valve Replacement for Bicuspid Aortic Valve Stenosis: A Practical Operative Overview. Circ Cardiovasc Interv. 2021 Jul;14(7):e009827. doi: 10.1161/CIRCINTERVENTIONS.120.009827. Epub 2021 Jun 16. PMID: 34130478; PMCID: PMC10332651*)

group exhibited a slightly higher mean post-TAVI gradient of about 2 mmHg [22].

14.5 Device Selection

TAVI for BAV requires careful selection of the THV system. Currently, two major types of THVs are used in BAV patients: balloon-expandable (BE) valves and self-expanding (SE) valves. Each type has distinct advantages and challenges based on BAV anatomy [23] (Fig. 14.5).

– Balloon-Expandable Valves: The Sapien 3 valve (Edwards Lifesciences) is widely used in BAV TAVI. Its external sealing skirt improves sealing in complex anatomies, such

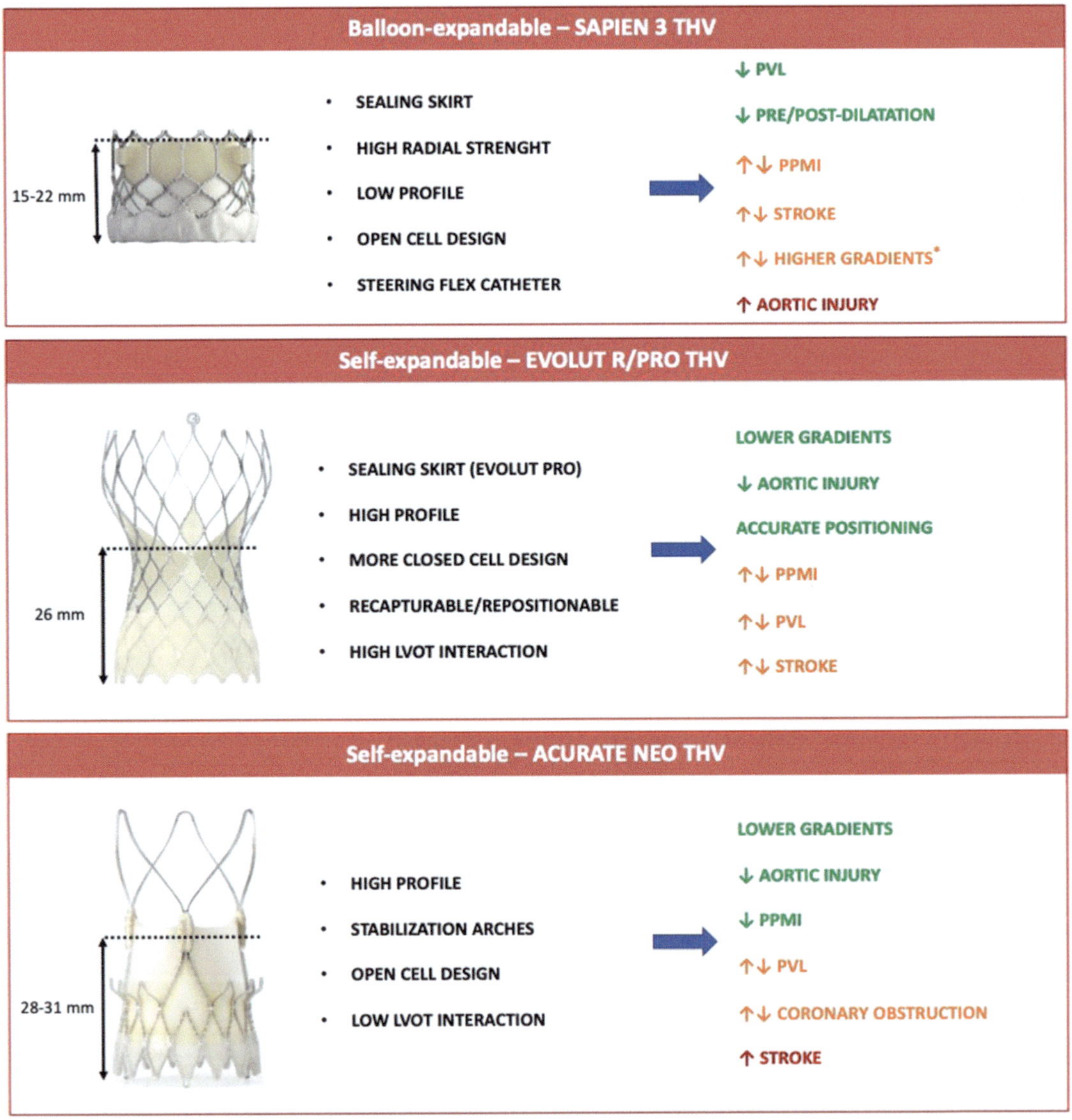

Fig. 14.5 New iterations of THVs with relative pros and cons for bicuspid anatomy. *THV transcatheter heart valve, PVL* paravalvular leak, *PPMI* permanent pacemaker implantation, *LVOT* left ventricle outflow tract. (*Modified from: Tarantini G, Fabris T. Transcatheter Aortic Valve Replacement for Bicuspid Aortic Valve Stenosis: A Practical Operative Overview. Circ Cardiovasc Interv. 2021 Jul;14(7):e009827. doi: 10.1161/CIRCINTERVENTIONS.120.009827. Epub 2021 Jun 16. PMID: 34130478; PMCID: PMC10332651*)

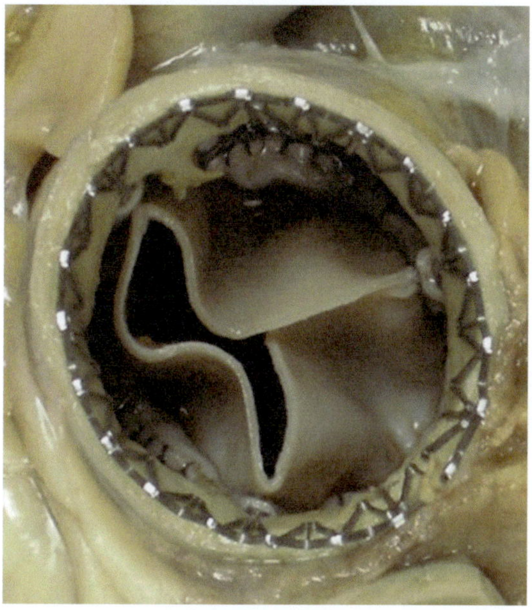

Fig. 14.6 Balloon-expandable Sapien valve in situ after deployment. From Thiene G, Basso C, Rizzo S, Della Barbera M, Valente M, Bortolotti U (eds). Pathology of Cardiac Valve Disease. 2023. Springer, Cham. ISBN: 978-3-031-35497-7

as BAV, and reduces the incidence of PVL. The high radial strength of the Sapien 3 valve is advantageous in heavily calcified BAVs, helping prevent under-expansion (Fig. 14.6). The flexibility of the delivery system offers advantages in cases of a horizontal aorta, as well as it can minimize the risk of aortic injury by reducing interaction with the malacic wall of a dilated aorta. Conversely, the rigid structure of balloon-expandable valves may heighten the risk of barotrauma-induced annular rupture or aortic injury, particularly in patients with significant calcification or severe aortopathy.

– Self-Expanding Valves: SEV, such as the Evolut R/Pro (Medtronic), are also commonly used in BAV patients. These valves offer flexibility in anatomies with irregular or elliptical annuli due to their ability to expand gradually. The supra-annular valve design provides excellent hemodynamic performance, even when the valve is under-expanded at the inflow portion. Additionally, the repositionability of some self-expanding valves allows for greater precision during deployment, reducing the risk of malposition and complications. However, the reduced radial force may result in THV underexpansion and an increased risk of PVL, particularly in anatomically constrained settings.

Recently, the AD HOC registry retrospectively evaluated the performance of modern-generation THVs in type 1 BAV stenosis [24]. After propensity-matching analysis, the authors reported that, compared to SEVs, BEVs were associated with less permanent pacemaker implantation (PPMI) and moderate or greater PVL but with more severe patient-prosthesis mismatch, resulting in an overall higher rate of 30-day device success and early safety [24]. However, in the absence of randomized comparisons among different types of THVs in BAV, a definitive recommendation on the preferable THV cannot be formulated. Rather, a patient-tailored device selection, based on different THV features, geometries, and interactions with the surrounding anatomy, is recommended [10].

14.6 Valve Implantation Techniques

Based on the three configurations of the aortic root for THV sizing, the operator can choose between two different landing zones (Fig. 14.7): (1) in codominant and VBR-dominant configurations, the VBR serves as the appropriate landing

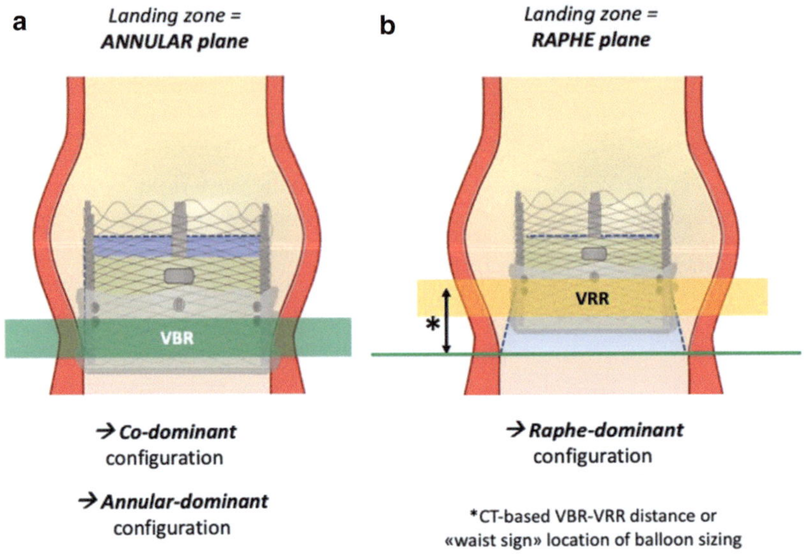

Fig. 14.7 Transcatheter heart valve (THV) positioning in bicuspid aortic valve. Illustration showing standard THV implantation at VBR in case of Co-dominant or Annular-dominant shapes (**a**), and optional higher THV implantation at VRR in case of Raphe-dominant shape (**b**). VBR indicates virtual basal ring; and VRR, virtual raphe ring. (*From: Tarantini G, Fabris T. Transcatheter Aortic Valve Replacement for Bicuspid Aortic Valve Stenosis: A Practical Operative Overview. Circ Cardiovasc Interv. 2021 Jul;14(7):e009827. doi: 10.1161/CIRCINTERVENTIONS.120.009827. Epub 2021 Jun 16. PMID: 34130478; PMCID: PMC10332651*)

zone, and (2) in VRR-dominant cases, a supra-annular landing zone should be considered, considering the characteristics of the raphe.

When the VRR is chosen as the landing zone, the THV should be implanted at a height that ensures optimal supra-annular sealing [19] (Fig. 14.8). Although a higher prosthesis implantation may theoretically reduce the risk of conduction disturbances by minimizing interaction with the conduction pathway, it also carries potential risks, such as device migration or coronary obstruction, depending on the anatomy of the aortic root [25].

Additionally, determining the ideal angiographic projection for THV deployment in BAV can be challenging, as the valve leaflets often appear irregular on fluoroscopy. Therefore, correct coaxial alignment of the THV frame should rely on a combination of CT-derived orthogonal views and valve orientation during fluoroscopy [10].

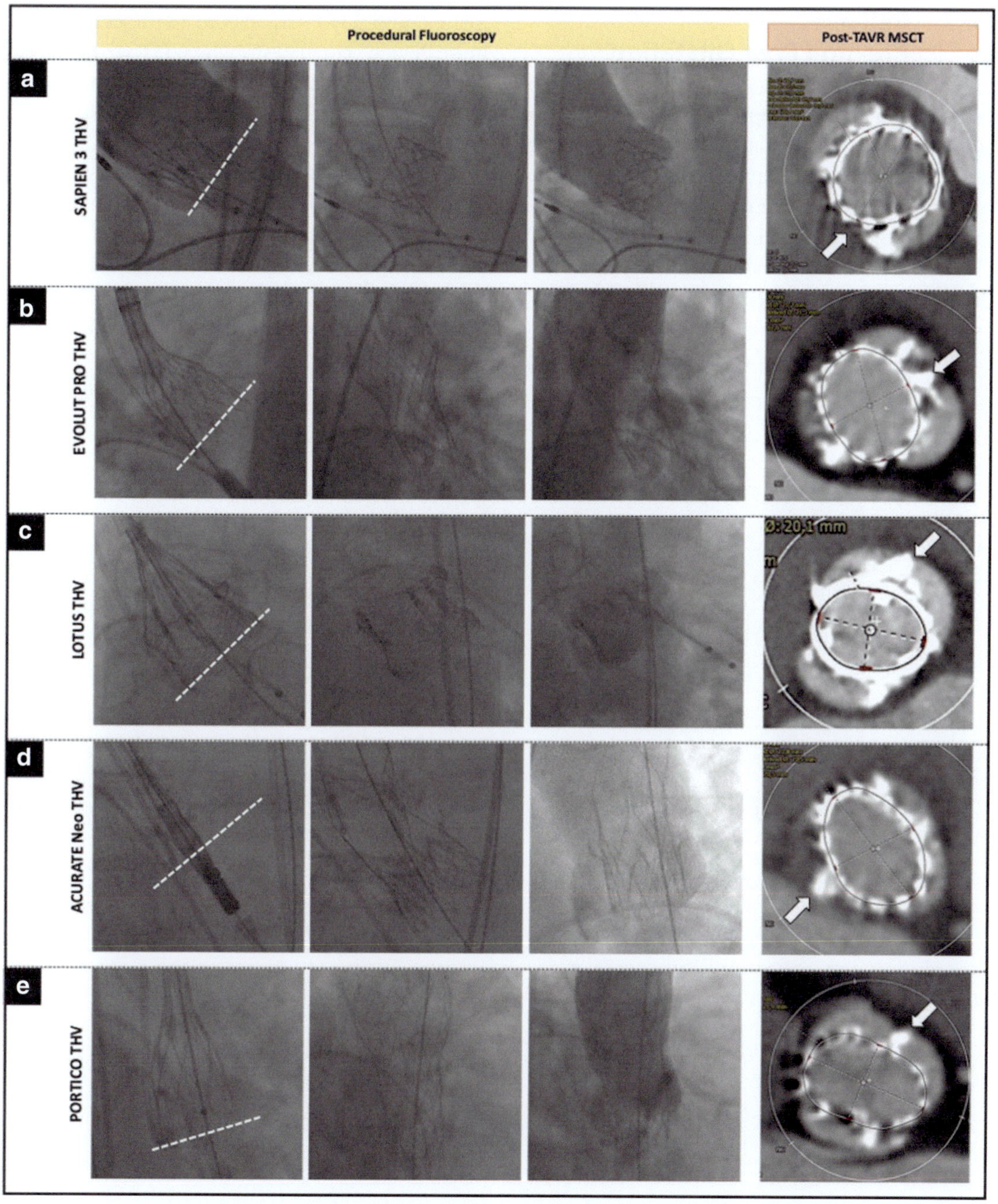

Fig. 14.8 THV implantation height in bicuspid aortic valve. Explicative cases of final high implantation with optimal supra-annular sealing in BAV raphe-type with VRR-dominant pattern: Sapien 3 THV (**a**); CoreValve Evolut PRO THV (**b**); Lotus valve (**c**); Acurate Neo THV (**d**); Portico THV (**e**). Dashed line indicates the plane of VBR. Arrow indicates the position of raphe. *THV* transcatheter aortic valve replacement, *BAV* bicuspid aortic valve, *VRR* virtual raphe ring. (*From: Tarantini G, Fabris T. Transcatheter Aortic Valve Replacement for Bicuspid Aortic Valve Stenosis: A Practical Operative Overview. Circ Cardiovasc Interv. 2021 Jul;14(7):e009827. doi: 10.1161/CIRCINTERVENTIONS.120.009827. Epub 2021 Jun 16. PMID: 34130478; PMCID: PMC10332651*)

14.7 Balloon Pre- and Post-Dilatation

Balloon aortic valvuloplasty can be a useful adjunct to MSCT scans for THV sizing in selected cases, as it helps reveal the mechanical properties of the annular and supra-annular structures by simulating the interaction with the THV [10, 26]. Specifically, balloon aortic pre-dilation offers several advantages: (1) it aids in crossing, correctly positioning, and expanding the THV; (2) it tests the raphe's resistance, providing insight into THV expansion; (3) it helps determine device size in cases where CT sizing is borderline or uncertain; and (4) it helps anticipate the risk of residual PVL and coronary obstruction through concomitant supra-annular aortography.

Regarding balloon sizing, the presence of a waist (indentation) above the annulus with minimal contrast regurgitation suggests that the supra-annular structure may provide sufficient anchoring force for the THV, making it a suitable level for device sizing and implantation (VRR-dominant) (Fig. 14.9).

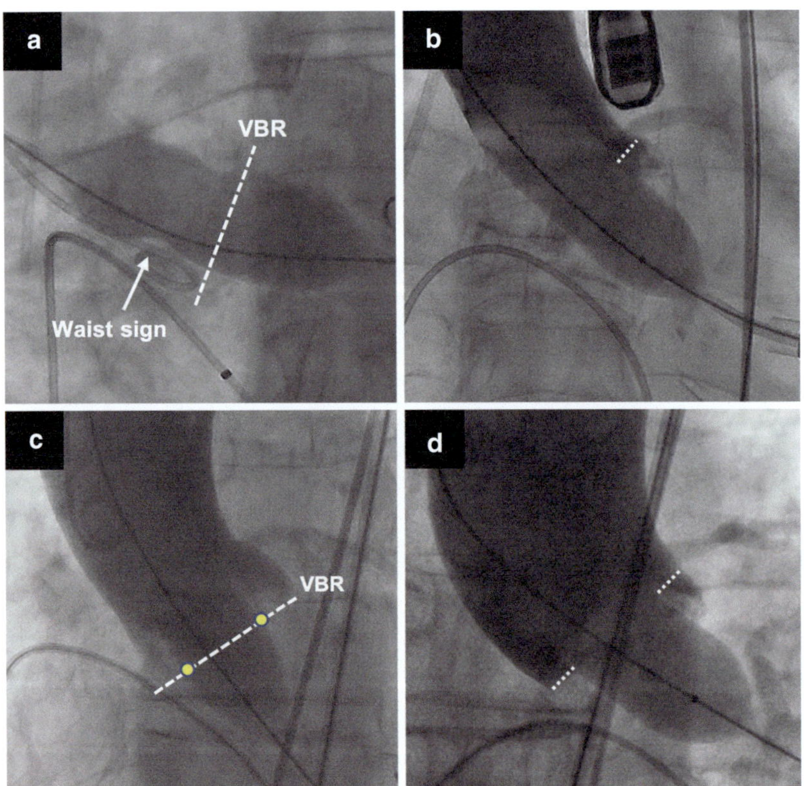

Fig. 14.9 Balloon sizing technique in bicuspid aortic valve. Supra-annular indentations of balloon sizing, suggesting a VRR-based sizing approach (**a**). Ante-raphe expansion of balloon sizing in bicommissural raphe-type, indicating a low-compliance of the raphe (**b**). Balloon sizing reaching the hinge points (yellow points) of VBR without supra-annular waist, suggesting a VBR-based sizing approach (**c**). Symmetrical expansion of balloon sizing in bicommissural raphe-type, predicting full and symmetrical prosthesis expansion (**d**). *VBR* virtual basal ring, *VRR* virtual raphe ring (*From: Tarantini G, Fabris T. Transcatheter Aortic Valve Replacement for Bicuspid Aortic Valve Stenosis: A Practical Operative Overview. Circ Cardiovasc Interv. 2021 Jul;14(7):e009827. doi: 10.1161/CIRCINTERVENTIONS.120.009827. Epub 2021 Jun 16. PMID: 34130478; PMCID: PMC10332651*)

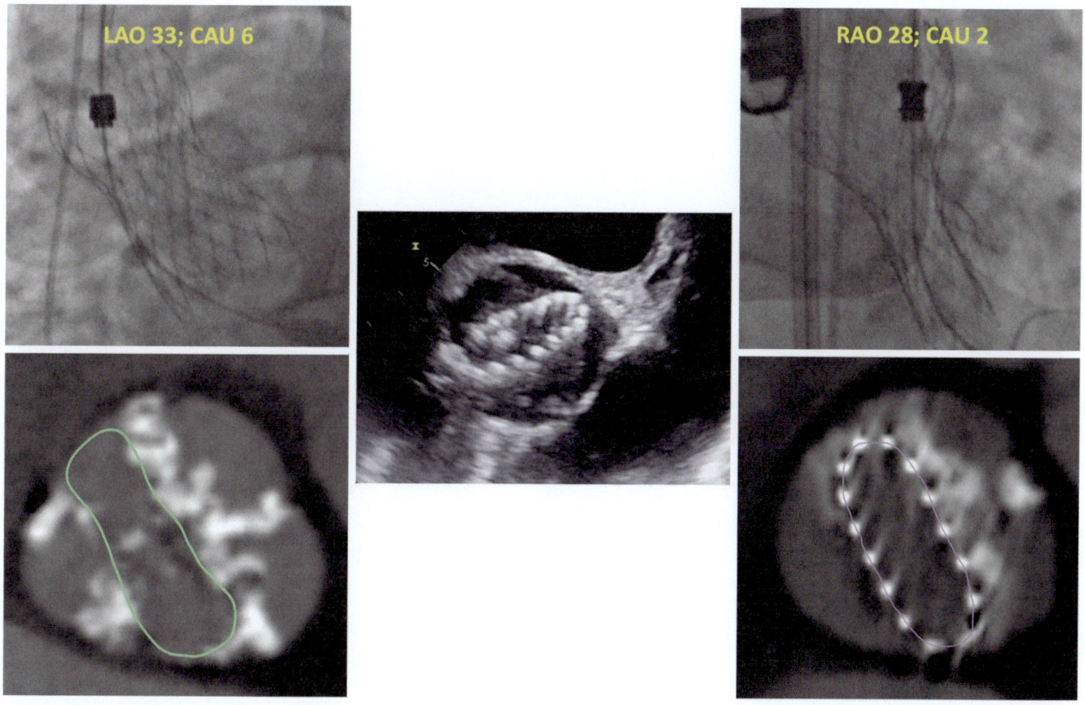

Fig. 14.10 Explicative case of asymmetrical THV expansion. Self-expanding Corevalve Evolut R 34 mm THV implanted in a type 1 BAV with calcific raphe between right and left cusp. On the top, fluoroscopic views in co-planar (left) and cusp-overlap (right) showing significant under-expansion of the THV along the ante-raphe space. On the bottom, pre- and post-TAVI MSCT reconstructions showing the asymmetrical expansion due to the presence of a calcific raphe, with good correlation between the perimeter of ante-raphe space and THV frame. *THV* transcatheter heart valve, *BAV* bicuspid aortic valve

While post-dilatation can help address THV distortion in BAV patients, it should be reserved for cases with significant residual PVL or transvalvular gradients due to the associated risk of balloon-related complications. An additional orthogonal view during valve deployment is recommended to detect potential stent frame under-expansion, which may necessitate invasive evaluation of transprosthetic gradients (Fig. 14.10). If post-dilatation is required, the final balloon size should consider balloon compliance, adverse aortic root characteristics (e.g., calcified left ventricular outflow tract), and the smallest diameter at either the VBR or VRR. The long-term effects of BAV-related THV distortion on structural valve degeneration, bioprosthetic failure, leaflet thrombosis, and risk of endocarditis remain under investigation [27, 28] (Figs. 14.11 and 14.12).

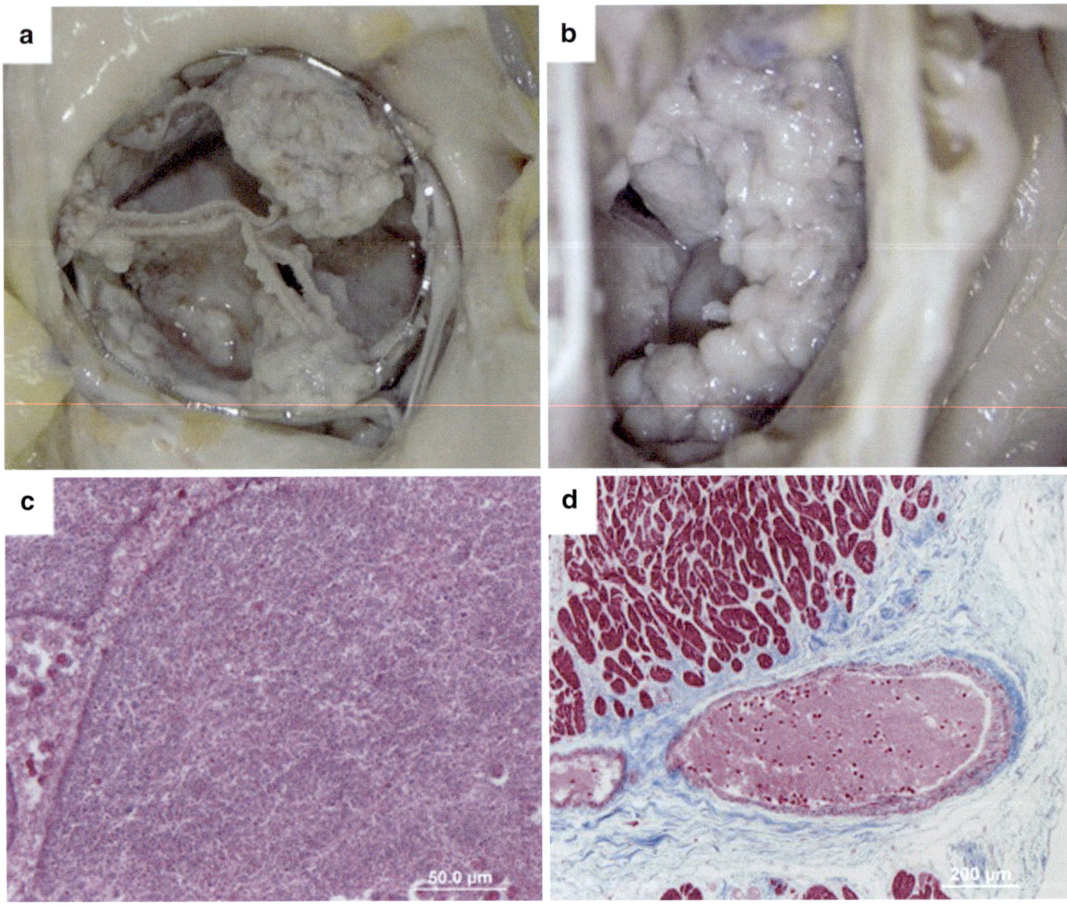

Fig. 14.11 (**a**, **b**) TAVI showing massive thrombotic vegetations blocking the valve leaflets. On histological examination, the thrombotic material consists of mixture of fibrin and platelet aggregates (**c**). Coronary artery embolization occurred (**d**). (**c**) Hematoxylin-eosin stain. (**d**) Azan Mallory stain. (*From Thiene G, Basso C, Rizzo S, Della Barbera M, Valente M, Bortolotti U (eds). Pathology of Cardiac Valve Disease. 2023. Springer, Cham. ISBN: 978-3-031-35497-7*)

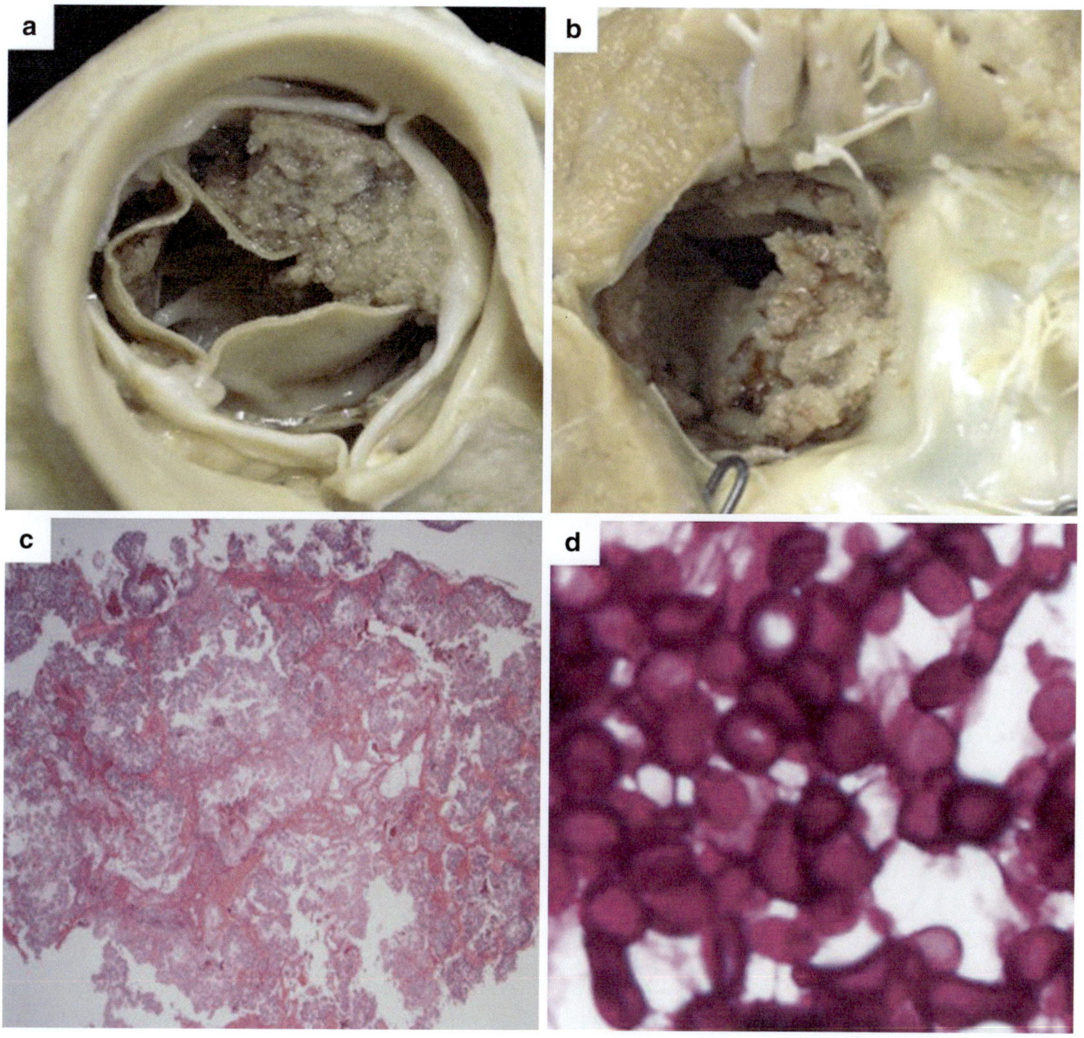

Fig. 14.12 Infective fungal endocarditis in TAVI. (**a**, **b**) At gross examination, note a prosthetic valve stenosis by polypous and friable vegetations that block a pericardial cusp and extends along the mitroaortic fibrous continuity. (**c**, **d**) Histologic examination shows clusters of Candida-type yeasts, hyphae, and pseudohyphae entrapped, within a fibrin network platelets, red cells, and granulocytes ((**c**), HE; (**d**), PAS; insert, Grocott stains). (*From Santos M et al., Eur Heart J. 2011. From Thiene G, Basso C, Rizzo S, Della Barbera M, Valente M, Bortolotti U (eds). Pathology of Cardiac Valve Disease. 2023. Springer, Cham. ISBN: 978-3-031-35497-7*)

14.8 Coronary Access and Commissural Alignment

Accessing the coronary arteries after TAVI can be challenging or even impossible if the THV commissures are positioned close to the coronary ostia. However, achieving patient-specific commissural alignment during the deployment of tall-frame self-expanding THVs can reduce the risk of difficult coronary access to less than 5% [29, 30]. Despite this, commissural alignment alone may not always result in optimal THV positioning for coronary engagement, especially in case of BAV anatomy, as other factors can influence access [31].

One key factor is the symmetry of the native aortic valve cusps, where any asymmetry can impair proper commissural alignment. This is because the angle between the two commissures of a THV is always fixed at 120°. The ALIGN-

TAVR Consortium recently defined native cusp symmetry based on the intercommissural angle of the largest cusp [32]. Valves with an angle between 120° and 125° are considered symmetric, while angles exceeding 125° are classified as mildly to severely asymmetric, with severity increasing by each 5° increment [32]. In a study using cardiac CT to evaluate 200 tricuspid and 200 bicuspid aortic valves, severe cusp asymmetry was observed more frequently in bicuspid valves (52.5%) than in tricuspid valves (2.5%), with the non-coronary cusp being the most commonly dominant cusp [33].

Coronary ostium eccentricity is another factor that can impact the risk of overlap between the coronary ostia and the new commissures after TAVI, even with optimal commissural alignment. The ALIGN-TAVR Consortium proposed a definition for coronary ostium eccentricity, classifying it based on the angle between the coronary ostium and the center of its respective cusp: 0°–10° indicates a centered ostium, 10°–20° signifies mild eccentricity, 20°–30° represents moderate eccentricity, and angles over 30° indicate severe eccentricity [32].

In cases with significant cusp asymmetry or coronary ostium eccentricity like in BAV, using coronary ostia overlap view during TAVI instead of the traditional cusp overlap view may be beneficial [34] (Fig. 14.13).

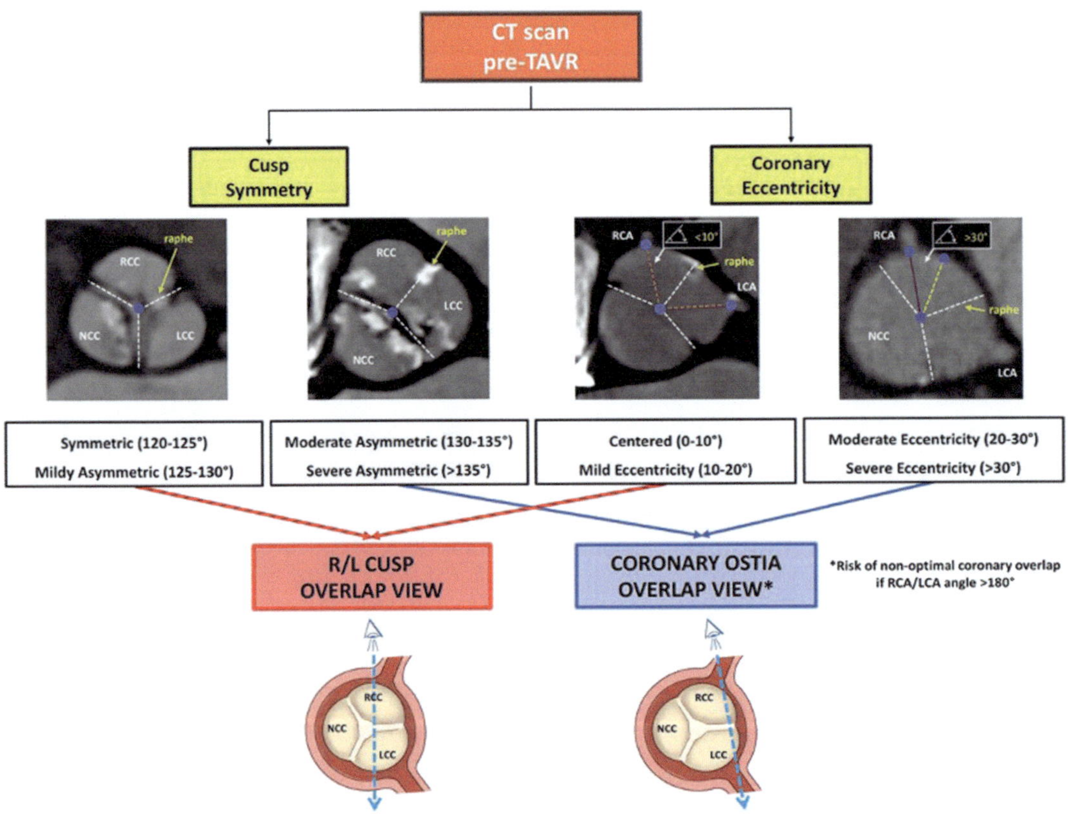

Fig. 14.13 Illustration of the flow algorithm when evaluating pre-procedural CT scan to customize THV alignment in bicuspid aortic valve. *CT* computed tomography; *THV* transcatheter heart valve, *BAV* bicuspid aortic valve, *TAVR* transcatheter aortic valve replacement, *R* right, *L* left, *RCA* right coronary artery, *LCA* left coronary artery. (*From: Tarantini G, Fabris T, Nai Fovino L, Cardaioli F, Pergola V, Montonati C, Rodinò G, Cabrelle G, Massussi M, Scotti A, Zuccarelli V, Sciarretta T, Masiero G, Gregori D, Napodano M, Fraccaro C, Continisio S, Iliceto S. Definition of trAnscatheter heart Valve orIeNtation in biCuspId aortic valve: The DA VINCI pilot study. Front Cardiovasc Med. 2022 Dec 12;9:1056496. doi: 10.3389/fcvm.2022.1056496. PMID: 36578836; PMCID: PMC9790995*)

14.9 Post-TAVI Computed Tomography

The complex anatomy of BAV and the increased risk of certain complications make post-TAVI CT an essential tool for a comprehensive assessment of procedural success and long-term valve performance. In fact, it allows for precise evaluation of valve expansion, positioning, and function, as well as the early detection of complications such as PVL, coronary obstruction, and leaflet thrombosis. Below are the main indications and utilities of post-TAVI CT in BAV patients:

- Assessment of prosthesis expansion and positioning: post-TAVI CT is valuable for confirming whether the THV is fully expanded and properly positioned, especially given the challenges posed by BAV anatomy, such as asymmetric calcification and elliptical annuli. Incomplete or asymmetric expansion can affect valve function and durability.
- Evaluation of PVL: Although echocardiography is the standard for detecting PVL, CT offers additional insights, particularly in identifying the anatomical causes of residual leak, such as under-expanded areas or gaps between the prosthesis and native annulus.
- Assessment of leaflet mobility and thrombosis: Reduced leaflet mobility and thrombosis are concerns following TAVI. Post-TAVI CT can detect subclinical leaflet thrombosis, which may not be immediately apparent on echocardiography. Identifying leaflet thrombosis early is critical for preventing potential valve dysfunction and embolic events.
- Planning for future interventions: As BAV patients are generally younger, they may require additional interventions for index THV failure. Understanding the current valve structure (e.g., expansion, alignment to coronaries, implantantion depth, etc.) and its relationship with surrounding aortic root structure is key to evaluate the feasibility and plan redo-TAVI procedure [35].

14.10 Clinical Outcomes

Due to the complex anatomical challenge that may pose significant short- and long-term risks for TAVI, BAV patients have traditionally been excluded from major randomized clinical trials comparing TAVI and SAVR [3–6]. The most reliable data on this subject come from two propensity-matched studies that reported comparable mortality and stroke rates up to 1 year [36, 37]. However, SAVR was associated with a higher incidence of acute kidney injury and new-onset atrial fibrillation, while TAVI showed a greater need for PPMI [36, 37].

Direct comparisons of outcomes between TAVI in BAV and TAV patients are largely based on retrospective registry data, with the strongest evidence derived from propensity-matched analyses and meta-analyses. Early experiences with first-generation THVs in high-risk BAV patients demonstrated comparable mortality rates to TAV patients, but BAV patients had lower device success rates, a higher need for surgical conversion, and more frequent significant PVL [12, 38–41]. However, the development of newer-generation THVs, improved operator skills, and the routine use of advanced imaging for preprocedural planning have markedly enhanced TAVI outcomes [38–41]. Recently, two propensity-matched analyses examined intermediate-risk patients from the TVT Registry who were treated with BEVs like Sapien 3 [42], and SEVs such as Evolut R/PRO [43]. These studies found similar one-year mortality rates between BAV and TAV patients [42, 43]. However, in BAV patients treated with BEVs, there was a higher rate of conversion to surgery, and those treated with SEVs had an increased incidence of significant PVL and required more reinterventions [42, 43]. Recent evidence highlights favorable clinical outcomes following TAVI in selected low-risk patients. A propensity-matched analysis from the Evolut Low Risk Bicuspid study found no significant differences in all-cause mortality, stroke, or echocardiographic outcomes at 1 year between BAV and TAV patients [44]. Similarly, a propensity-matched analysis from the PARTNER 3 Bicuspid Registry reported comparable rates for the composite endpoint of death, stroke, and cardiovascular rehospitalization at 1 year [45].

Outcomes after TAVI in BAV patients appear to be heavily influenced by the specific anatomical characteristics of the valve. Yoon et al. identified both a calcified raphe and extensive leaflet calcification as independent predictors of procedural complications and two-year all-cause mortality [46]. Patients exhibiting these morphological features had a higher likelihood of aortic root injury, moderate-to-severe PVL, and increased 30-day mortality rates [46].

While improvements in the valve sealing mechanisms in newer-generation devices have substantially reduced the incidence of PVL in BAV patients, bringing it down to just 2.7% [38], the issue of PPMI has not seen similar improvements. Data from the Bivolut X Registry [21] and the TAVR Low Risk Bicuspid Study [45] indicated significantly elevated rates of PPMI in BAV patients treated with SEVs compared to TAV patients. The recent adoption of the cusp overlap technique for SEV implantation in BAV patients offers a promising approach to reducing PPMI rates [47].

The risk of periprocedural stroke during TAVI is another area of concern, particularly due to the potential embolization of small debris during valve implantation. High levels of valve calcification and additional procedural steps, such as pre- and post-dilatation, further elevate this risk (Fig. 14.14). According to the TVT Registry, the incidence of cerebral ischemic events was not significantly different between BAV and TAV patients treated with newer BEV devices [42]. However, other studies have suggested an increased stroke risk in BAV patients [48]. The recent PROTECTED TAVR trial examined the role of cerebral embolic protection devices (CEPD) in reducing this risk but found no significant differences in the incidence of periprocedural stroke or transient ischemic attack with the use of CEPD [49]. Nevertheless, given the potentially disabling consequences of periprocedural stroke, systematic use of CEPD may be considered in younger BAV patients.

Another key concern in BAV patients undergoing TAVI is the risk of THV underexpansion in anatomically challenging valves, which can neg-

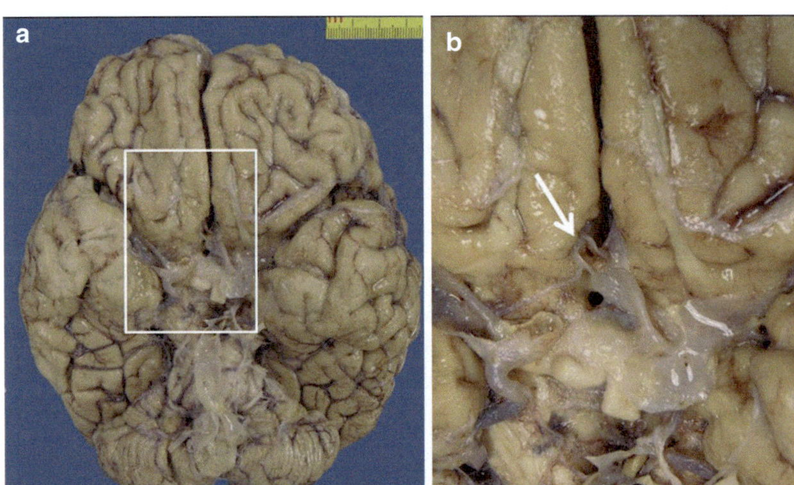

Fig. 14.14 Embolism of the right anterior cerebral artery due to detachment of a calcific nodule from a native aortic cusp, occurred during Sapien bioprosthetic valve deployment with ballooning. (**a**) Panoramic view. (**b**) Closeup: note the calcific embolic fragment occluding the right anterior cerebral artery. Arrow indicates the calcific embolism. (*From Thiene G, Basso C, Rizzo S, Della Barbera M, Valente M, Bortolotti U (eds). Pathology of Cardiac Valve Disease. 2023. Springer, Cham. ISBN: 978-3-031-35497-7*)

atively impact THV durability and increase the likelihood of leaflet thrombosis. This risk is especially pronounced in younger patients, where accelerated prosthetic degeneration could pose long-term issues. Long-term data on TAVI outcomes in BAV patients are still scarce but eagerly anticipated to better understand the durability and safety of this approach in younger, low-risk populations.

14.11 Future Directions

As TAVI continues to evolve, the indications for its use in BAV stenosis are expanding. The introduction of new-generation THVs has improved outcomes in BAV patients, but further research is needed to optimize device selection, procedural techniques, and patient selection criteria. Large-scale prospective studies with core laboratory CT scan analyses will delineate the specific BAV subtypes that benefit most from TAVI [50].

14.12 Conclusion

TAVI has emerged as a viable treatment option for BAV stenosis, particularly in elderly and high-risk patients. While early outcomes were suboptimal due to the anatomical complexities of BAV, advances in imaging, device technology, and procedural techniques have improved the safety and efficacy of TAVI in this population. However, careful patient selection and meticulous procedural planning are crucial for optimizing outcomes.

References

1. Siu SC, Silversides CK. Bicuspid aortic valve disease. J Am Coll Cardiol. 2010;55:2789–800.
2. Michelena HI, Desjardins VA, Avierinos JF, et al. Natural history of asymptomatic patients with normally functioning or minimally dysfunctional bicuspid aortic valve in the community. Circulation. 2008;117:2776–84.
3. Vahanian A, Beyersdorf F, Praz F, Milojevic M, Baldus S, Bauersachs J, Capodanno D, Conradi L, De Bonis M, De Paulis R, Delgado V, Freemantle N, Gilard M, Haugaa KH, Jeppsson A, Jüni P, Pierard L, Prendergast BD, Sádaba JR, Tribouilloy C, Wojakowski W, ESC/EACTS Scientific Document Group, ESC Scientific Document Group. 2021 ESC/EACTS guidelines for the management of valvular heart disease. Eur Heart J. 2021:ehab395. https://doi.org/10.1093/eurheartj/ehab395.
4. Otto CM, Nishimura RA, Bonow RO, Carabello BA, Erwin JP 3rd, Gentile F, Jneid H, Krieger EV, Mack M, McLeod C, O'Gara PT, Rigolin VH, Sundt TM 3rd, Thompson A, Toly C. 2020 ACC/AHA guideline for the management of patients with valvular heart disease: executive summary: a report of the American College of Cardiology/American Heart Association Joint Committee on Clinical Practice Guidelines. J Am Coll Cardiol. 2021;77(4):450–500. https://doi.org/10.1016/j.jacc.2020.11.035. Epub 2020 Dec 17. Erratum in: J Am Coll Cardiol. 2021 Mar 9;77(9):1276
5. Leon MB, Smith CR, Mack MJ, Makkar RR, Svensson LG, Kodali SK, Thourani VH, Tuzcu EM, Miller DC, Herrmann HC, Doshi D, Cohen DJ, Pichard AD, Kapadia S, Dewey T, Babaliaros V, Szeto WY, Williams MR, Kereiakes D, Zajarias A, Greason KL, Whisenant BK, Hodson RW, Moses JW, Trento A, Brown DL, Fearon WF, Pibarot P, Hahn RT, Jaber WA, Anderson WN, Alu MC, Webb JG, PARTNER 2 Investigators. Transcatheter or surgical aortic-valve replacement in intermediate-risk patients. N Engl J Med. 2016;374(17):1609–20. https://doi.org/10.1056/NEJMoa1514616. Epub 2016 Apr 2
6. Reardon MJ, Van Mieghem NM, Popma JJ, Kleiman NS, Søndergaard L, Mumtaz M, Adams DH, Deeb GM, Maini B, Gada H, Chetcuti S, Gleason T, Heiser J, Lange R, Merhi W, Oh JK, Olsen PS, Piazza N, Williams M, Windecker S, Yakubov SJ, Grube E, Makkar R, Lee JS, Conte J, Vang E, Nguyen H, Chang Y, Mugglin AS, Serruys PW, Kappetein AP, SURTAVI Investigators. Surgical or transcatheter aortic-valve replacement in intermediate-risk patients. N Engl J Med. 2017;376(14):1321–31. https://doi.org/10.1056/NEJMoa1700456. Epub 2017 Mar 17
7. Mack MJ, Leon MB, Thourani VH, Makkar R, Kodali SK, Russo M, Kapadia SR, Malaisrie SC, Cohen DJ, Pibarot P, Leipsic J, Hahn RT, Blanke P, Williams MR, McCabe JM, Brown DL, Babaliaros V, Goldman S, Szeto WY, Genereux P, Pershad A, Pocock SJ, Alu MC, Webb JG, Smith CR, PARTNER 3 Investigators. Transcatheter aortic-valve replacement with a balloon-expandable valve in low-risk patients. N Engl J Med. 2019;380(18):1695–705. https://doi.org/10.1056/NEJMoa1814052. Epub 2019 Mar 16
8. Popma JJ, Deeb GM, Yakubov SJ, Mumtaz M, Gada H, O'Hair D, Bajwa T, Heiser JC, Merhi W, Kleiman NS, Askew J, Sorajja P, Rovin J, Chetcuti SJ, Adams DH, Teirstein PS, Zorn GL 3rd, Forrest JK, Tchétché D, Resar J, Walton A, Piazza N, Ramlawi B, Robinson N, Petrossian G, Gleason TG, Oh JK, Boulware MJ, Qiao H, Mugglin AS, Reardon MJ, Evolut Low Risk Trial Investigators. Transcatheter aortic-valve replacement with a self-expanding valve in low-risk patients.

N Engl J Med. 2019;380(18):1706–15. https://doi.org/10.1056/NEJMoa1816885. Epub 2019 Mar 16

9. Tarantini G, Nai Fovino L, Gersh BJ. Transcatheter aortic valve implantation in lower-risk patients: what is the perspective? Eur Heart J. 2018;39:658–66.
10. Tarantini G, Fabris T. Transcatheter aortic valve replacement for bicuspid aortic valve stenosis: a practical operative overview. Circ Cardiovasc Interv. 2021;14(7):e009827. https://doi.org/10.1161/CIRCINTERVENTIONS.120.009827. Epub 2021 Jun 16. PMID: 34130478; PMCID: PMC10332651
11. Sievers HH, Schmidtke C. A classification system for the bicuspid aortic valve from 304 surgical specimens. J Thorac Cardiovasc Surg. 2007;133(5):1226–33. https://doi.org/10.1016/j.jtcvs.2007.01.039.
12. Jilaihawi H, Chen M, Webb J, Himbert D, Ruiz CE, Rodés-Cabau J, Pache G, Colombo A, Nickenig G, Lee M, et al. A bicuspid aortic valve imaging classification for the TAVR era. JACC Cardiovasc Imaging. 2016;9:1145–58.
13. Blanke P, Weir-McCall JR, Achenbach S, Delgado V, Hausleiter J, Jilaihawi H, Marwan M, Nørgaard BL, Piazza N, Schoenhagen P, Leipsic JA. Computed tomography imaging in the context of transcatheter aortic valve implantation (TAVI)/transcatheter aortic valve replacement (TAVR): an expert consensus document of the society of cardiovascular computed tomography. JACC Cardiovasc Imaging. 2019;12(1):1–24. https://doi.org/10.1016/j.jcmg.2018.12.003.
14. Frangieh AH, Michel J, Deutsch O, Joner M, Pellegrini C, Rheude T, Bleiziffer S, Kasel AM. Aortic annulus sizing in stenotic bicommissural non-raphe-type bicuspid aortic valves: reconstructing a three-dimensional structure using only two hinge points. Clin Res Cardiol. 2019;108:6–15.
15. Piazza N, de Jaegere P, Schultz C, Becker AE, Serruys PW, Anderson RH. Anatomy of the aortic valvar complex and its implications for transcatheter implantation of the aortic valve. Circ Cardiovasc Interv. 2008;1:74–81.
16. Tchetche D, de Biase C, van Gils L, Parma R, Ochala A, Lefevre T, Hovasse T, De Backer O, Sondergaard L, Bleiziffer S, Lange R, Kornowski R, Landes U, Norgaard BL, Biasco L, Philippart R, Molina-Martin de Nicolas J, Mylotte D, Lemee C, Dumonteil N, Van Mieghem NM. Bicuspid aortic valve anatomy and relationship with devices: the BAVARD multicenter registry. Circ Cardiovasc Interv. 2019;12(1):e007107. https://doi.org/10.1161/CIRCINTERVENTIONS.118.007107.
17. Iannopollo G, Romano V, Buzzatti N, Ancona M, Ferri L, Russo F, et al. Supra-annular sizing of transcatheter aortic valve prostheses in raphe-type bicuspid aortic valve disease: the LIRA method. Int J Cardiol. 2020;317:144–51. https://doi.org/10.1016/j.ijcard.2020.05.076.
18. Petronio AS, Angelillis M, De Backer O, Giannini C, Costa G, Fiorina C, et al. Bicuspid aortic valve sizing for transcatheter aortic valve implantation: development and validation of an algorithm based on multislice computed tomography. J Cardiovasc Comput Tomogr. 2020;14:452–61. https://doi.org/10.1016/j.jcct.2020.01.007.
19. Blackman D, Gabbieri D, Del Blanco BG, Kempfert J, Laine M, Mascherbauer J, Parma R, Tchétché D. Expert consensus on sizing and positioning of SAPIEN 3/ultra in bicuspid aortic valves. Cardiol Ther. 2021;10(2):277–88. https://doi.org/10.1007/s40119-021-00223-9. Epub 2021 Jun 3. PMID: 34081277; PMCID: PMC8555019
20. Kim WK, Renker M, Rolf A, Fischer-Rasokat U, Wiedemeyer J, Doss M, Möllmann H, Walther T, Nef H, Hamm CW, Liebetrau C. Annular versus supra-annular sizing for TAVI in bicuspid aortic valve stenosis. EuroIntervention. 2019;15(3):e231–8. https://doi.org/10.4244/EIJ-D-19-00236.
21. Tchétché D, Ziviello F, De Biase C, De Backer O, Hovasse T, Leroux L, Petronio AS, Saint-Etienne C, Teles RC, Modine T, Sudre A, Teiger E, Mylotte D, Souteyrand G, Piazza N, Casassus F, Sondergaard L, Angelillis M, Nolasco T, Siddiqui S, Kardys I, Dumonteil N, Van Mieghem NM. Transcatheter aortic valve implantation with the Evolut platform for bicuspid aortic valve stenosis: the international, multicentre, prospective BIVOLUTX registry. EuroIntervention. 2023;19(6):502–11. https://doi.org/10.4244/EIJ-D-23-00021. PMID: 37203860; PMCID: PMC10436069
22. Buono A, De Biase C, Fabris T, Bellamoli M, Kim WK, Montarello N, Costa G, Zito A, Alfadhel M, Koren O, Fezzi S, Bellini B, Massussi M, Scotti A, Bai L, Costa G, Mazzapicchi A, Giacomin E, Gorla R, Latini A, Fraccaro C, Sondergaard L, Strazzieri O, Boiago M, Busco M, Charitos E, Orbach A, Messina A, Bettari L, Navazio E, Paglianiti DA, Nagasaka T, Napodano M, Villa E, Angelillis M, Ielasi A, Landes U, Brambilla N, Bedogni F, Mangieri A, Saia F, Favero L, Chen M, Adamo M, Latib A, Petronio AS, Montorfano M, Makkar RR, Mylotte D, Blackman DJ, Barbanti M, De Backer O, Tchètchè D, Maffeo D, Tarantini G. CharActeristics, sizing anD outcomes of stenotic, tapered, rapHe-type bicuspid aOrtic valves treated with trans-catheter device implantation: insights the AD HOC registry. Int J Cardiol. 2024;417:132569. https://doi.org/10.1016/j.ijcard.2024.132569. Epub 2024 Sep 19
23. Mangieri A, Tchetchè D, Kim WK, Pagnesi M, Sinning JM, Landes U, Kornowski R, De Backer O, Nickenig G, Ielasi A, De Biase C, Søndergaard L, De Marco F, Montorfano M, Chiarito M, Regazzoli D, Stefanini G, Presbitero P, Toggweiler S, Tamburino C, Immè S, Tarantini G, Sievert H, Schäfer U, Kempfert J, Wöehrle J, Gallo F, Laricchia A, Latib A, Giannini F, Colombo A. Balloon versus self-expandable valve for the treatment of bicuspid aortic valve stenosis: insights from the BEAT International Collaborative Registrys. Circ Cardiovasc Interv. 2020;13(7):e008714. https://doi.org/10.1161/CIRCINTERVENTIONS.119.008714. Epub 2020 Jul 10

24. Buono A, Zito A, Kim WK, Fabris T, De Biase C, Bellamoli M, Montarello N, Costa G, Alfadhel M, Koren O, Fezzi S, Bellini B, Massussi M, Scotti A, Bai L, Costa G, Mazzapicchi A, Giacomin E, Gorla R, Hug K, Briguori C, Bettari L, Messina A, Boiago M, Villa E, Renker M, Garcia Gomez M, Fraccaro C, De Rosa ML, Patel V, Trani C, De Carlo M, Laterra G, Latini A, Pellegrini D, Ielasi A, Orbach A, Landes U, Rheude T, Testa L, Amat Santos I, Mangieri A, Saia F, Favero L, Chen M, Adamo M, Latib A, Sonia Petronio A, Montorfano M, Makkar RR, Mylotte D, Blackman DJ, Burzotta F, Barbanti M, De Backer O, Tchètchè D, Maffeo D, Tarantini G. Balloon-expandable vs self-expanding valves for transcatheter treatment of Sievers type 1 bicuspid aortic stenosis. JACC Cardiovasc Interv. 2024.:S1936-8798(24)01028-8; https://doi.org/10.1016/j.jcin.2024.07.031. Epub ahead of print
25. Giannini F, Tzanis G, Gallo F, Mangieri A, Laricchia A, Mazzavillani M, Fisicaro A, Gallone G, Colombo A. Looking for optimal implantation height in bicuspid aortic valve patients undergoing transcatheter aortic valve implantation: a case series. Cardiovasc Revasc Med. 2020;21:25–7.
26. Liu X, He Y, Zhu Q, Gao F, He W, Yu L, Zhou Q, Kong M, Wang J. Supra-annular structure assessment for self-expanding transcatheter heart valve size selection in patients with bicuspid aortic valve. Catheter Cardiovasc Interv. 2018;91:986–94.
27. Bailey J, Curzen N, Bressloff NW. The impact of imperfect frame deployment and rotational orientation on stress within the prosthetic leaflets during transcatheter aortic valve implantation. J Biomech. 2017;53:22–8.
28. Kawamori H, Yoon SH, Chakravarty T, Maeno Y, Kashif M, Israr S, Abramowitz Y, Mangat G, Miyasaka M, Rami T, et al. Computed tomography characteristics of the aortic valve and the geometry of SAPIEN 3 transcatheter heart valve in patients with bicuspid aortic valve disease. Eur Heart J Cardiovasc Imaging. 2018;19:1408–18.
29. Nai Fovino L, Scotti A, Massussi M, Cardaioli F, Rodinò G, Matsuda Y, Pavei A, Masiero G, Napodano M, Fraccaro C, et al. Coronary angiography after transcatheter aortic valve replacement (TAVR) to evaluate the risk of coronary access impairment after TAVR-in-TAVR. J Am Heart Assoc. 2020;9:e016446.
30. Tang GHL, Zaid S, Fuchs A, Yamabe T, Yazdchi F, Gupta E, Ahmad H, Kofoed KF, Goldberg JB, Undemir C, et al. Alignment of transcatheter aortic-valve neo-commissures (ALIGN TAVR): impact on final valve orientation and coronary artery overlap. JACC Cardiovasc Interv. 2020;13:1030–42.
31. Tarantini G, Fabris T, Nai FL. TAVR-in-TAVR and coronary access: importance of preprocedural planning. EuroIntervention. 2020;16:e129–32.
32. Tang GHL, Amat-Santos IJ, De Backer O, Avvedimento M, Redondo A, Barbanti M, Costa G, Tchétché D, Eltchaninoff H, Kim WK, Zaid S, Tarantini G, Søndergaard L. Rationale, definitions, techniques, and outcomes of commissural alignment in TAVR: from the ALIGN-TAVR consortium. JACC Cardiovasc Interv. 2022;15(15):1497–518. https://doi.org/10.1016/j.jcin.2022.06.001.
33. Wang L, Ming Wang L, Chen W, Chen X. Bicuspid aortic valve: a review of its genetics and clinical significance. J Heart Valve Dis. 2016;25(5):568–73.
34. Tarantini G, Fabris T, Nai Fovino L, Cardaioli F, Pergola V, Montonati C, Rodinò G, Cabrelle G, Massussi M, Scotti A, Zuccarelli V, Sciarretta T, Masiero G, Gregori D, Napodano M, Fraccaro C, Continisio S, Iliceto S. Definition of trAnscatheter heart valve orIeNtation in biCuspId aortic valve: the DA VINCI pilot study. Front Cardiovasc Med. 2022;9:1056496. https://doi.org/10.3389/fcvm.2022.1056496. PMID: 36578836; PMCID: PMC9790995
35. Tarantini G, Sathananthan J, Fabris T, Landes U, Bapat VN, Khan JM, Nai Fovino L, Zaid S, Van Mieghem NM, Latib A, Waksman R, De Backer O, Rogers T, Søndergaard L, Tang GHL. Transcatheter aortic valve replacement in failed transcatheter bioprosthetic valves. JACC Cardiovasc Interv. 2022;15(18):1777–93. https://doi.org/10.1016/j.jcin.2022.07.035.
36. Mentias A, Sarrazin MV, Desai MY, Saad M, Horwitz PA, Kapadia S, Girotra S. Transcatheter versus surgical aortic valve replacement in patients with bicuspid aortic valve stenosis. J Am Coll Cardiol. 2020;75:2518–9.
37. Elbadawi A, Saad M, Elgendy IY, Barssoum K, Omer MA, Soliman A, Almahmoud MF, Ogunbayo GO, Mentias A, Gilani S, Jneid H, Aronow HD, Kleiman N, Abbott JD. Temporal trends and outcomes of transcatheter versus surgical aortic valve replacement for bicuspid aortic valve stenosis. JACC Cardiovasc Interv. 2019;12(18):1811–22. https://doi.org/10.1016/j.jcin.2019.06.037.
38. Yoon S-H, Lefèvre T, Ahn J-M, Perlman GY, Dvir D, Latib A, Barbanti M, Deuschl F, De Backer O, Blanke P, Modine T. Transcatheter aortic valve replacement with early- and new-generation devices in bicuspid aortic valve stenosis. J Am Coll Cardiol. 2016;68:1195–205.
39. Perlman GY, Blanke P, Dvir D, Pache G, Modine T, Barbanti M, Holy EW, Treede H, Ruile P, Neumann F-J, et al. Bicuspid aortic valve stenosis: favorable early outcomes with a next-generation transcatheter heart valve in a multicenter study. JACC Cardiovasc Interv. 2016;9:817–24.
40. Yoon S-H, Bleiziffer S, De Backer O, Delgado V, Arai T, Ziegelmueller J, Barbanti M, Sharma R, Perlman GY, Khalique OK, et al. Outcomes in transcatheter aortic valve replacement for bicuspid versus tricuspid aortic valve stenosis. J Am Coll Cardiol. 2017;69:2579–89.
41. Ueshima D, Nai Fovino L, Brener S, Fabris T, Scotti A, Barioli A, Giacoppo D, Pavei A, Fraccaro C, Napodano M, Tarantini G. Transcatheter aortic valve replacement for bicuspid aortic valve steno-

sis with first- and new-generation bioprostheses: a systematic review and meta-analysis. Int J Cardiol. 2020;298:76–82.

42. Makkar RR, Yoon SH, Leon MB, Chakravarty T, Rinaldi M, Shah PB, Skipper ER, Thourani VH, Babaliaros V, Cheng W, et al. Association between transcatheter aortic valve replacement for bicuspid vs tricuspid aortic stenosis and mortality or stroke. JAMA. 2019;321:2193–202. https://doi.org/10.1001/jama.2019.7108.
43. Forrest JK, Kaple RK, Ramlawi B, Gleason TG, Meduri CU, Yakubov SJ, Jilaihawi H, Liu F, Reardon MJ. Transcatheter aortic valve replacement in bicuspid versus tricuspid aortic valves from the STS/ACC TVT registry. JACC Cardiovasc Interv. 2020;13:1749–59. https://doi.org/10.1016/j.jcin.2020.03.022.
44. Deeb GM, Reardon MJ, Ramlawi B, Yakubov SJ, Chetcuti SJ, Kleiman NS, Mangi AA, Zahr F, Song HK, Gada H, Mumtaz M, Heiser J, Merhi W, Murrah CP, Noel T, Kirshner M, Byrne T, Ito S, Huang J, Forrest JK. Propensity-matched 1-year outcomes following transcatheter aortic valve replacement in low-risk bicuspid and tricuspid patients. JACC Cardiovasc Interv. 2022;15(5):511–22. https://doi.org/10.1016/j.jcin.2021.10.027.
45. Williams MR, Jilaihawi H, Makkar R, O'Neill WW, Guyton R, Malaisrie SC, Brown DL, Blanke P, Leipsic JA, Pibarot P, Hahn RT, Leon MB, Cohen DJ, Bax JJ, Kodali SK, Mack MJ, Lu M, Webb JG. The PARTNER 3 bicuspid registry for transcatheter aortic valve replacement in low-surgical-risk patients. JACC Cardiovasc Interv. 2022;15(5):523–32. https://doi.org/10.1016/j.jcin.2022.01.279.
46. Yoon SH, Kim WK, Dhoble A, Milhorini Pio S, Babaliaros V, Jilaihawi H, Pilgrim T, De Backer O, Bleiziffer S, Vincent F, Shmidt T, Butter C, Kamioka N, Eschenbach L, Renker M, Asami M, Lazkani M, Fujita B, Birs A, Barbanti M, Pershad A, Landes U, Oldemeyer B, Kitamura M, Oakley L, Ochiai T, Chakravarty T, Nakamura M, Ruile P, Deuschl F, Berman D, Modine T, Ensminger S, Kornowski R, Lange R, McCabe JM, Williams MR, Whisenant B, Delgado V, Windecker S, Van Belle E, Sondergaard L, Chevalier B, Mack M, Bax JJ, Leon MB, Makkar RR. Bicuspid aortic valve stenosis transcatheter aortic valve replacement registry investigators. Bicuspid aortic valve morphology and outcomes after transcatheter aortic valve replacement. J Am Coll Cardiol. 2020;76(9):1018–30. https://doi.org/10.1016/j.jacc.2020.07.005.
47. Aljabbary T, Wijeysundera H, Radhakrishnan S. Cusp overlap method for self-expanding transcatheter aortic valve replacement. Can J Cardiol. 2020;36:S32–3.
48. Montalto C, Sticchi A, Crimi G, Laricchia A, Khokhar AA, Giannini F, Reimers B, Colombo A, Latib A, Waksman R, Mangieri A. Outcomes after transcatheter aortic valve replacement in bicuspid versus tricuspid anatomy: a systematic review and meta-analysis. JACC Cardiovasc Interv. 2021;14(19):2144–55. https://doi.org/10.1016/j.jcin.2021.07.052.
49. Kapadia SR, Makkar R, Leon M, Abdel Wahab M, Waggoner T, Massberg S, Rottbauer W, Horr S, Sondergaard L, Karha J, Gooley R, Satler L, Stoler RC, Messé SR, Baron SJ, Seeger J, Kodali S, Krishnaswamy A, Thourani VH, Harrington K, Pocock S, Modolo R, Allocco DJ, Meredith IT, Linke A, PROTECTED TAVR Investigators. Cerebral embolic protection during transcatheter aortic-valve replacement. N Engl J Med. 2022;387(14):1253–63. https://doi.org/10.1056/NEJMoa2204961. Epub 2022 Sep 17
50. Tarantini G, Fabris T, Cardaioli F, Nai FL. Coronary access after transcatheter aortic valve replacement in patients with bicuspid aortic valve: lights and shades. JACC Cardiovasc Interv. 2019;12:1190–1.

Further Reading

Abecasis J, Maltês S, Santos RR, Lopes P, Manso RT, Gil V, Cardim N, Ramos S, Félix A. Subendocardial "ischemic-like" state in patients with severe aortic stenosis: insights from myocardial histopathology and ultrastructure. Cardiovasc Pathol. 2024;69:107589.

Abott ME. Coarctation of the aorta of adult type; statistical study and historical retrospect of 200 recorded cases with autopsy; of stenosis or obliteration of descending arch in subjects above age of two years. Am Heart J. 1928;3:574.

Aicher D, Kunihara T, Abou Issa O, Brittner B, Graber S, Schafers HJ. Valve configuration determines long-term results after repair of the bicuspid aortic valve. Circulation. 2011;123:178–85.

Al-Hussaini A. Pregnancy and aortic dissections. Eur Heart J. 2020;41(44):4243–4. https://doi.org/10.1093/eurheartj/ehaa754. PMID: 33295620.

Alsoufi B, Borger MA, Armstrong S, Maganti M, David TE. Results of valve preservation and repair for bicuspid aortic valve insufficiency. J Heart Valve Dis. 2005;14:752–9.

Angelini A, Ho SY, Anderson RH, et al. The morphology of the normal aortic valve as compared with the aortic valve having two leaflets. J Thorac Cardiovasc Surg. 1989;98:362–7.

Bacchion F, Cukon S, Rizzoli G, Gerosa G, Daliento L, Thiene G, Basso C. Infective endocarditis in bicuspid aortic valve: atrioventricular block as sign of perivalvular abscess. Cardiovasc Pathol. 2007;16(4):252–5.

Bambury MK, Cosgrove DM, Thomas JD, et al. Hemodynamic stability during 17 years of the Carpentier-Edwards aortic pericardial bioprosthesis. Ann Thorac Surg. 2002;73:1460–5.

Basso C, Frescura C, Corrado D, Muriago M, Angelini A, Daliento L, Thiene G. Congenital heart disease and sudden death in the young. Hum Pathol. 1995;26:1065–72.

Basso C, Boschello M, Perrone C, Mecenero A, Cera A, Bicego D, Thiene G, De Dominicis E. An echocardiographic survey of primary school children for bicuspid aortic valve. Am J Cardiol. 2004;93(5):661–3. https://doi.org/10.1016/j.amjcard.2003.11.031. PMID: 14996606.

Bauer M, Gliech V, Siniawiski H, Hetzer L. Configuration of ascending aorta in patients with bicuspid and tricuspid aortic valve disease undergoing aortic valve replacement with or without reduction aortoplasty. J Heart Valve Dis. 2006;15:594–600.

Bentall H, De Bono A. A technique for complete replacement of the ascending aorta. Thorax. 1968;23(4):338–9.

Beppu S, Suzuki S, Matsuda H, Ohmori F, Nagata S, Miyatake K. Rapidity of progression of aortic stenosis in patients with congenital bicuspid aortic valves. Am J Cardiol. 1993;71:322–7.

Beroukhim RS, Kruzick TL, Taylor AL, Gao D, Yetman AT. Progression of aortic dilation in children with a functionally normal bicuspid aortic valve. Am J Cardiol. 2006;98(6):828–30.

Beyer SE, Dicks AB, Shainker SA, Feinberg L, Schermerhorn ML, Secemsky EA, Carroll BJ. Pregnancy-associated arterial dissections: a nationwide cohort study. Eur Heart J. 2020;41(44):4234–42. https://doi.org/10.1093/eurheartj/ehaa497. Erratum in: Eur Heart J 2021 Jul 31;42(29):2863. PMID: 32728725.

Bohbot Y, Denev S, Benvenga RM, Philip M, Michelena HI, Citro R, Habib G, Tribouilloy C. Characteristics and prognosis of isolated aortic valve infective endocarditis in patients with bicuspid aortic valves: a propensity matched study. Front Cardiovasc Med. 2023;10:1304957.

Bonderman D, Gharehbaghi-Schell E, Wollenek G, et al. Mechanism underlying aortic dilatation in congenital aortic valve malformation. Circulation. 1999;99:2138–43.

Boodhwani M, de Kerchove L, Glineur D, et al. Repair of regurgitant bicuspid aortic valves: a systematic approach. J Thorac Cardiovasc Surg. 2010;140:276–84.

Borger MA, Preston M, Ivanov J, et al. Should the ascending aorta be replaced more frequently in patients with bicuspid aortic valve disease? J Thorac Cardiovasc Surg. 2004;128:677–83.

Boyer JK, Gutierrez F, Braverman AC. Approach to the dilated aortic root. Curr Opin Cardiol. 2004;19:563–9.

G. Thiene et al. (eds.), *Bicuspid Aortic Valve Syndrome*, https://doi.org/10.1007/978-3-031-85589-4

Boyum J, Fellinger EK, Schmoker JD, et al. Matrix metalloproteinase activity in thoracic aortic aneurysms associated with bicuspid and tricuspid aortic valves. J Thorac Cardiovasc Surg. 2004;127:686–91.

Brandenburg RO, Tajik AJ, Edwards WD, et al. Accuracy of 2 dimensional echocardiographic diagnosis of congenitally bicuspid aortic valve: echocardiographic-anatomic correlation in 115 patients. Am J Cardiol. 1983;51:1469–73.

Braverman AC, Güven H, Beardslee MA, Makan M, Kates AM, Moon MR. The bicuspid aortic valve. Curr Probl Cardiol. 2005;30(9):470–522. https://doi.org/10.1016/j.cpcardiol.2005.06.002. PMID: 16129122.

Bray JJH, Freer R, Pitcher A, Kharbanda R. Family screening for bicuspid aortic valve and aortic dilatation: a meta-analysis. Eur Heart J. 2023;44(33):3152–64.

Broberg CS, Therrien J. Understanding and treating aortopathy in bicuspid aortic valve. Trends Cardiovasc Med. 2015;25(5):445–51. https://doi.org/10.1016/j.tcm.2014.12.006. Epub 2014 Dec 17. PMID: 25640310.

Brown CB, Baldwin HS. Neural crest contribution to the cardiovascular system. Adv Exp Med Biol. 2006;589:134–54.

Brown C, Sane DC, Kitzman DW. Bicuspid aortic valves in monozygotic twins. Echocardiography. 2003;20:183–4.

Bulut HI, Arjomandi Rad A, Syrengela A-A, Ttofi I, Djordjevic J, Kaur R, Keiralla A, Krasopoulos G. A comprehensive review of management strategies for bicuspid aortic valve (BAV): exploring epidemiology, aetiology, aortopathy, and interventions in light of recent guidelines. J Cardiovasc Dev Dis. 2023;10(9):398. https://doi.org/10.3390/jcdd10090398.

Casselman FP, Giinov A, Akhrass R, Kasirajan V, Blackstone E, Cosgrove D. Intermediate-term durability of bicuspid aortic valve repair for prolapsing leaflet. Eur J Cardiothorac Surg. 1999;15:302–8.

Castonguay MC, Burner KD, Edwards WD, Baddour LM, Maleszewski JJ. Surgical pathology of native valve endocarditis in 310 specimens from 287 patients (1985-2004). Cardiovasc Pathol. 2013;22(1):19–27.

Chan KL, Stinson WA, Veinot JP. Reliability of transthoracic echocardiography in the assessment of aortic valve morphology: pathological correlation in 178 patients. Can J Cardiol. 1999;15:48–52.

Chatrath N, Westaby J, Finocchiaro G, Sharma S, Esteban MT, Papadakis M, Sheppard MN. The role of the bicuspid aortic valve in sudden cardiac death-findings at cardiac autopsy. Cardiovasc Pathol. 2023;65:107527.

Chiam P, Chao V, et al. Percutaneous transcatheter heart valve implantation in a bicuspid aortic valve. JACC. 2010;3:559–61.

Ciotti GR, Vlahos AP, Silverman NH. Morphology and function of the bicuspid aortic valve with and without coarctation of the aorta in the young. Am J Cardiol. 2006;98:1096–102.

Clarke DR, Bishop DA. Congenital malformation of the aortic valve and left ventricular outflow tract. In: Baue AE, editor. Glenn's thoracic and cardiovascular surgery. 6th ed. Eat Norwalk, CT: Appleton & Lange; 1996. p. 1221–42.

Clementi M, Notari L, Borghi A, Tenconi R. Familial congenital bicuspid aortic valve: a disorder of uncertain inheritance. Am J Med Genet. 1996;62:336–8.

Cosgrove DM, Rosenkranz ER, Hendren WG, Barlett JC, Stewart WJ. Valvuloplasty for aortic insufficiency. J Thorac Cardiovasc Surg. 1991;102:571–7.

Cotrufo M, Della Corte A. The association of bicuspid aortic valve disease with asymmetric dilatation of the tubular ascending aorta: identification of a definite syndrome. J Cardiovasc Med (Hargestown). 2009;10:291–7.

Cotrufo M, Della Corte A, De Santo LS, et al. Different patterns of extracellular matrix protein expression in the convexity and concavity of the dilated aorta with bicuspid aortic valve: preliminary results. J Thorac Cardiovasc Surg. 2005;130:504–11.

Cripe L, Andelfinger G, at al. Bicuspid aortic valve is heritable. J Am Coll Cardiol. 2004;44:138–43.

Crociani P. Antibiotic prophylaxis of infective endocarditis in dental procedures: adverse reactions. Cardiologia Ambulatoriale. 2024;1:111–20.

Dalla Corte A, De Santo LS, Montagnani S, et al. Spatial patterns of matrix protein expression in dilated ascending aorta with aortic regurgitation: congenital bicuspid valve versus Marfan's syndrome. J Heart Valve Dis. 2006;15:20–7; discusion 27.

Dalla Corte A, Quarto C, Bancone C, et al. Spatiotemporal patterns of smooth muscle cell changes in ascending aortic dilatation with bicuspid and tricuspid aortic valve stenosis: focus on cell-matrix signaling. J Thorac Cardiovasc Surg. 2008;135:8–18,18 el–2.

Dargis N, Lamontagne M, Gaudreault N, Sbarra L, Henry C, Pibarot P, Mathieu P, Bossé Y. Identification of gender-specific genetic variants in patients with bicuspid aortic valve. Am J Cardiol. 2016;117(3):420–6.

David TE, Omran A, Ivanov J, et al. Dilatation of the pulmonary autograft after the Ross procedure. J Thorac Cardiovasc Surg. 2000;119:210–20.

David TE, Ivanov J, Armstrong S, et al. Late result of heart valve replacement with the Hancock II bioprosthesis. J Thorac Cardiovasc Surg. 2001;121:268–77.

David TE, Feindel CM, Webb GD, Colman JM, Armstrong S, Maganti M. Long term results of aortic valve-sparing operations for aortic root aneurysm. J Thorac Cardiovasc Surg. 2006;132(2):347–54.

Davies RR, Gallo A, Coady MA, et al. Novel measurement of relative aortic size predicts rupture of thoracic aortic aneurysms. Ann Thorac Surg. 2006;81:169–77.

De Backer J, Prakash SK. Refining the course: long-term outcome in patients with bicuspid aortic valve. Eur Heart J. 2023;44(43):4563–5. https://doi.org/10.1093/eurheartj/ehad621. PMID: 37793128.

De Kerchove L, Rubay J, Pasquet A, Poncelet A, Ovaert C, Pirotte M, et al. Ross operation in the adult: long-term

outcomes after root replacement and inclusion techniques. Ann Thorac Surg. 2009;87:95–102.
De Kerchove L, Boodhwani M, Glineur D, et al. Valve-sparing root replacement with reimplantation technique to increase the durability of bicuspid aortic valve repair. J Thorac Cardiovasc Surg. 2011;142(9):1430–8.
De Martino A, Morganti R, Falcetta G, Scioti G, Milano AD, Pucci A, Bortolotti U. Acute aortic dissection and pregnancy: review and meta-analysis of incidence, presentation, and pathologic substrates. J Card Surg. 2019;34(12):1591–7. https://doi.org/10.1111/jocs.14305. Epub 2019 Nov 5. PMID: 31794127.
De Reijer PM, Sallee D III, van der Velden P, et al. Haemodynamic predictors of aortic dilatation in bicuspid aortic valve by velocity-encoded cardiovascular magnetic resonance. J Cardiovasc Magn Reson. 2010;12:4.
De Sa M, Moshkovitz Y, Butany J, et al. Histologic abnormalities of the ascending aorta and pulmonary trunk in patients with bicuspid aortic valve disease: clinical relevance to the Ross procedure. J Thorac Cardiovasc Surg. 1999;118:588–94.
Della Barbera M, Valente M, Basso C, Thiene G. Morphologic studies of cell endogenous repopulation in decellularized aortic and pulmonary homografts implanted in sheep. Cardiovasc Pathol. 2015;24(2):102–9. https://doi.org/10.1016/j.carpath.2014.10.001. Epub 2014 Oct 16. Erratum in: Cardiovasc Pathol. 2015 May-Jun;24(3):198. PMID: 25541180.
Dellgren G, Feindel CM, Bos J, et al. Aortic valve replacement with the Toronto SPV: long term clinical and hemodynamic result. Eur J Cardiothorac Surg. 2002a;21:698–702.
Dellgren G, David TE, Raanani E, et al. Late hemodynamic and clinical outcomes of aortic valve replacement with Carpentier-Edwards Perimount pericardial bioprosthesis. J Thorac Cardiovasc Surg. 2002b;124:146–54.
Duran AC, et al. Bicuspid aortic valves in hearts with other congenital heart disease. J Heart Valve Dis. 1995;4:581–90.
Elkins RC, Thompson DM, Lane MM, Elkins CC, Peyton MD. Ross operation: 16-year experience. J Thorac Cardiovasc Surg. 2008;136:623–30, 630.e1–5.
Emanuel R, Withers R, O'Brien K, et al. Congenitally bicuspid aortic valve. Clinicogenetic study of 41 families. Br Heart J. 1978;40:1402–7.
Erbel R, Aboyans V, Boileau C, Bossone E, Di Bartolomeo R, Eggebrecht H, Evangelista A, Falk V, Frank H, Gaemperli O, Grabenwoger M, Haverich A, Iung B, John Manolis A, Meijboom F, Nienaber CA, Roffi M, Rousseau H, Sechtem U, Sirnes PA, von Allmen RS, Vrints CJ, Authors/Task Force Members. Corrigendum to: 2014 ESC guidelines on the diagnosis and treatment of aortic diseases. Eur Heart J. 2015;36(41):2779. https://doi.org/10.1093/eurheartj/ehv178. Epub 2015 Jun 29. Erratum for: Eur Heart J. 2014 Nov 1;35(41):2873–926.
Etz CD, Homann TM, Silovitz D, et al. Long-term survival after the Bentall Procedure in 206 patients with bicuspid aortic valve. Ann Thorac Surg. 2007;84:1186–94.
Evangelista A, Michelena HI. Family screening for bicuspid aortic valve: indicated, but easy to implement? Eur Heart J. 2023;44(33):3165–7.
Fedak PW, Verma S, David TE, et al. Clinical and pathophysiological implications of a bicuspid aortic valve. Circulation. 2002;106:9000–4.
Fedak PWM, de Sa MPL, Verma S, et al. Vascular matrix remodelling in patients with bicuspid aortic valve malformation: implication for aortic dilatation. J Thorac Cardiovasc Surg. 2003;126:797–806.
Fenoglio JJ, McAlister HA, Decastro CM, et al. Congenital bicuspid aortic valve after age 20. Am J Cardiol. 1997;39:164–9.
Fernandes SM, Sanders SP, Khairy P, et al. Morphology of bicuspid aortic valve in children and adolescents. J Am Coll Cardiol. 2004;44:1648–51.
Fernandez B, Fernandez MC, Duran AC, et al. Anatomy and formation of congenital bicuspid and quadricuspid pulmonary valves in Syrian hamsters. Anat Rec. 1998;250:70–9.
Friedman T, Mani A, Elefteriaded JA. Bicuspid aortic valve: clinical approach and scientific review of a common clinical entity. Expert Rev Cardiovasc Ther. 2008;6:235–48.
Garg V, Muth AN, Ransom JF, et al. Mutations in NOTCH1 cause aortic valve disease. Nature. 2005;437:270–4.
Gehlen J, Stundl A, Debiec R, Fontana F, Krane M, Sharipova D, Nelson CP, Al-Kassou B, Giel AS, Sinning JM, Bruenger CMH, Zelck CF, Koebbe LL, Braund PS, Webb TR, Hetherington S, Ensminger S, Fujita B, Mohamed SA, Shrestha M, Krueger H, Siepe M, Kari FA, Nordbeck P, Buravezky L, Kelm M, Veulemans V, Adam M, Baldus S, Laugwitz KL, Haas Y, Karck M, Mehlhorn U, Conzelmann LO, Breitenbach I, Lebherz C, Urbanski P, Kim WK, Kandels J, Ellinghaus D, Nowak-Goettl U, Hoffmann P, Wirth F, Doppler S, Lahm H, Dreßen M, von Scheidt M, Knoll K, Kessler T, Hengstenberg C, Schunkert H, Nickenig G, Nöthen MM, Bolger AP, Abdelilah-Seyfried S, Samani NJ, Erdmann J, Trenkwalder T, Schumacher J. Elucidation of the genetic causes of bicuspid aortic valve disease. Cardiovasc Res. 2023;119(3):857–66.
Gerosa G, Pontarollo S, Iliceto S, Di Marco F. An alternative technique for aortic root remodeling in patients with bicuspid aortic valve. J Thorac Cardiovasc Surg. 2007;133:249–50.
Giusti S, Cocco P, Thiene G. The bicuspid aortic valve: a "minor" congenital cardiopathy at risk of serious complications. G Ital Cardiol. 1991;21:189–201.
Glick BN, Roberts WC. Congenitally bicuspid aortic valve in multiple family members. Am J Cardiol. 1994;73:400–4.
Glotzbach JP, Hanson HA, Tonna JE, Horns JJ, McCarty Allen C, Presson AP, Griffin CL, Zak M, Sharma V, Tristani-Firouzi M, Selzman CH. Familial asso-

ciations of prevalence and cause-specific mortality for thoracic aortic disease and bicuspid aortic valve in a large-population database. Circulation. 2023;148(8):637–47.

Grewal N, Girdauskas E, Idhrees M, Velayudhan B, Klautz R, Driessen A, Poelmann RE. Structural abnormalities in the non-dilated ascending aortic wall of bicuspid aortic valve patients. Cardiovasc Pathol. 2023;62:107478.

Guiney TE, Davies MJ, Parker DJ, et al. The etiology and course of isolated severe aortic regurgitation. Br Heart J. 1987;58:358–68.

Guntheroth WG, Spiers PS. Does aortic root dilatation with bicuspid aortic valves occur as a primary tissue abnormality or as a relatively benign poststenotic phenomenon? Am J Cardiol. 2005;95:820.

Guo DC, Pannu H, Tran Fadulu V, et al. Mutations in smooth muscle alpha-actin (ACTA2) lead to thoracic aortic aneurysm and dissections. Nat Genet. 2007;39:1488–93.

Gurvitz M, Chang RK, Drant S, Allada V. Frequency of aortic root dilation in children with a bicuspid valve. Am J Cardiol. 2004;94:1337–40.

Halushka MK, Angelini A, Bartoloni G, Basso C, Batoroeva L, Bruneval P, Buja LM, Butany J, d'Amati G, Fallon JT, Gallagher PJ, Gittenberger-de Groot AC, Gouveia RH, Kholova I, Kelly KL, Leone O, Litovsky SH, Maleszewski JJ, Miller DV, Mitchell RN, Preston SD, Pucci A, Radio SJ, Rodriguez ER, Sheppard MN, Stone JR, Suvarna SK, Tan CD, Thiene G, Veinot JP, van der Wal AC. Consensus statement on surgical pathology of the aorta from the Society for Cardiovascular Pathology and the Association for European Cardiovascular Pathology: II. Noninflammatory degenerative diseases—nomenclature and diagnostic criteria. Cardiovasc Pathol. 2016;25(3):247–57.

Hayashida K, Bouvier E, Lefèvre T, Chevalier B, Hovasse T, Romano M, Garot P, Watanabe Y, Farge A, Donzeau-Gouge P, Cormier B, Morice MC. Transcatheter aortic valve implantation for patients with severe bicuspid aortic valve stenosis. Circ Cardiovasc Interv. 2013;6(3):284–91.

Haydar HS, He G-W, Hovaguimian H, McIrvin DM, King DH, Starr A. Valve repair for aortic insufficiency: surgical classification an techniques. Eur J Cardiothorac Surg. 1997;11:258–65.

Hope MD, Hope TA, Meadows AK, et al. Bicuspi aortic valve: four-dimensional MR evaluation of ascending aortic systolic flow patterns. Radiology. 2010;255:53–61.

Horstkotte D, Follath F, Gutschik E, et al. Guidelines on prevention, diagnosis and treatment of infective endocarditis executive summary; the task force on infective endocarditis of the European Society of Cardiology. Eur Heart J. 2004;25(3):267–76.

Hultgren R, Sakalihasan N. Do we need new thresholds for surgical repair in patients with ascending thoracic aortic aneurysm disease? Eur Heart J. 2023;44(43):4589–91.

Huntington K, Hunter AG, Chan KL. A prospective study to assess the frequency of familial clustering of congenital bicuspid aortic valve. J Am Coll Cardiol. 1997;30:1809–12.

Hutchins GM, Nazarian IH, Bulkley BH. Association of left dominant coronary arterial system with congenital bicuspid aortic valve. Am J Cardiol. 1978;42:57–9.

Immer FF, Bansi AG, Immer-Bansi AS, McDougall J, Zehr KJ, Schaff HV, Carrel TP. Aortic dissection in pregnancy: analysis of risk factors and outcome. Ann Thorac Surg. 2003;76(1):309–14. https://doi.org/10.1016/s0003-4975(03)00169-3. PMID: 12842575.

Jain D, Dietz HC, Oswald GL, Maleszewski JJ, Halushka MK. Causes and histopathology of ascending aortic disease in children and young adults. Cardiovasc Pathol. 2011;20(1):15–25. https://doi.org/10.1016/j.carpath.2009.09.008. PMID: 19926309; PMCID: PMC3046386.

Jamieson WR, Burr LH, Miyagishima RT, Germann E, Macnab JS, Stanford E, et al. Carpentier-Edwards supra-annular aortic porcine bioprosthesis: clinical performance over 20 years. J Thorac Cardiovasc Surg. 2005;130:994–1000.

Kappetein AP, Glittemberger-de Groot AC, Zwinderman AH, et al. The neural crest as a possible pathogenetic factor in coarctation of the aorta and bicuspid aortic valve. J Thorac Cardiovasc Surg. 1991;102:830–6.

Laenens D, Stassen J, Galloo X, Ewe SH, Singh GK, Ammanullah MR, Hirasawa K, Sia CH, Butcher SC, Chew NWS, Kong WKF, Poh KK, Ding ZP, Ajmone Marsan N, Bax JJ. The impact of atrial fibrillation on prognosis in aortic stenosis. Eur Heart J Qual Care Clin Outcomes. 2023;9(8):778–84.

Laforest B, Andelfinger G, Nemer M. Loss of Gata5 in mice leads to bicuspid aortic valve. J Clin Invest. 2011;121(7):2876–87.

Larson EW, Edwards WD. Risk factors for aortic dissection: a necropsy study of 161 cases. Am J Cardiol. 1984;53:849–55.

Le Maire SA, Wang X, Wilks JA, et al. Matrix metalloproteinase in ascending aortic aneurism: bicuspid versus trileaflet aortic valves. J Surg Res. 2005;123:40–8.

Leon MB, Smith CR, Mack M, et al. Transcatheter aortic-valve implantation for aortic stenosis in patients who cannot undergo surgery. N Engl J Med. 2010;363:1597–607.

Lewin MB, Otto CM. The bicuspid aortic valve: adverse outcomes from infancy to old age. Circulation. 2005;111:832–4.

Manalo-Estrella P, Barker AE. Histopathologic findings in human aortic media associated with pregnancy. Arch Pathol. 1967;83(4):336–41.

Mancuso D, Basso C, Cardiaioli P, et al. Clefted bicuspid aortic valve. Cardiovasc Pathol. 2002;11:217–20.

Markl M, Draney MT, Hope MD, et al. Time-resolved 3-dimensional velocity mapping in the thoracic aorta: visualization of 3-directional blood flow patterns in healthy volunteers and patients. J Comput Assist Tomogr. 2004;28:459–68.

Martin LJ, Ramachandran V, Cripe LH, et al. Evidence in favor of linkage to human chromosomal regions 18q, 5q and 13q for bicuspid aortic valve and associated cardiovascular malformations. Hum Genet. 2007;121:275–84.

Matthias Bechtel JF, Noack F, Sayk F, Erasmi AW, Bartels C, Sievers HH. Histopathological grading of ascending aortic aneurysm: comparison of patients with bicuspid versus tricuspid aortic valve. J Heart Valve Dis. 2003;12(1):54–61.

Matura LA, Ho VB, Rosing DR, Bondi CA. Aortic dilatation and dissection in Turner syndrome. Circulation. 2007;116:1663–70.

McKusick VA. Association of congenital bicuspid aortic valve and erdheim's cystic medial necrosis. Lancet. 1972;1(7758):1026–7.

Michelena HI. The bicuspid aortic valve aortopathy mystery continues: are we that mediocre? Trends Cardiovasc Med. 2015;25(5):452–5.

Michelena HI, Desjardins VA, Avierinos JF, Russo A, Nkomo VT, Sundt TM, Pellikka PA, Tajik AJ, Enriquez-Sarano M. Natural history of asymptomatic patients with normally functioning or minimally dysfunctional bicuspid aortic valve in the community. Circulation. 2008;117(21):2776–84.

Michelena H, Khanna A, Mahoney D, et al. Incidence of aortic complications in patients with bicuspid aortic valves. JAMA. 2011;306(10):1104–13.

Milleron O, Eliahou L, Jondeau G. There is insufficient evidence to lower the threshold for prophylactic aortic surgery. Eur Heart J. 2024;45(8):632.

Mills P, Leech G, Davies M, Leathan A. The natural history of a non-stenotic bicuspid aortic valve. Br Heart J. 1978;40(9):951–7.

Mohamed SA, Aherrahrou Z, Liptau H, et al. Novel missense mutations (p. T596 and p. 1797H) in NOTCH1 in patients with bicuspid aortic valve. Biochem Biophys Res Commun. 2006;345:1460–165.

Moidl R, Moritz A, Simon P, Kupilik N, Wolner E, Mohl W. Echocardiographic result after repair of incompetent bicuspid aortic valves. Ann Thorac Surg. 1995;60:669–72.

Monda E, Fusco A, Melis D, et al. Clinical significance of family history and bicuspid aortic valve in children and young adult patients with Marfan syndrome. Cardiol Young. 2020;30(5):663–7.

Moore P, Egito E, Mowrey H, Perry SB, Lock JE, Keane JF. Midterm result of balloon dilatation of congenital aortic stenosis: predictors of success. J Am Coll Cardiol. 1996;27:1257–63.

Murphy ES, Rosch J, Rahimtoola S. The frequency and significance of coronary arterial dominance in isolated aortic stenosis. Am J Cardiol. 1977;39:505–9.

Nanda NC, Graniac R, Manning J, Mahoney EB, Lipchik EO, De Weese JA. Echocardiographic recognition of the congenital bicuspid aortic valve. Circulation. 1974;49:870–5.

Nappi F, Avtaar Singh SS, de Siena PM. Bicuspid aortic valve in children and young adults for cardiologists and cardiac surgeons: state-of-the-art of literature review. J Cardiovasc Dev Dis. 2024;11(10):317. https://doi.org/10.3390/jcdd11100317. PMID: 39452287; PMCID: PMC11509083.

Nataatmadjia M, West M, West J, et al. Abnormal extracellular matrix protein transport associated with increased apoptosis of vascular smooth muscle cells in Marfan syndrome and bicuspid aortic valve thoracic aortic aneurysm. Circulation. 2003;108(Suppl II):II-329–34.

Navaro GM, Mishra M, Griffin BP. Incidence and echocardiographic features of congenital unicuspid aortic valve in a adult population. J Heart Valves Dis. 2003;12:674–8.

Nishimura RA, Carabello BA, Faxon DP, et al. Focused update incorporated into the ACC/AHA 2006 guidelines for the management of patients with valvular heart disease. A report of the American College of Cardiology/American Heart Association Task Force on practice guidelines (Writing Committee to revise the 1998 guidelines for the management of patients with valvular heart disease). JACC. 2008a;52:1–142.

Nishimura RA, Carabello BA, Faxon DP, et al. ACC/AHA 2008 guideline update on valvular heart disease: focused update on infective endocarditis: a report of the American College of Cardiology/America Heart Association Task Force on practice guidelines. J Am Coll Cardiol. 2008b;52:676–85.

Nistri S, Sorbo MD, Marin M, Palisi M, Scognamiglio R, Thiene G. Aortic root dilatation in young men with normally functioning bicuspid aortic valves. Heart. 1999;82:19–22.

Nistri S, Sorbo MD, Basso C, Thiene G. Bicuspid aortic valve: abnormal aortic elastic properties. J Heart Valve Dis. 2002;11(3):369–73; discussion 373–4. PMID: 12056729.

Nistri S, Basso C, Marzari C, Mormino P, Thiene G. Frequency of bicuspid aortic valve in young male conscripts by echocardiogram. Am J Cardiol. 2005;96(5):718–21. https://doi.org/10.1016/j.amjcard.2005.04.051. PMID: 16125502.

Nistri S, Grande-Allen J, Noale M, Basso C, Siviero P, Maggi S, Crepaldi G, Thiene G. Aortic elasticity and size in bicuspid aortic valve syndrome. Eur Heart J. 2008;29(4):472–9. https://doi.org/10.1093/eurheartj/ehm528. Epub 2007 Dec 20. PMID: 18096569.

Nistri S, Basso C, Thiene G. La valvola aortica bicuspide [Bicuspid aortic valve]. G Ital Cardiol (Rome). 2012a;13(1):25–37.

Nistri S, Porciani MC, Attanasio M, Abbate R, Gensini GF, Pepe G. Association of Marfan syndrome and bicuspid aortic valve: frequency and outcome. Int J Cardiol. 2012b;155(2):324–5.

O'Sullivan CJ, Windecker S. Implications of bicuspid aortic valves for transcatheter aortic valve implantation. Circ Cardiovasc Interv. 2013;6(3):204–6.

Oliver JM, Gallego P, Gonzales A, et al. Risk factors for aortic complications in adults with coarctation of the aorta. J Am Coll Cardiol. 2004;44:1641–7.

Olson LJ, Subramanian R, Edwards WD. Surgical pathology of pure aortic insufficiency: a study of 225 cases. Mayo Clin Proc. 1984;59:835–41.

Osler W. The bicuspid condition of the aortic valve. Trans Assos Am Physicians. 1986;2:185–92.

Passik CS, Ackermann DM, Pluth JK, Edwards WD. Temporal changes in the causes of aortic stenosis: a surgical pathologic study of 646 cases. Mayo Clin Proc. 1987;62:119–23.

Pereira SC, Abrantes AL, António PS, Morais P, Sousa C, David C, Pinto FJ, Almeida AG, Caldeira D. Infective endocarditis risk in patients with bicuspid aortic valve: systematic review and meta-analysis. Int J Cardiol Heart Vasc. 2023;47:101249.

Pileggi S, De Chiara B, Magnoli M, Franzosi MG, Merlanti B, Bianchini F, Moreo A, Romeo G, Russo CF, Rizzo S, Basso C, Martinelli L, Maseri A, VAR Study Group. Sequencing of NOTCH1 gene in an Italian population with bicuspid aortic valve: preliminary results from the GISSI OUTLIERS VAR study. Gene. 2019;715:143970. https://doi.org/10.1016/j.gene.2019.143970. Epub 2019 Jul 19. PMID: 31330235.

Pollari F, Berretta P, Albertini A, Carrel T, Teoh K, Meuris B, Villa E, Kappert U, Andreas M, Solinas M, Misfeld M, Savini C, Fiore A, Shrestha M, Santarpino G, Martinelli GL, Mignosa C, Glauber M, Yan T, Fischlein T, Di Eusanio M. Pacemaker after sutureless and rapid-deployment prostheses: a progress report from the SURD-IR. Thorac Cardiovasc Surg. 2023;71(7):557–65.

Prakash SK, Michelena HI, Milewicz DM. Bicuspid aortic valve and thoracic aortic disease: further evidence of clinically silent but deadly risk to family members of affected individuals. Circulation. 2023;148(8):648–50.

Pucci A, Rossetti M, Lenzi C, Buja ML. The cardiovascular pathologist in the aortic team. Cardiovasc Pathol. 2024;72:107649.

Razzolini R, Longhi S, Tarantini G, Rizzo S, Napodano M, Abate E, Fraccaro C, Thiene G, Iliceto S, Gerosa G, Basso C. Relation of aortic valve weight to severity of aortic stenosis. Am J Cardiol. 2011;107(5):741–6. https://doi.org/10.1016/j.amjcard.2010.10.052. Epub 2011 Jan 19. PMID: 21247543.

Rizzo S, Basso C, Thiene G. Aorta bicuspide dall'anatomia patologica alla clinica. Cardiol Sci. 2024;12:8–14.

Roberts WC. The congenitally aortic bicuspid valve. A study of 85 autopsy cases. Am J Cardiol. 1970a;26:72–83.

Roberts WC. The structure of the aortic valve in clinically isolated aortic stenosis. Circulation. 1970b;42:91–7.

Roberts WC. Prophylactic replacement of a dilated ascending aorta at the time of aortic valve replacement of a dysfunctioning congenitally unicuspid or bicuspid aortic valve. Am J Cardiol. 2011;108(9):1371–2.

Roberts WC, Ko JM. Frequency by decades of unicuspid, bicuspid, and tricuspid aortic valves in adults having isolated aortic valve replacement for aortic stenosis, with or without associated aortic regurgitation. Circulation. 2005;111:920–5.

Roberts WC, Morrow AG, McInthosh CL, et al. Congenitally bicuspid aortic valve causing severe pure aortic regurgitation without superimposed infective endocarditis. Am J Cardiol. 1981;47:206–9.

Robicsek F, Thubrikar MJ, Cook JW, Fowler B. The congenitally bicuspid aortic valve: how does it function? Why does it fail? Ann Thorac Surg. 2004;77(1):177–85.

Rosenfeld HM, Landzberg MJ, Perry SB, Colan SD, Keane JF, Lock JE. Balloon aortic valvuloplasty in the young adult with congenital aortic stenosis. Am J Cardiol. 1994;73:1112–7.

Russo CF, Mazzetti S, Garatti A, et al. Aortic complications after bicuspid aortic valve replacement: long term results. Ann Thorac Surg. 2002;74:S1773–6, discussion S1792–9.

Sabet HY, Edwards WD, Tazelaar HD, et al. Congenitally bicuspid aortic valves: a surgical pathology study of 542 cases and a literature review of 2715 additional cases. Mayo Clin Proc. 1999;74:14–26.

Salem DN, Stein PD, Al-Ahmad A, Bussey HI, Horstkotte D, Miller N, et al. Antithrombotic therapy in valvular heart disease-native and prosthetic: the seventh ACCP conference on antithrombotic and thrombolytic therapy. Chest. 2004;126(3 Suppl):457S–82S.

Sans-Coma V, Fernandez B, Duran AC, Thiene G, Arque JM, Munoz-Chapuli R, Cardo M. Fusion of valve cushions as a key factor in the formation of congenital bicuspid aortic valve in Syrian hamsters. Anat Rec. 1996;244:490–8.

Sans-Coma V, Fernandez M, Fernandez B, et al. Genetically alike Syrian hamsters display both bifoliate and trifoliate aortic valves. J Anat. 2012;220(1):92–101. https://doi.org/10.1111/j.1469-7580.

Sareyyupoglu B, Suri RM, Shaff HV, Dearani JA, Daly RC, et al. Survival and reoperation risk following bicuspid aortic valve-sparing root replacement. J Heart Valve Dis. 2009;18(1):1–8.

Schaefer BM, Lewin MB, Stout KK, et al. Usefulness of bicuspid aortic valve phenotype to predict elastic properties of the ascending aorta. Am J Cardiol. 2007;99:686–90.

Schäfers H-J. Chapter 1—Anatomy. In: Current treatment of aortic regurgitation. Bremen: UNI-MED Verlag; 2013. ISBN: 9783837414066.

Schmid FX, Bielemberg K, Schneider A, et al. Ascending aortic aneurysm associated with bicuspid and tricuspid aortic valve: involvement and clinical relevance of smooth muscle cell apoptosis and expression of cell death-initiating proteins. Eur J Cardiothorac Surg. 2003;23:537–43.

Schofer N, Blankenberg S. Transfemoral or surgical aortic valve replacement in young low-risk patients with tricuspid and bicuspid aortic valve stenosis. Eur Heart J. 2024;45(37):3815–7.

Shafers HJ, Aicher D, Langer F, Lausberg HF. Preservation of bicuspid aortic valve. Ann Thorac Surg. 2007;83(2):S740–5.

Sieber-Blum M. Cardiac neural crest stem cells. Anat Rec A Discov Mol Cell Evol Biol. 2004;276(1):34–42.

Sievers HH, Schmidtke C. A classification system for the bicuspid aortic valve from 304 surgical specimens. J Thorac Cardiovasc Surg. 2007;133(5):1226–33.

Siguero-Álvarez M, Salguero-Jiménez A, Grego-Bessa J, de la Barrera J, MacGrogan D, Prados B, Sánchez-Sáez F, Piñeiro-Sabarís R, Felipe-Medina N, Torroja C, Gómez MJ, Sabater-Molina M, Escribá R, Richaud-Patin I, Iglesias-García O, Sbroggio M, Callejas S, O'Regan DP, McGurk KA, Dopazo A, Giovinazzo G, Ibañez B, Monserrat L, Pérez-Pomares JM, Sánchez-Cabo F, Pendas AM, Raya A, Gimeno-Blanes JR, de la Pompa JL. A human hereditary cardiomyopathy shares a genetic substrate with bicuspid aortic valve. Circulation. 2023;147(1):47–65.

Siu SC, Silversides CK. Bicuspid aortic valve disease. J Am Coll Cardiol. 2010;55:2789–800.

Small AM, Peloso GM, Linefsky J, Aragam J, Galloway A, Tanukonda V, Wang LC, Yu Z, Sunitha Selvaraj M, Farber-Eger EH, Baker MT, Setia-Verma S, Lee SSK, Preuss M, Ritchie MD, Damrauer SM, Rader DJ, Wells QS, Loos R, Lubitz SA, Thanassoulis G, Cho K, Wilson PWF, VA Million Veteran Program, Natarajan P, O'Donnell CJ. Multiancestry genome-wide association study of aortic stenosis identifies multiple novel loci in the million veteran program. Circulation. 2023;147(12):942–55.

Soto-Navarrete MT, López-Unzu MÁ, Durán AC, Fernández B. Embryonic development of bicuspid aortic valves. Prog Cardiovasc Dis. 2020;63(4):407–18.

Stone JR, Bruneval P, Angelini A, Bartoloni G, Basso C, Batoroeva L, Buja LM, Butany J, d'Amati G, Fallon JT, Gittenberger-de Groot AC, Gouveia RH, Halushka MK, Kelly KL, Kholova I, Leone O, Litovsky SH, Maleszewski JJ, Miller DV, Mitchell RN, Preston SD, Pucci A, Radio SJ, Rodriguez ER, Sheppard MN, Suvarna SK, Tan CD, Thiene G, van der Wal AC, Veinot JP. Consensus statement on surgical pathology of the aorta from the Society for Cardiovascular Pathology and the Association for European Cardiovascular Pathology: I. Inflammatory diseases. Cardiovasc Pathol. 2015;24(5):267–78. https://doi.org/10.1016/j.carpath.2015.05.001. Epub 2015 May 16. PMID: 26051917.

Svensson LG, Kim KH, Lytle BW, Cosgrove DM. Relationship of aortic cross-sectional area to height ratio and risk of aortic dissection in patients with bicuspid aortic valves. J Thorac Cardiovasc Surg. 2003;126:892–3.

Takamura K, Okishima T, Ohlo S, et al. Association of cephalic neural crest cells with cardiovascular development, particularly that of the semilunar valves. Anat Embryol. 1990;182:263–70.

Takkemberg JJ, Klieverik LM, Schoof PH, van Suylen RJ, van Herwerden LA, Zondervan PE, et al. The Ross procedure: a systematic review and meta-analysis. Circulation. 2009;119:222–8.

Thiene G, Basso C. Le aortopatie. Cardiol Sci. 2006a;4:267–75.

Thiene G, Basso C. Pathology and pathogenesis of infective endocarditis in native heart valves. Cardiovasc Pathol. 2006b;15(5):256–63.

Thiene G, Corrado D, Rigato I, Basso C. Why and how to support screening strategies to prevent sudden death in athletes. Cell Tissue Res. 2012;348(2):315–8. https://doi.org/10.1007/s00441-012-1398-4. Epub 2012 Mar 29. PMID: 22457200.

Thiene G, Basso C, Della Barbera M. Pathology of the aorta and aorta as homograft. J Cardiovasc Dev Dis. 2021;8(7):76. https://doi.org/10.3390/jcdd8070076. PMID: 34209632; PMCID: PMC8304113.

Thiene G, Rizzo S, Basso C. Bicuspid aortic valve: the most frequent and not so benign congenital heart disease. Cardiovasc Pathol. 2024;70:107604.

Totaro P, Caimi A, Formenton G, Musto M, Schembri M, Morganti S, Pelenghi S, Auricchio F. Bicuspid valve aortopathy: is it reasonable to define a different surgical cutoff based on different aortic wall mechanical properties compared to those of the tricuspid valve? J Cardiovasc Dev Dis. 2024;11(10):312.

Tzemos N, Therrien J, Yip J, et al. Outcomes in adults with bicuspid aortic valves. JAMA. 2008;300:1317–25.

Tzemos N, Lyseggen E, Silversides C, et al. Endothelial function, carotid-femoral stiffness, and plasma matrix metalloproteinase-2 in men with bicuspid aortic valve and dilated aorta. J Am Coll Cardiol. 2010;55(7):660–8.

Vahanian A, Alfieri OR, Al-Attar N, et al. Transcatheter valve implantation for patients with aortic stenosis: a position statement from the European Association of Cardio-Thoracic Surgery (EACTS) and the European Society of Cardiology (ESC), in collaboration with the European Association of Percutaneous Cardiovascular Interventions (EAPCI). Eur J Cardiothorac Surg. 2008;34(1):1–8.

Vowels TJ, Gonzalez-Stawinski GV, Ko JM, Trachiotis GD, Roberts BJ, Roberts CS, Roberts WC. Anomalous cord from the raphe of a congenitally bicuspid aortic valve to the aortic wall producing either acute or chronic aortic regurgitation. J Am Coll Cardiol. 2014;63(2):153–7.

Vriz O, Aboyans V, D'Andrea A, Ferrara F, Acri E, Limongelli G, Della Corte A, Driussi C, Bettio M, Pluchinotta FR, Citro R, Russo MG, Isselbacher E, Bossone E. Normal values of aortic root dimensions in healthy adults. Am J Cardiol. 2014;114(6):921–7.

Wallby L, Janerot-Sjoberg B, Steffensen T, Broqvist M. T lymphocyte infiltration in non rheumatic aortic stenosis: a comparative descriptive study between tricuspid and bicuspid aortic valves. Heart. 2002;88:348–51.

Ward C. Clinical significance of the bicuspid aortic valve. Heart. 2000;83:81–5.

Warnes CA, Williams RG, Bashore TM, et al. ACC/AHA 2008 guidelines for the management of adults with congenital heart disease: a report of the American College of Cardiology/American Heart Association Task Force on Practice Guidelines (Writing Committee to develop guidelines on the management of adults with congenital heart disease). J Am Coll Cardiol. 2008;52:e1–121.

Wedin JO, Vedin O, Rodin S, Simonson OE, Hörsne Malmborg J, Pallin J, James SK, Flachskampf FA, Ståhle E, Grinnemo KH. Patients with bicuspid aortic stenosis demonstrate adverse left ventricular remodeling and impaired cardiac function before surgery with increased risk of postoperative heart failure. Circulation. 2022;146(17):1310–22.

Wigstrom L, Sjoqvist L, Wranne B. Temporally resolved 3D phase-contrast imaging. Magn Reson Med. 1996;36:800–3.

Wijesinghe N, Ye J, Rodés-Cabau J, Cheung A, Velianou JL, Natarajan MK, Dumont E, Nietlispach F, Gurvitch R, Wood DA, Tay E, Webb JG. Transcatheter aortic valve implantation in patients with bicuspid aortic valve stenosis. JACC Cardiovasc Interv. 2010;3(11):1122–5.

Wu J, Zafar MA, Liu Y, Chen JF, Li Y, Ziganshin BA, Ellauzi H, Mukherjee SK, Rizzo JA, Elefteriades JA. Fate of the unoperated ascending thoracic aortic aneurysm: three-decade experience from the Aortic Institute at Yale University. Eur Heart J. 2023;44(43):4579–88.

Yang LT, Ye Z, Wajih Ullah M, Maleszewski JJ, Scott CG, Padang R, Pislaru SV, Nkomo VT, Mankad SV, Pellikka PA, Oh JK, Roger VL, Enriquez-Sarano M, Michelena HI. Bicuspid aortic valve: long-term morbidity and mortality. Eur Heart J. 2023;44(43):4549–62.

Yener N, Oktar GL, Erer D, et al. Bicuspid aortic valve. Ann Thorac Cardiovasc Surg. 2002;8:264–7.

Zhang Y, Choi BH, Chee HK, Kim JS, Ko SM. Aortic valve dysfunction and aortopathy based on the presence of raphe in patients with bicuspid aortic valve disease. J Cardiovasc Dev Dis. 2023;10(9):372.

Zhu JM, Ma WG, Peterss S, Wang LF, Qiao ZY, Ziganshin BA, Zheng J, Liu YM, Elefteriades JA, Sun LZ. Aortic dissection in pregnancy: management strategy and outcomes. Ann Thorac Surg. 2017;103(4):1199–206. https://doi.org/10.1016/j.athoracsur.2016.08.089. Epub 2016 Nov 5. PMID: 27825688.

MIX
Papier aus verantwortungsvollen Quellen
Paper from responsible sources
FSC® C105338

If you have any concerns about our products,
you can contact us on
ProductSafety@springernature.com

In case Publisher is established outside the EU,
the EU authorized representative is:
Springer Nature Customer Service Center GmbH
Europaplatz 3, 69115 Heidelberg, Germany

Printed by Libri Plureos GmbH
in Hamburg, Germany